AF815401

METHODS IN MOLECULAR BIOLOGY

Series Editor
John M. Walker
School of Life and Medical Sciences
University of Hertfordshire
Hatfield, Hertfordshire, AL10 9AB, UK

For further volumes:
http://www.springer.com/series/7651

Nucleic Acid Crystallography

Methods and Protocols

Edited by

Eric Ennifar

CNRS "Architecture et Réactivité de l'ARN", Institut de Biologie Moléculaire et Cellulaire, Strasbourg, France

Humana Press

Editor
Eric Ennifar
CNRS "Architecture et Réactivité de l'ARN"
Institut de Biologie Moléculaire et Cellulaire
Strasbourg, France

ISSN 1064-3745 ISSN 1940-6029 (electronic)
Methods in Molecular Biology
ISBN 978-1-4939-2762-3 ISBN 978-1-4939-2763-0 (eBook)
DOI 10.1007/978-1-4939-2763-0

Library of Congress Control Number: 2015941273

Springer New York Heidelberg Dordrecht London
© Springer Science+Business Media New York 2016
This work is subject to copyright. All rights are reserved by the Publisher, whether the whole or part of the material is concerned, specifically the rights of translation, reprinting, reuse of illustrations, recitation, broadcasting, reproduction on microfilms or in any other physical way, and transmission or information storage and retrieval, electronic adaptation, computer software, or by similar or dissimilar methodology now known or hereafter developed.
The use of general descriptive names, registered names, trademarks, service marks, etc. in this publication does not imply, even in the absence of a specific statement, that such names are exempt from the relevant protective laws and regulations and therefore free for general use.
The publisher, the authors and the editors are safe to assume that the advice and information in this book are believed to be true and accurate at the date of publication. Neither the publisher nor the authors or the editors give a warranty, express or implied, with respect to the material contained herein or for any errors or omissions that may have been made.

Printed on acid-free paper

Humana Press is a brand of Springer
Springer Science+Business Media LLC New York is part of Springer Science+Business Media (www.springer.com)

Preface

Knowledge of three-dimensional structures of biological macromolecules is essential for a complete understanding of their functions and interactions. X-ray crystallography is the most widely used technique in structural biology and can provide highly detailed structures of macromolecular molecules without any size limit. Among the 93,700 crystal structures that have been deposited to date (January 2015) in the Protein Data Bank, less than 1,600 are RNA- or DNA-only structures and about 4,800 protein/nucleic acid complexes. Consequently, biological X-ray crystallography has long been synonymous with protein crystallography, and most of the tools in macromolecular crystallography were primarily developed for protein structure determination. In the past two decades, considerable progresses have been made in nucleic acid crystallography, with improvements in nucleic acid synthesis, purification, crystallization, and structure determination. This led to a rapid growth in the number of RNA or DNA structures and nucleic acid/protein complexes, and several very large RNA structures (such as self-splicing introns) or ribonucleoprotein particles (such as the ribosome) were solved. This volume includes comprehensive and up-to-date coverage of all nucleic-acid-specific steps used in X-ray crystallography, from macromolecule production to structure determination. The protocols presented in this book are aimed at researchers and students interested into structural biology of DNA or RNA, alone or in complex with proteins or ligands. Chapters dedicated to RNA preparation and crystallogenesis should be of particular interest for beginners, while those focussed on data collection, phasing, and refinement will be particularly useful to people with a good level of expertise. Several functional case studies are presented in the last part of this book.

Finally, I would like to thank all the authors who contributed to this book by providing high-quality chapters.

Strasbourg, France ***Eric Ennifar***

Contents

Contributors

SULTAN AGALAROV • *Institute of Protein Research, Russian Academy of Sciences, Moscow, Russia*
JENNIFER ARCHER • *Department of Chemistry, University of Central Florida, Orlando, FL, USA*
PASCAL AUFFINGER • *Architecture et Réactivité de l'ARN, Institut de Biologie Moléculaire et Cellulaire, UPR 9002 CNRS/Université de Strasbourg, Strasbourg, France*
ROBERT T. BATEY • *Department of Chemistry and Biochemistry, University of Colorado at Boulder, Boulder, CO, USA*
ULRICH BAUMANN • *Institute of Biochemistry, University of Cologne, Cologne, Germany*
FRANCK BRACHET • *Laboratoire de Cristallographie et RMN Biologiques, CNRS/Paris Sorbonne Cité, Paris, France*
GÉRARD BRICOGNE • *Global Phasing Ltd., Cambridge, UK*
TOSHIYUKI CHATAKE • *Research Reactor Institute, Kyoto University, Osaka, Japan*
FANG-CHIEH CHOU • *Department of Biochemistry, Stanford University, Stanford, CA, USA*
THIBAULT CREPIN • *UVHCI, Université de Grenoble Alpes/EMBL/CNRS, Saint-Martin-d'Hères, France*
LUIGI D'ASCENZO • *Institut de Biologie Moléculaire et Cellulaire, UPR 9002 CNRS/Université de Strasbourg, Strasbourg, France*
RHIJU DAS • *Department of Biochemistry, Stanford University, Stanford, CA, USA*
CLÉMENT DÉGUT • *Laboratoire de Cristallographie et RMN Biologiques, CNRS/Paris Sorbonne Cité, Paris, France*
SERGEY M. DIBROV • *Department of Chemistry and Biochemistry, University of California, La Jolla, San Diego, CA, USA*
KAY DIEDERICHS • *Department of Biology, Universität Konstanz, Konstanz, Germany*
PHILIPPE DUMAS • *Institut de Biologie Moléculaire et Cellulaire, UPR 9002 CNRS/Université de Strasbourg, Strasbourg, France*
NATHANIEL ECHOLS • *Physical Biosciences Division, Lawrence Berkeley National Laboratory, Berkeley, CA, USA*
MARTIN EGLI • *Department of Biochemistry, Vanderbilt University School of Medicine, Nashville, TN, USA*
ERIC ENNIFAR • *"Architecture et Réactivité de l'ARN", Institut de Biologie Moléculaire et Cellulaire, Strasbourg, France*
ADRIAN R. FERRÉ-D'AMARÉ • *Laboratory of RNA Biophysics and Cellular Physiology, National Heart, Lung and Blood Institute, Bethesda, MD, USA*
AARON D. FINKE • *Paul Scherrer Institute, Swiss Light Source, Villigen, Switzerland*
ANG GAO • *Department of Biochemistry and Molecular Pharmacology, New York University School of Medicine, New York, NY, USA*
THOMAS HERMANN • *Department of Chemistry and Biochemistry, University of California, La Jolla, San Diego, CA, USA*
ZHEN HUANG • *Department of Chemistry, Georgia State University, Atlanta, GA, USA; Department of Biology, Georgia State University, Atlanta, GA, USA*

SIBO JIANG • *Department of Chemistry, Georgia State University, Atlanta, GA, USA; Department of Biology, Georgia State University, Atlanta, GA, USA*
JEFFREY S. KIEFT • *Department of Biochemistry and Molecular Genetics, Howard Hughes Medical Institute, University of Colorado Denver School of Medicine, Aurora, CO, USA*
JIRO KONDO • *Department of Materials and Life Sciences, Faculty of Science and Technology, Sophia University, Tokyo, Japan*
MARCO MARCIA • *European Molecular Biology Laboratory, Grenoble Outstation, Grenoble Cedex, France*
BENOÎT MASQUIDA • *Génomique Microbiologie et Génétique Moléculaire, Institut de Physiologie et de Chimie Biologique, UMR 7156 CNRS/Université de Strasbourg, Strasbourg, France*
MÉLANIE MEYER • *Département de Biologie Structurale et Intégrative, IGBMC, Université de Strasbourg/CNRS/INSERM, Illkirch, France*
JOELLE MEZHER • *Institut de Biologie Moléculaire et Cellulaire, UPR 9002 CNRS/Université de Strasbourg, Strasbourg, France*
ALEXANDRE MONOD • *UVHCI, Université de Grenoble Alpes/EMBL/CNRS, Paris, France*
MAX H. NANAO • *European Molecular Biology Laboratory, Grenoble, France*
VINCENT OLIÉRIC • *Paul Scherrer Institute, Swiss Light Source, Villigen, Switzerland*
PRADEEP S. PALLAN • *Department of Biochemistry, Vanderbilt University School of Medicine, Nashville, TN, USA*
EZEQUIEL PANEPUCCI • *Paul Scherrer Institute, Swiss Light Source, Villigen, Switzerland*
ALLA PESELIS • *Department of Biochemistry and Molecular Pharmacology, New York University School of Medicine, New York, NY, USA*
ALEXANDER SERGANOV • *Department of Biochemistry and Molecular Pharmacology, New York University School of Medicine, New York, USA*
EILEEN SHERMAN • *University of Central Florida, Orlando, FL, USA*
HUIYAN SUN • *Department of Chemistry, Georgia State University, Atlanta, GA, USA; Department of Biology, Georgia State University, Atlanta, GA, USA*
THOMAS C. TERWILLIGER • *Bioscience Division, Los Alamos National Laboratory, Los Alamos, NM, USA*
CARINE TISNÉ • *Laboratoire de Cristallographie et RMN Biologiques, CNRS/Paris Sorbonne Cité, Paris, France*
ALEXANDRE URZHUMTSEV • *Centre for Integrative Biology, IGBMC, Université de Strasbourg/CNRS/INSERM, Illkirch, France*
LUDMILA URZHUMTSEVA • *Institut de Biologie Moléculaire et Cellulaire, UPR 9002 CNRS/Université de Strasbourg, Strasbourg, France*
NIKITA VASILYEV • *Department of Biochemistry and Molecular Pharmacology, New York University School of Medicine, New York, NY, USA*
CYRIELLE DA VEIGA • *Institut de Biologie Moléculaire et Cellulaire, UPR 9002 CNRS/Université de Strasbourg, Strasbourg, France*
CLEMENS VONRHEIN • *Global Phasing Ltd., Cambridge, UK*
MEITIAN WANG • *Paul Scherrer Institute, Swiss Light Source, Villigen, Switzerland*
ERIC WESTHOF • *Institut de Biologie Moléculaire et Cellulaire, UPR 9002 CNRS/Université de Strasbourg, Strasbourg, France*
JING-DONG YE • *Department of Chemistry, University of Central Florida, Orlando, FL, USA*

MARAT YUSUPOV • *Département de Biologie et de Génomique Structurales, IGBMC, Université de Strasbourg/CNRS/INSERM, Illkirch, France*
GULNARA YUSUPOVA • *Département de Biologie et de Génomique Structurales, IGBMC, Université de Strasbourg/CNRS/INSERM, Illkirch, France*
CHLOE ZUBIETA • *Structural Biology Group, European Synchrotron Radiation Facility, Grenoble, France; CNRS, Univ. Grenoble Alpes, CEA, DSV, INRA, iRTSV, Laboratoire de Physiologie Cellulaire Végétale, UMR, Grenoble, France*

Part I

Introduction

Chapter 1

Perspectives and Pitfalls in Nucleic Acids Crystallography

Eric Westhof

Abstract

X-ray crystallography offers precious and striking knowledge on biomolecular architectures. Although safeguards do exist to guarantee the accuracy of the structures deposited in databases, they are not always applied, leading to the spread of inaccurate data. The importance of validation reports in the publication process is emphasized.

Key words Nucleic acids, Crystallography, Model quality, RNA, DNA, B factors, Validation

1 Introduction

The vast majority of our structural knowledge of nucleic acids, as it is for any biological molecule, is derived from X-ray diffraction data of crystals. The accuracy of the final structural model results from a complex set of factors. *First* and foremost, the preparation and purification of chemically pure RNA molecules constitute major experimental bottlenecks. Except for transfer RNAs or ribosomes, not too many RNA molecules or RNA–protein complexes can be obtained in sufficient quantities by extraction and purification from cultivated cells. Fortunately, chemical techniques and methods based on molecular biology tools have progressed enormously in the last 15 years. *Secondly*, the preparation of crystals diffracting at the highest possible resolution, in order to derive reliable structural models, is a major experimental challenge extremely demanding in patient manpower. *Thirdly*, data collection, treatment, the solution of the phasing problem, and the refinement form the core of the crystallographic work. From the early times of visual estimation of diffracted data obtained using copper anodes to present-day "zero-noise photon counting detectors" installed at novel generation synchrotron lines, the quality of the measured data and the proper choice of their mathematical treatment were both essential to derive accurate and cogent crystallographic structures that are biologically pertinent.

Eric Ennifar (ed.), *Nucleic Acid Crystallography: Methods and Protocols*, Methods in Molecular Biology, vol. 1320,
DOI 10.1007/978-1-4939-2763-0_1, © Springer Science+Business Media New York 2016

Finally, the biological relevance of the derived structural models has to be assessed and discussed. Highly chemically precise and accurate structures can still be of doubtful biological relevance. In this respect, the telling saga of the hammerhead ribozyme should be kept in mind. Those four main aspects of nucleic acid crystallography are addressed in this book.

The first eight articles are dedicated to RNA preparation and crystallogenesis. *Alexander Serganov*, who crystallized and solved so many beautiful structures of riboswitches, starts with a protocol for the preparation of short oligonucleotides with modified 5′-ends often a must for RNA–protein complexes (with *Nikita Vasilyev*). It follows with a detailed description on how to prepare crystals of riboswitches with and without ligands (with *Alla Peselis* and *Ang Gao*). Transfer RNA was the first large RNA molecule to be crystallized (tRNA-Phe) from yeast because it could easily be separated from a mixture of tRNAs owing to the large modified base at position 37 in the anticodon loop (the wybutine or Y base). *Clément Dégut, Alexandre Monod, Franck Brachet, Thibaut Crépin*, and *Carine Tisné* describe original methods for producing and purifying large quantities of tRNAs for crystallization. In a next chapter, *Mélanie Meyer* and *Benoît Masquida* describe how to use polyacrylamide gel electrophoresis for preparing milligram amounts of large (around 200 nt) RNAs. Over the years, lots of efforts have been put in the development of systematic approaches for assisting both crystallogenesis of RNA molecules and the solution to the phase problem of RNA crystals. A very successful method has been developed by *Adrian Ferré-D'Amaré,* while he was in *Jennifer Doudna's* laboratory, the use of the U1A protein. Here, *Adrian Ferré-D'Amaré* describes the improved and very practical method. This chapter is followed by a more recent method developed by *Jing-Dong Ye* while he was in *Piccirilli's* laboratory, the use of Fab fragments from phage display against a target RNA. With *Eileen Sherman* and *Jennifer Archer*, *Jing-Dong Ye* outlines very precisely the successive steps leading to the selection and co-crystallization of RNA molecules with a Fab assisting the crystallization and facilitating the resolution of the crystal structure. *Martin Egli* and *Pradeep S. Pallan* present the sole chapter on a DNA system, clearly a sign of present RNA-dominated times. They describe how the co-crystallization of DNA dodecamers with non-cleaving RNAse H enzymes allows for the determination of sequences that do not crystallize by themselves. Finally in this first set of chapters, *Cyrielle da Veiga, Joelle Mezher, Philippe Dumas* and *Eric Ennifar* explain how a powerful biophysical method, isothermal titration calorimetry, by monitoring RNA/ligand binding parameters can increase the rate of success of crystallization experiments.

The next nine chapters form the core of the crystallographic work, once crystals have been obtained successfully, and they are devoted to data collection, data treatment, the solution of the phase problem, and the final structural refinement. The first two chapters are of utmost importance because they deal with how the diffraction data are collected and computationally treated in order to obtain the best data for deriving accurate atomic models and precise phases by single- or multi-wavelength anomalous dispersion. Both chapters should constitute required reading by any crystallographer collecting data, either student or more senior. Indeed, evolving modern techniques and tools certainly allow for very rapid and simplified data collection protocols, but are still very demanding in high-level knowledge of data collection strategies in order to prevent model errors and loss of data resolution. The chapter written by *Kay Diederichs*, although focused on the widely used package XDS, contains a very general and rich presentation of data collection and data treatment. The second chapter, by *Aaron D. Finke, Ezequiel Panepucci, Meitian Wang, Vincent Olieric, Clemens Vonrhein*, and *Gérard Bricogne*, reviews all the fundamental aspects covering collection of anomalous data for phasing. In addition, this chapter provides advice helping quick and appropriate decision making while at the synchrotron where time is counted. The replacement of an oxygen atom by a sulfur atom (as in selenomethionine) has been extremely successful for solving the phase problem in protein crystallography. *Huiyan Sun, Sibo Jiang*, and *Zhen Huang* describe the synthesis of 2-Se-uridine containing RNAs to help in phasing. *Chloe Zubieta* and *Max H. Nanao* detail the process, called radiation damage induced phasing (or RIP), a surprising approach of transforming a drawback of X-ray irradiation into a method for solving the phase problem. In a next chapter, *Robert T. Batey* and *Jeffrey S. Kieft* describe the method they pioneered, the directed soaking of hexamine cations (iridium or cobalt) into RNA crystals containing a GoU wobble motif. *Marco Marcia* expounds very clearly and pragmatically the molecular replacement method for solving the phase problem. This very powerful method is going to take more and more importance owing to the increasing number of RNA crystal structures being regularly solved. The chapter by *Alexandre Urzhumtsev, Ludmila Urzhumtseva*, and *Ulrich Baumann* addresses a characteristic feature of nucleic acids, helical symmetry. The crystal packing is generally such that pseudo infinite helices appear, resulting in severe difficulties by standard molecular replacement methods. The authors describe various stratagems to turn a nucleic acid structural drawback into a positive contribution to the phase problem. A first crystallographic model fitting in the density map is rarely perfect and error-free in geometry, stereochemistry and without steric clashes. At this stage, the long process of refinement starts under the control of various tools monitoring geometry and fit into the density maps. *Fang-Chieh Chou, Nathaniel Echols, Thomas C. Terwilliger*, and *Rhiju Das* describe the very useful pipeline ERRASER (Enumerative Real-Space Refinement ASsisted by

Electron density under Rosetta) that couples reciprocal space in Phenix with real space refinement using Phenix. In a final chapter of this set on data collection, phasing, and refinement, *Toshiyuki Chatake* unfolds the steps necessary to perform neutron diffraction analysis of nucleic acid crystals. Neutron crystallography is a powerful tool for determining fine details in the hydration shells around nucleic acids, but also establishing correct enzymatic reactions.

In a last set of four chapters, various aspects related to the biology of nucleic acids are exposed. Three chapters are concerned with ribosomal translation and, among those, two are concerned with drug binding to RNA fragments. *Sultan Agalarov*, *Marat Yusupov*, and *Gulnara Yusupova* describe the in vitro reconstitution from free ribosomal RNA and ribosomal proteins into functional 30S subunit particles for structural and functional studies. In a following chapter, *Jiro Kondo* convincingly argues for the use of model RNA oligomers for systematic and thorough analysis of antibiotic binding to ribosomal fragments. In a similar vein, *Sergey M. Dibrov* and *Thomas Hermann* show how an important viral translation inhibitor of HCV could be structurally characterized by crystallizing it with a subdomain of the Internal Ribosome Entry Site of HCV. Finally, *Luigi D'Ascenzo* and *Pascal Auffinger* discuss the identification of ions in crystal structures, focusing especially on anions around the negatively charged nucleic acids. Several of these anions binding to nucleic acids are clearly due to the buffers used in crystallization conditions and, thus, not necessarily biologically relevant. But, at least, they should not be confused with other types of ions usually more linked to nucleic acids (like, for example, magnesium ions). They end up their chapter with a set of precious guidelines for avoiding solvent identification errors.

X-ray crystallography is a beautiful and successful tool for determining biomolecular structures and architectures. Its success is, undoubtedly, due to the fact that crystallography is firmly and solidly grounded in physics and chemistry. In addition, the extraordinary developments in data collection and computer power allow for complete mathematical and statistical treatments at all stages of the crystallographic process. However, despite the numerous mathematical or chemical safeguards and warnings available, errors in stereochemistry, in RNA folds or in solvent or ligand identifications do occur and are, unfortunately, rarely corrected afterwards. In the well-established protein crystallography field, several papers have appeared describing fatal mistakes or mishandling of data reporting, together with ways how to prevent and correct such errors [1–6]. Almost all, if not all, remarks, recommendations and suggestions made in those papers apply also to the field of nucleic acids crystallography. Very valuable tools for validations [7] and corrections [3] of nucleic acid structures do exist. The Protein Data Bank offers also various validation tools with compelling metrics [2]. As a personal note, while a post-doctoral fellow in

Madison, Wisconsin, in the laboratory of M. Sundaralingam (who with pencil and paper could draw in pre-ORTEP times nucleotides in a unit cell [8], but who unfortunately disappeared during the devastating South-East Asian tsunami 10 years ago), like other fellows in the laboratory, I had to type in all the crystal parameters and coordinates in order to check for stereochemistry, chirality and contacts for every single crystal structure of a nucleic acid component being submitted or published. Those were the days when computer-based databases, like the Cambridge Data Base [9] or the Protein Data Base [10], started to emerge fully. Interestingly, in the latter paper, a footnote warns "The Bank will assume no responsibility for checkout or correction of errors in such deposited programs." Owing to the amazing interest and usefulness of structural databanks, a similar stand is not held anymore. And, in 1990, Brändén and Jones [5] wrote "It is the crystallographer's responsibility to make sure that incorrect protein structures do not reach the literature." Should the problems with bad protein structures be now duplicated for the structures of RNA? One can certainly argue that there is information in data with low signal to noise. But, can one really be confident in a RNA structure at 3.7 Å resolution with average B-factors around 200 Å^2, bad clash score, and poor PDB validation metrics? Only a systematic analysis of the various regions of the RNA molecule in electron density maps would allow a knowledgeable interested person to reach a personal opinion. Several of the chapters in this book address this difficult question and various protocols, programs and tools for assessing the quality of the data, the refinement, and the final structure are presented in detail. However, as set forth by several authors [1, 2], referees and journal editors have also a major role to play in order to prevent incorrect structures to reach the literature and, almost worse, depository databases. Referees should request complete and detailed statistics tables, validation reports and quality indicators, coordinates, and electron-density maps and editors as well as authors should comply with such requests despite fierce competition for publication.

References

1. Brown EN, Ramaswamy S (2007) Quality of protein crystal structures. Acta Cryst D 63:941–950
2. Read RJ, Adams PD, Arendall WB 3rd, Brunger AT, Emsley P, Joosten RP, Kleywegt GJ, Krissinel EB, Lutteke T, Otwinowski Z, Perrakis A, Richardson JS, Sheffler WH, Smith JL, Tickle IJ, Vriend G, Zwart PH (2011) A new generation of crystallographic validation tools for the protein data bank. Structure 19:1395–1412
3. Chou FC, Sripakdeevong P, Dibrov SM, Hermann T, Das R (2013) Correcting pervasive errors in RNA crystallography through enumerative structure prediction. Nat Methods 10:74–76
4. Wlodawer A, Minor W, Dauter Z, Jaskolski M (2013) Protein crystallography for aspiring crystallographers or how to avoid pitfalls and traps in macromolecular structure determination. FEBS J 280:5705–5736

5. Brändén C-I, Jones TA (1990) Between objectivity and subjectivity. Nature 343:687–689
6. Dauter Z, Wlodawer A, Minor W, Jaskolski M, Rupp B (2014) Avoidable errors in deposited macromolecular structures: an impediment to efficient data mining. IUCrJ 1:179–193
7. Davis IW, Leaver-Fay A, Chen VB, Block JN, Kapral GJ, Wang X, Murray LW, Arendall WB 3rd, Snoeyink J, Richardson JS, Richardson DC (2007) MolProbity: all-atom contacts and structure validation for proteins and nucleic acids. Nucleic Acids Res 35:W375–W383
8. Sundaralingam M, Jensen LH (1965) Stereochemistry of nucleic acid constituents. I. Refinement of the structure of cytidylic acid b. J Mol Biol 13:914–929
9. Kennard O, Allen FH, Brice MD, Hummelink TWA, Motherwell WDS, Roidgers JR, Watson DG (1977) Computer based systems for the retrieval of data:crystallographyt. Pure Appl Chem 49:1807–1816
10. Bernstein FC, Koetzle TF, Williams GJ, Meyer EF Jr, Brice MD, Rodgers JR, Kennard O, Shimanouchi T, Tasumi M (1977) The protein data bank: a computer-based archival file for macromolecular structures. J Mol Biol 112:535–542

Part II

RNA Preparation and Crystallogenesis

Chapter 2

Preparation of Short 5′-Triphosphorylated Oligoribonucleotides for Crystallographic and Biochemical Studies

Nikita Vasilyev and Alexander Serganov

Abstract

RNA molecules participate in virtually all cellular processes ranging from transfer of hereditary information to gene expression control. In cells, many RNAs form specific interactions with proteins often using short nucleotide sequences for protein recognition. Biochemical and structural studies of such RNA–protein complexes demand preparation of short RNAs. Although short RNAs can be synthesized chemically, certain proteins require monophosphate or triphosphate moieties on the 5′ end of RNA. Given high cost of chemical triphosphorylation, broad application of such RNAs is impractical. In vitro transcription of RNA by DNA-dependent bacteriophage T7 RNA polymerase provides an alternative option to prepare short RNAs with different phosphorylation states as well as modifications on the 5′ terminus. Here we outline the in vitro transcription methodology employed to prepare ≤5-mer oligoribonucleotide for structural and biochemical applications. The chapter describes the principles of construct design, in vitro transcription and RNA purification applied for characterization of a protein that targets the 5′ end of RNA.

Key words Triphosphorylated RNA, 5′ Modification, Crystallization, Ion-exchange chromatography, In vitro transcription

1 Introduction

RNA–protein interactions are fundamental to core cellular processes. Recent technological advances have allowed genome-wide sequencing of multiple mRNA targets for RNA-binding proteins [1, 2]. Together with mRNA–protein binding data and variations in specificity of "canonical" RNA binding domains, these results suggest that many mRNA-binding proteins recognize protein-specific sets of mRNA sites. These sets could be considered an "mRNA recognition code" that defines interactions and function of mRNA-binding proteins [3]. Many RNA-binding proteins recognize specifically or semi-specifically short RNA sequences. Therefore, biochemical and crystallographic studies of RNA-binding proteins would greatly benefit from the technological

Eric Ennifar (ed.), *Nucleic Acid Crystallography: Methods and Protocols*, Methods in Molecular Biology, vol. 1320,
DOI 10.1007/978-1-4939-2763-0_2, © Springer Science+Business Media New York 2016

capability to synthesize short RNAs. Interactions with RNA for a number of important RNA-binding proteins, for instance, Argonaut [4], RIG-I [5, 6], and others [7], depend on the specific readout of the phosphorylation state of the RNA termini. Certain proteins, for instance, bacterial pyrophosphatase RppH [8], utilize triphosphorylated RNAs as reaction substrates. Studies of such proteins demands short RNAs with different number of phosphates on the 5′ terminus as well as RNAs with phosphate analogs. Here, we describe an inexpensive biochemical method for producing short (≤5-mer) phosphorylated RNA oligonucleotides at the scale sufficient for the majority of research applications.

The proposed approach for synthesis of short oligoribonucleotides is based on the in vitro transcription with bacteriophage T7 DNA-dependent RNA polymerase [9]. In vitro transcription with phage polymerases can produce both short and long RNAs and is currently the method of choice for making long (>100 nucleotides, nts) RNAs. Chemical synthesis is often used for preparing non-phosphorylated short and medium size (2–30 nts) RNAs. The chemical synthesis is relatively inexpensive for short non-phosphorylated RNAs, however addition of a monophosphate and triphosphate to the 5′ end inquire additional costs, which in the case of triphosphorylation are so high that extensive application of such RNAs for biochemical and crystallographic studies becomes impractical. In contrast to the chemical synthesis, RNAs prepared by in vitro transcription contain a natural 5′-triphosphorylated terminus and the transcription reactions can be also tweaked to begin with monophosphorylated, diphosphorylated, or modified triphosphorylated termini.

The method presented here takes into account two characteristics of the T7 RNA polymerase. First, the polymerase is able to transcribe in vitro from both supercoiled and linear DNA templates and is highly effective with double- and single-stranded synthetic templates containing a double-stranded T7 promoter. Therefore, a DNA template can be easily prepared by annealing two chemically synthesized DNA oligonucleotides, which reconstitutes the T7 promoter, followed by the target RNA sequence. Second, T7 polymerase can prematurely terminate transcription in the presence of an incomplete set of NTPs. If one or two out of four NTPs are not present during reaction, polymerase cannot complete RNA transcription and releases so-called "abortive" transcripts [10, 11].

Unfortunately, the run-off T7 polymerase-driven transcription produces "slippage" RNAs which contain extra non-template nucleotides [12], "turn-around" transcripts [13], and other aberrant products ([14–16], and references within). Several strategies have been used to minimize synthesis of aberrant RNAs with defects on the 3′ end of RNA, for instance, by preparing partially 2′-*O*-methyl-modified DNA template [17], by generating a 5′

overhang in the end of the template DNA strand [18] or by placing a self-cleaving ribozyme downstream of the desired RNA [19–21]. Defects on the 5′ end of RNA can be alleviated by reducing the number of 5′ cytosines in the template coding strand [15] or by incorporating the ribozyme upstream of the target RNA [19]. Thus, template DNA sequences should be carefully designed to minimize production of the anomalous RNA species and transcribed RNAs must be further purified. Addition of noncoding nucleotides to the 5′ end of a DNA template after the coding region appeared to improve the yield of the target oligoribonucleotide and minimize aberrant transcripts [11, 12]. Therefore, in our preparations, we routinely use coding DNA strands that contain 2–5 extra nucleotides which do not occur in the coding region and therefore would cause T7 RNA polymerase to abort transcription if the corresponding complementary nucleotides are omitted from the transcription mixture. In such transcription mixtures, the target RNA typically accounts for the vast majority of RNA products and therefore can be easily purified by anion-exchange chromatography.

Our method was developed for biochemical and crystallographic characterization of *Escherichia coli* RppH [8], the enzyme responsible for initiation of the 5′-dependent mRNA decays by cleaving off a pyrophosphate moiety from the 5′ end of natural triphosphorylated mRNAs. The technique was optimized for the synthesis of the 2–5 nt long RNAs with natural or modified triphosphorylated as well as diphosphorylated 5′ ends. Transcription reactions described here are typically carried out at the semipreparative scale which can yield sufficient amount of RNA for successful applications in both biochemical and crystallographic studies of proteins and other macromolecules.

2 Materials

2.1 RNA Preparation

2.1.1 DNA Templates

1. Chemically synthesized, deprotected, and desalted DNA oligonucleotides (*see* **Note 1**) at 100 pmol/μL concentration: “TopA,” a top strand of the T7 Class II ϕ2.5 promoter, 5′-TAATACGACTCACTATT; “TopB” a top strand of the Class III promoter, 5′-TAATACGACTCACTATA; “Bottom” (BotA and BotB types) oligonucleotides containing a complementary or bottom strand of the T7 promoter with an RNA coding sequence depicted by a stretch of Ns, 3′-ATTATGCTGAGTGATAANNNNN for Class II ϕ2.5 promoter (BotA oligonucleotides), or 3′-ATTATGCT GAGTGATATNNNNNN for Class III promoter (BotB oligonucleotides).
2. Diethyl pyrocarbonate (DEPC)-treated water (*see* **Note 2**).

3. TE buffer: 10 mM Tris–HCl, pH 8.0, 1 mM ethylenediaminetetraacetic acid (EDTA) (*see* **Note 3**).
4. 0.2 mL PCR tubes.
5. Thermal cycler.

2.2 In vitro Transcription

1. Transcription mixture components: 1 M Tris–HCl, pH 8.0, 1 M dithiothreitol (DTT), 250 mM spermidine–HCl, 100 mM solutions of each ribonucleotide triphosphate (ATP, GTP, CTP, and UTP), 1 M $MgCl_2$, and T7 RNA polymerase (6 mg/mL) (*see* **Note 4**). Nucleotides with diphosphates or monophosphates (ADP, AMP, etc.) or their analogs with a modified 5′ end (AMPcPP, where c is a methylene bridge between phosphates, Jena Biosciences, Jena, Germany) can be used for certain transcription reactions.
2. 0.5 M EDTA.
3. Water bath.
4. Bench top centrifuge for 1.5 mL Eppendorf and 50 mL tubes.

2.3 RNA Chromatography

1. 5 mL HiTrap Q HP column (GE Healthcare, Pittsburgh, PA, USA).
2. Mono Q 5/50 (1 mL) column (GE Healthcare, Pittsburgh, PA, USA).
3. ÄKTApurifier chromatography system (GE Healthcare, Pittsburgh, PA, USA).
4. Column cleaning solutions: 0.5 M NaOH and 2 M NaCl.
5. Buffer A: 20 mM Tris–HCl, pH 8.0.
6. Buffer B: 20 mM Tris–HCl, pH 8.0, 1 M NaCl.
7. 100 % ethanol.
8. 80 % (v/v) ethanol.
9. 5 M NaCl.
10. 0.22 μm syringe filters.
11. 10 mL syringes.

3 Methods

3.1 Construct Design

DNA templates for T7 RNA polymerase are designed to contain a double-stranded 17 base pair T7 promoter (ϕ2.5 class II promoter for initiation with A or class III promoter for initiation with G, Fig. 1a) [22] with a downstream single-stranded extension encoding for the target oligoribonucleotide. The coding region typically contains a 2–5 nt sequence (*see* **Note 5**) complementary to the target RNA, followed by a "stalling" nucleotide absent in the coding region. Omission of the complementary nucleotide for

A

Class II ϕ2.5 promoter

TopA 5′-**TAATACGACTCACTATT**-3′

|||||||||||||||||

BotA# 3′-**ATTATGCTGAGTGATAA**TNNNN-5′

pppANNNN

Class III promoter

TopB 5′-**TAATACGACTCACTATA**-3′

|||||||||||||||||

BotB# 3′-**ATTATGCTGAGTGATAT**CNNNN-5′

pppGNNNN

B

BotA1 3′-**ATTATGCTGAGTGATAA**TCAGA-5′

ATP+GTP pppAG

ATP+GTP+UTP pppAGU

BotA2 3′-**ATTATGCTGAGTGATAA**TTAGA-5′

ATP pppAA

ATP+UTP pppAAU

BotA3 3′-**ATTATGCTGAGTGATAA**TAAGA-5′

ATP+UTP pppAUU

BotA4 3′-**ATTATGCTGAGTGATAA**TGACA-5′

ATP+CTP pppAC

ATP+CTP+UTP pppACU

BotA5 3′-**ATTATGCTGAGTGATAA**TACAA-5′

ATP+UTP pppAU

BotB1 3′-**ATTATGCTGAGTGATAT**CCTGA-5′

GTP pppGG

GTP+ATP pppGGA

Fig. 1 DNA templates and corresponding RNA oligonucleotides. (**a**) General schematics of DNA templates with class II (*top*) and class III (*bottom*) T7 RNA polymerase promoters. A DNA template is composed of two uneven DNA oligonucleotides: T7 promoter strand (*top most*) is encoded by the TopA oligonucleotide while the complementary promoter strand (shown below) with the RNA coding sequence (ANNNN) is encoded by the BotA oligonucleotide. Promoter sequences are in *bold*. The name of each DNA oligonucleotide refers to the type of the promoter: A for class II ϕ2.5 and B for class III, respectively. *Bottom* oligonucleotides are also numbered. (**b**) Examples of the *bottom* oligonucleotides and corresponding RNAs, shown with NTP mixtures used in transcription. *Underlined* nucleotides indicate positions where T7 RNA polymerase aborts transcription in the absence of complementary NTP

the stalling nucleotide from the transcription mixture causes T7 RNA polymerase to stall and abort transcription at this site. Examples of DNA templates and corresponding oligoribonucleotides are depicted in Fig. 1b with stalling nucleotides in the template strand underlined. For instance, to synthesize pppApG, the DNA template should be prepared by annealing TopA and BotA1 oligonucleotides, while transcription reaction should contain ATP and GTP. To make pppApGpU from the same template, transcription reaction should contain ATP, GTP, and UTP.

To synthesize, ppcpApG, transcription mixture should contain AMPcPP and GTP. To synthesize pppGpG, the DNA template should be prepared with TopB and BotB1 oligonucleotides and transcription reaction should be carried out with GTP.

3.2 RNA Oligonucleotides Preparation

3.2.1 DNA Template Preparation

1. Mix TopA and one of the BotA oligonucleotides diluted in water at the final concentration 10 μM in 100 μL volume. For example, to synthesize dinucleotide pppApG, mix TopA and BotA1 DNA oligonucleotides (Fig. 1).
2. Anneal oligonucleotides to form a DNA duplex in 0.2 mL PCR tube in thermal cycler by heating at 98 °C for 2 min, cooling to 37 °C at 1 °C/s, and incubating samples at 37 °C for 10 min.

3.2.2 In Vitro Transcription

A typical transcription reaction is carried out in 1 mL volume of the mixture that contains 100 mM Tris–HCl, pH 8.0, 40 mM DTT, 2 mM spermidine, 15–16 mM total NTPs, 1 μM DNA template, 20 mM $MgCl_2$, and 50 μg/mL of T7 RNA polymerase. The mixture should be preheated at 37 °C prior to the addition of the DNA template and polymerase. Transcription of each DNA template should be routinely optimized for template and polymerase concentrations in the analytical scale (25 μL) reactions. Concentrations of individual NTPs should be adjusted according to the RNA oligonucleotide sequence. For example, to synthesize the dinucleotide pppApG, use 8 mM of each ATP and GTP; for pppApGpU, use 5 mM of each ATP, GTP, and UTP; and for pppApGpA—10 mM ATP and 5 mM GTP. To prepare RNA with a diphosphate or monophosphate on the 5′ end, conduct transcription reactions with the corresponding NDP or NMP. To obtain RNA with a modification in the 5′-triphosphate moiety, transcription should be carried out with the corresponding modified precursor (NMPcPP, NMPPcP, etc). Make sure that T7 RNA polymerase can initiate transcription from a nucleotide analog (*see* **Note 6**). Incubate reactions at 37 °C for 2–4 h. Reactions should get cloudy from the precipitation of magnesium pyrophosphate. After incubation, add 100 μL of 0.5 M EDTA and dissolve the precipitate by vigorous vortexing (*see* **Note 7**). At this point, transcription mixtures can be frozen and stored at −20 °C prior to purification.

3.2.3 RNA Chromatography

1. Thaw reactions, dilute them with 5 mL Buffer A, and filter through 0.22 μm syringe filter.
2. Separate a desired oligoribonucleotide from unincorporated NTPs and other reaction products using anion exchange chromatography on 5 mL HiTrap Q HP column (*see* **Note 8**). Load an RNA sample on the column pre-equilibrated with 25 mL Buffer A. Wash away unbound material with 20 mL

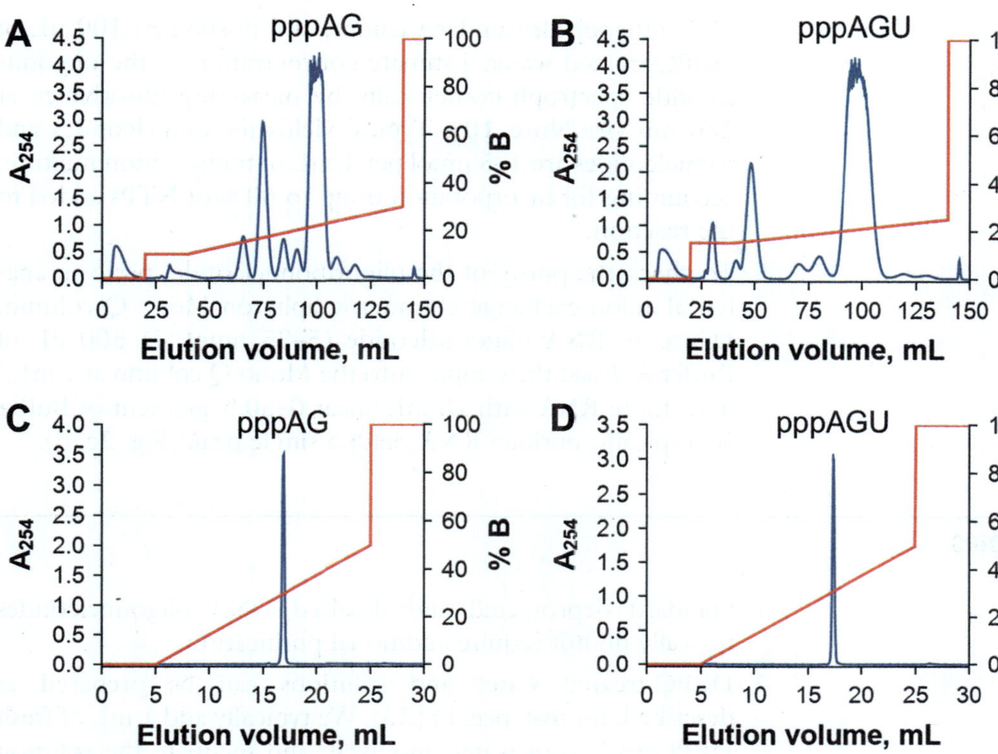

Fig. 2 Purification and analysis of RNA oligonucleotides using anion exchange chromatography. Purification of pppAG (**a**) and pppAGU (**b**) RNAs was performed on 5 mL HiTrap Q HP column using linear gradient 10–30 % (**a**) or 15–25 % (**b**) of buffer B over 100 mL. Analysis of purified pppAG (**c**) and pppAGU (**d**) was performed by chromatography on 1 mL Mono Q column with gradient elution of 0–50 % buffer B over 20 mL. Elution profiles (*blue lines*) are shown with theoretical gradients (*red lines*)

Buffer A. Increase concentration of Buffer B to 10 % and continue washing the column with another 20 mL of buffer to remove weakly bound nucleotides and reaction products. Elute an RNA oligonucleotide by a 100-mL linear 10–30 % gradient of Buffer B (*see* **Note 9**). Collect 2 mL fractions. Remove DNA template from the column with a step elution using 15 mL of 100 % Buffer B. Keep flow rate at 1 mL/min during all chromatographic steps. Monitor elution at 254 nm. Before purification of another RNA oligonucleotide, clean the column with 0.5 M NaOH and 2 M NaCl according to the manufacturer instructions.

3. The target RNA oligonucleotide is usually eluted as the largest peak during the gradient step (Fig. 2a, b). Combine fractions containing the RNA in 50 mL tube. Add 5 M NaCl to the final concentration of 0.3 M. Add 3 volumes of ethanol and mix the solution well. Incubate at −20 °C overnight.
4. Collect precipitated RNA by centrifugation at ≥12,000 × *g* for 30 min at +4 °C. Discard the supernatant. Wash pellet with

80 % ethanol, dry under vacuum, and dissolve in 100 μL of DEPC-treated water. Estimate concentration of the oligonucleotide spectrophotometrically by measuring absorbance at 260 nm (*see* **Note 10**). Typical yields for dinucleotides and trinucleotides are 1–5 μmol per 1 mL of transcription mixture, accounting for incorporation of up to 60 % of NTPs added to the reaction.

5. To check the purity of the oligoribonucleotide, perform analytical anion-exchange chromatography on Mono Q column. Dilute an RNA oligonucleotide (5–25 nmol) in 500 μL of Buffer A. Load the sample onto the Mono Q column at 1 mL/min. Elute RNA with 20 mL linear 0–50 % gradient of Buffer B. Typically, purified RNA yields a single peak (Fig. 2c, d).

4 Notes

1. Standard deprotected and desalted DNA oligonucleotides typically do not require additional purification.
2. DEPC-treated water and solutions can be prepared as described, for instance, in [23]. We typically add 1 mL of fresh DEPC to 2 L of water, mix well, and incubate the solution overnight at 37 °C. DEPC inactivates ribonucleases by irreversible modification of catalytic residues. DEPC is then decomposed by autoclaving the solution at 121 °C for 1 h.
3. All solutions must be prepared with RNase-free chemicals and DEPC-treated water. All glassware should be heated at 160 °C for at least 1 h to remove RNases and all plasticware must be RNase-free.
4. We routinely use T7 RNA polymerase purified from *E. coli* using the plasmid described in [24]. Alternatively, the polymerase can be purchased from different suppliers.
5. Synthesis of longer RNA oligonucleotides (>6–8 nt) is less effective, possibly due to stabilization of the hybrid between RNA and template DNA. The run-off transcription by the method described by Milligan et al. seems to be more suitable for preparation of long oligoribonucleotides [9].
6. Modified NTPs as well as NDPs and NMPs can be accommodated by T7 RNA polymerase to initiate transcription. In addition to regular NTPs (ATP and GTP), we have successfully used ADP and AMPcPP as initiating nucleotides for transcription of short oligoribonucleotides.
7. Although pyrophosphate precipitate can be alternatively removed by centrifugation at high speed for ~30 min at 4 °C, the precipitate may be formed during chromatography and clog the column. Therefore, it is strongly recommended to

dissolve the precipitate with EDTA or alternatively remove it by pyrophosphatase treatment.

8. Single-nucleotide resolution of RNA can be achieved with alternative high-resolution anion exchange columns such as Resource Q and Mono Q columns (GE Healthcare, Pittsburgh, PA, USA).
9. To improve resolution, parameters of elution can be adjusted for each particular oligonucleotide. Elution with 20× column volume gradient of 100–300 mM NaCl at pH 8.0 normally resolves mononucleotides, dinucleotides, and trinucleotides. Typically resolution can be improved by a longer gradient with a smaller change in the NaCl concentration. Resolution also depends on pH of buffer. Finely tuned parameters of gradient elution may resolve oligonucleotides not only according to their length but also according to their base composition (Fig. 3).
10. Extinction coefficient of RNA oligonucleotides can be calculated with, for instance, OligoAnalyzer (Integrated DNA Technologies, Coralville, IA, USA) or similar programs.

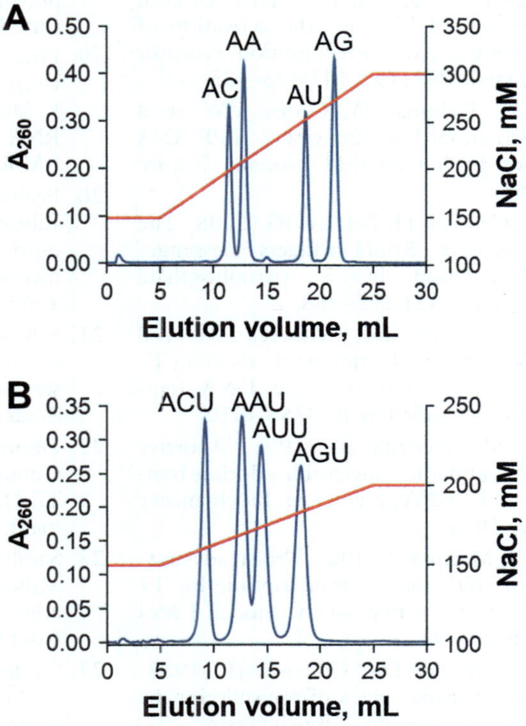

Fig. 3 Separation of triphosphorylated dinucleotides and trinucleotides according to their base composition. Four different RNAs were separated from a mixture on 1 mL Mono Q column. (**a**) Dinucleotides were separated using 150–300 mM NaCl gradient in a buffer containing 20 mM Tris–HCl, pH 9.0. (**b**) Trinucleotides were separated by 150–200 mM NaCl gradient in a buffer containing 20 mM Citrate-Na, pH 4.0

References

1. Baltz AG, Munschauer M, Schwanhäusser B et al (2012) The mRNA-bound proteome and its global occupancy profile on protein-coding transcripts. Mol Cell 46:674–690
2. Castello A, Fischer B, Eichelbaum K et al (2012) Insights into RNA biology from an atlas of mammalian mRNA-binding proteins. Cell 149:1393–1406
3. Serganov A, Patel DJ (2008) Towards deciphering the principles underlying an mRNA recognition code. Curr Opin Struct Biol 18:120–129
4. Ma J-B, Yuan Y-R, Meister G et al (2005) Structural basis for 5′-end-specific recognition of guide RNA by the *A. fulgidus* Piwi protein. Nature 434:666–670
5. Wang Y, Ludwig J, Schuberth C et al (2010) Structural and functional insights into 5′-ppp RNA pattern recognition by the innate immune receptor RIG-I. Nat Struct Mol Biol 17: 781–787
6. Kowalinski E, Lunardi T, McCarthy AA et al (2011) Structural basis for the activation of innate immune pattern-recognition receptor RIG-I by viral RNA. Cell 147:423–435
7. Abbas YM, Pichlmair A, Górna MW et al (2013) Structural basis for viral 5′-PPP-RNA recognition by human IFIT proteins. Nature 494:60–64
8. Deana A, Celesnik H, Belasco JG (2008) The bacterial enzyme RppH triggers messenger RNA degradation by 5′ pyrophosphate removal. Nature 451:355–358
9. Milligan JF, Groebe DR, Witherell GW et al (1987) Oligoribonucleotide synthesis using T7 RNA polymerase and synthetic DNA templates. Nucleic Acids Res 15:8783–8798
10. Moroney SE, Piccirilli JA (1991) Abortive products as initiating nucleotides during transcription by T7 RNA polymerase. Biochemistry 30:10343–10349
11. Kuzmine I, Martin CT (2001) Pre-steady-state kinetics of initiation of transcription by T7 RNA polymerase: a new kinetic model. J Mol Biol 305:559–566
12. Martin CT, Muller DK, Coleman JE (1988) Processivity in early stages of transcription by T7 RNA polymerase. Biochemistry 27: 3966–3974
13. Rong M, Durbin RK, McAllister WT (1998) Template strand switching by T7 RNA polymerase. J Biol Chem 273:10253–10260
14. Pleiss JA, Derrick ML, Uhlenbeck OC (1998) T7 RNA polymerase produces 5′ end heterogeneity during in vitro transcription from certain templates. RNA 4:1313–1317
15. Helm M, Brulé H, Giegé R et al (1999) More mistakes by T7 RNA polymerase at the 5′ ends of in vitro-transcribed RNAs. RNA 5:618–621
16. Nacheva GA, Berzal-Herranz A (2003) Preventing nondesired RNA-primed RNA extension catalyzed by T7 RNA polymerase. Eur J Biochem 270:1458–1465
17. Kao C, Zheng M, Rüdisser S (1999) A simple and efficient method to reduce nontemplated nucleotide addition at the 3′ terminus of RNAs transcribed by T7 RNA polymerase. RNA 5:1268–1272
18. Schenborn ET, Mierendorf RC (1985) A novel transcription property of SP6 and T7 RNA polymerases: dependence on template structure. Nucleic Acids Res 13:6223–6236
19. Price SR, Ito N, Oubridge C et al (1995) Crystallization of RNA-protein complexes. I. Methods for the large-scale preparation of RNA suitable for crystallographic studies. J Mol Biol 249:398–408
20. Fechter P, Rudinger J, Giegé R et al (1998) Ribozyme processed tRNA transcripts with unfriendly internal promoter for T7 RNA polymerase: production and activity. FEBS Lett 436:99–103
21. Schürer H, Lang K, Schuster J et al (2002) A universal method to produce in vitro transcripts with homogeneous 3′ ends. Nucleic Acids Res 30:e56
22. Dunn JJ, Studier FW, Gottesman M (1983) Complete nucleotide sequence of bacteriophage T7 DNA and the locations of T7 genetic elements. J Mol Biol 166:477–535
23. Sambrook J, Fritch EF, Maniatis T (1989) Molecular cloning: a laboratory manual, 2nd edn. Cold Spring Harbor Laboratory Press, Cold Spring Harbor, NY
24. Studier FW, Rosenberg AH, Dunn JJ et al (1990) Use of T7 RNA polymerase to direct expression of cloned genes. Methods Enzymol 185:60–89

Chapter 3

Preparation and Crystallization of Riboswitches

Alla Peselis, Ang Gao, and Alexander Serganov

Abstract

Recent studies have revealed that the majority of biological processes are controlled by noncoding RNAs. Among many classes of noncoding RNAs, metabolite-sensing segments of mRNAs called riboswitches are unique. Discovered over a decade ago in all three kingdoms of life, riboswitches specifically and directly interact with various metabolites and regulate expression of multiple genes, often associated with metabolism and transport of small molecules. Thus, riboswitches do not depend on proteins for binding to small molecules and play a role as both metabolite sensors and effectors of gene control. Riboswitches are typically located in the untranslated regions of mRNAs where they form alternative structures in the presence and absence of the ligand and modulate expression of genes through the formation of regulatory elements. To understand the mechanism of the riboswitch-driven gene control, it is important to elucidate how riboswitches interact with cognate and discriminate against non-cognate ligands. Here we outline the methodology to synthesize riboswitch RNAs and prepare riboswitch–ligand complexes for crystallographic and biochemical studies. The chapter describes how to design, prepare, and conduct crystallization screening of riboswitch–ligand complexes. The methodology was refined on crystallographic studies of several riboswitches and can be employed for other types of RNA molecules.

Key words B_{12} riboswitch, Fluoride riboswitch, RNA secondary structure, Crystallization

1 Introduction

Discovered in 2002 as regulators of vitamin-associated genes [1–3], riboswitches have quickly emerged as one of the most important feedback regulatory system in bacteria, with representatives found in eukaryotes and archaea [4]. The most important feature of riboswitches is their ability to control expression of genes through sensing and direct binding of various cellular metabolites. In contrast to many other regulatory circuits, riboswitches do not require proteins for sensing regulatory cues and therefore are considered noncoding RNA regulators of gene expression [5].

In bacteria, riboswitches typically reside in the 5′ untranslated regions of mRNAs encoding for genes associated with metabolism and the transport of cellular metabolites [6, 7]. The majority of riboswitches falls within a 35–250 nucleotide (nt) range and is

Eric Ennifar (ed.), *Nucleic Acid Crystallography: Methods and Protocols*, Methods in Molecular Biology, vol. 1320,
DOI 10.1007/978-1-4939-2763-0_3, © Springer Science+Business Media New York 2016

composed of two regions, an evolutionarily conserved metabolite-sensing or aptamer domain and a variable expression platform bearing regulatory signals. Regulatory switching depends on the presence of a cognate metabolite at the threshold concentration. Upon metabolite binding, the sensing domain adopts a stable metabolite-bound structure which triggers formation of a particular conformation in the downstream expression platform. If the cognate metabolite is not present in the cells, the sensing domain and expression platform form an alternative conformation in which both riboswitch regions are engaged in base pairing. The sequence of the expression platform defines the genetic outcome and the mechanism of the gene control (reviewed in [8]). Most bacterial riboswitches control transcription termination by forming a Rho-independent transcription terminator in the expression platform in response to metabolite binding. Many other bacterial riboswitches regulate access of the ribosome to the translation initiation signals thus impacting on translation of the message. Some riboswitches modulate gene expression through control of mRNA stability and translation via different mechanisms, such as mRNA self-cleavage and alternative splicing.

Riboswitches demonstrate highly specific binding to their cognate ligands, which involve inorganic ions, nucleobases and their derivatives, amino acids, sugars, and protein coenzymes (for review *see* [9]). The specificity of metabolite binding is entirely encoded in the three-dimensional structures of the metabolite-sensing domains of riboswitches. Given large differences in the chemical structure of the riboswitch ligands, these domains adopt distinct folds and often use unique combinations of structural principles for binding to their cognate ligands. Therefore, understanding the molecular mechanism for each riboswitch would require determination of the three-dimensional structure for individual riboswitches.

Structural studies require milligram quantities of riboswitch–ligand complexes, amounts that are not easily obtained. Despite feasibility of a chemical synthesis for RNA longer than 100 nts, the yield and costs of such synthesis makes it impractical for structural studies, which demand large quantities of multiple pure RNAs. Although we did use chemical synthesis for riboswitches in the range of 30–55 nts, many riboswitches are much longer and therefore cannot be synthesized chemically with reasonable costs and sufficient quantity. To prepare RNA for structural studies, we routinely carry out in vitro transcription with DNA-dependent bacteriophage T7 RNA polymerase [10] followed by two purification steps. Here we describe an updated RNA preparation and crystallization procedure based on the methodology developed earlier for preparation of large quantities of riboswitch RNAs for crystallographic studies [11]. The method is optimized for 30–250 nucleotides (nts) long RNAs and includes preparation of a DNA construct, in vitro transcription, purification of RNA by gel and ion-exchange chromatography, and the setup of crystallization screens. Our current methodology has been recently applied for crystallization of adenosylcobalamin (AdoCbl) [12] and fluoride riboswitches and can be employed for preparation of large quantities of other RNAs.

2 Materials

2.1 Construct Preparation

2.1.1 Construct Design

1. Benchtop incubator or water bath reaching 95 °C.
2. Thermal cycler.
3. iProof High Fidelity DNA Polymerase provided with 5× iProof HF Buffer (Bio-Rad, Hercules, CA, USA) or another high fidelity thermostable polymerase.
4. 10 mM deoxyribonucleotide triphosphate (dATP, dGTP, dCTP, and dTTP) mixture.
5. Desalted and deprotected DNA oligonucleotides dissolved in water.
6. PCR fragment gel extraction kit (Life Technologies, Grand Island, NY, USA).
7. 0.1 M $MgCl_2$.
8. Gel casting tray, gel chamber, and other items for agarose gels.
9. Ultrapure agarose.
10. 50× TAE buffer: 2 M Tris-acetate, pH 8.0, 50 mM ethylenediaminetetraacetic acid (EDTA).

2.1.2 Cloning

1. *Hind*III, 20,000 U/mL, and *Stu*I, 20,000 U/mL, restriction enzymes (New England Biolabs, Ipswich, MA, USA), supplied with 10× NE buffer 2.1 (NEB 2.1) and 10× CutSmart Buffer, respectively.
2. 2× Quick Ligation reaction buffer (New England Biolabs, Ipswich, MA, USA): 132 mM Tris–HCl, pH 7.6, 20 mM $MgCl_2$, 2 mM dithiothreitol (DTT), 2 mM ATP, 15 % Polyethylene glycol 6,000 (PEG 6,000).
3. Quick T4 DNA ligase (New England Biolabs, Ipswich, MA, USA).
4. *Escherichia coli* DH5α competent cells (Life Technologies, Grand Island, NY, USA).
5. 100 mg/mL ampicillin. Dissolve in 50 % ethanol and store at −20 °C.
6. DNA Clean kit (Zymo Research, Irvine, CA, USA).
7. Miniprep and Gigaprep plasmid DNA purification kits, PureLink Quick gel extraction kit (Life Technologies, Grand Island, NY, USA).

8. Luria Bertani (LB) broth/Amp medium and plates: 10 g tryptone, 5 g yeast extract, 10 g NaCl; dissolve in 1 L of water, autoclave for 30 min, and add ampicillin to 100 μg/mL. To prepare LB-agar for plates, add 4 g LB agar to 100 mL of water. Sterilize media by autoclaving.

2.2 RNA Preparation

2.2.1 DNA Template Preparation

1. Diethyl pyrocarbonate (DEPC)-treated water (*see* **Note 1**).
2. 1 M Tris–HCl, pH 8.0, (*see* **Note 2**).
3. Phenol–chloroform mixture (1:1): phenol, pH 8.0, (ACROS Organics, Geel, Belgium), chloroform (Fisher Scientific, Pittsburg, PA, USA).

2.2.2 In vitro Transcription

1. Transcription mixture components: 1 M Tris–HCl, pH 7.9, 1 M DTT, 250 mM spermidine–HCl, 100 mM of each ribonucleotide triphosphate (ATP, GTP, CTP, and UTP), 2 M $MgCl_2$, and 6 mg/mL T7 RNA polymerase.
2. Disposable filtration devices with a molecular weight cutoff of 5,000 Da (Corning, Tewksbury, MA, USA).

2.2.3 Polyacrylamide Gel Electrophoresis (PAGE)

1. RNAse ERASE (MP Biomedicals, Santa Ana, CA, USA).
2. 10× Tris/borate/EDTA (TBE) buffer: 890 mM Tris-borate, pH 8.3, 20 mM EDTA.
3. 2 L 40 % acrylamide–bis-acrylamide stock solution 29:1 (w/w): 27.5 g *N,N′*-methylenebisacrylamide and 772.4 g acrylamide (Sigma-Aldrich, St. Louis, MO, USA) dissolved in 2 L of DEPC water. Store in the dark.
4. Polyacrylamide gel (PAG) stock solution: 1× TBE buffer, 8 M urea (Sigma-Aldrich, St. Louis, MO), 20 % acrylamide–bis-acrylamide solution 29:1 (w/w). Store in the dark for up to 3 months.
5. TBE/urea solution: 1× TBE, 8 M urea solution. Store in the dark for up to 3 months.
6. Denaturing loading buffer: 1× TBE, 8 M urea, spatula tip of xylene cyanol and bromophenol blue.
7. 10 % ammonium persulfate (APS) solution. Store at +4 °C in the dark for no more than 4 weeks.
8. Tetramethylethylenediamine (TEMED).
9. Thin layer chromatography (TLC) plates, 20×20 cm (Selecto Scientific, Suwanee, GA, USA).
10. Mega-Gel high throughput vertical gel system with custom glass plates measuring 45×50 cm and 0.5 cm spacers (C.B.S. Scientific, Del Mar, CA, USA).
11. Disposable scalpels.

12. Elutrap electroelution systems, each with four Elutrap chambers, assembled with BT1 and BT2 membranes (Whatman, Dassel, Germany).
13. 20 mM Tris-acetate, pH 8.0.
14. Ethanol precipitation reagents: 3 M sodium acetate, pH 5.2, 100 % ethanol, 80 % (v/v) ethanol solution.
15. Handheld UV lamp.

2.2.4 RNA Chromatography

1. Mono Q 5/50 column (1 mL) and AKTA system (GE Healthcare, Piscataway, NJ, USA).
2. Cleaning solutions: 0.5 M NaOH, 10 % acetic acid.
3. Buffer A: 20 mM Tris–HCl, pH 7.0.
4. Buffer B: 20 mM Tris–HCl, pH 7.0, 1 M NaCl.
5. Column storage solution: 20 % ethanol.

2.3 Initial Crystallization Screening

1. Binding buffer: 50 mM Tris–HCl, pH 7.0, 50 mM K-acetate and 5 mM $MgCl_2$.
2. Mosquito liquid handler (TTP Labtech, Melbourn, UK).
3. 8-Position 2 μL micro-reservoir strips for Mosquito liquid handler (TTP Labtech, Melbourn, UK).
4. MRC 2 well crystallization plate (Hampton Research, Aliso Viejo, CA, USA).
5. 3 in. wide Crystal Clear sealing tape (Hampton Research, Aliso Viejo, CA, USA).
6. ClearSeal film (Hampton Research, Aliso Viejo, CA, USA).
7. Sealing Film Applicator (Hampton Research, Aliso Viejo, CA, USA).
8. Masterblock 96 Deep Well Polypropylene plate (Hampton Research, Aliso Viejo, CA, USA).
9. AlumaSeal II Sealing Film (Hampton Research, Aliso Viejo, CA, USA).
10. Commercial sparse matrix kits: Classics Suite, Classics II Suite, Nucleix Suite (Qiagen, Valencia, CA, USA), and kits from other suppliers.
11. Pre-greased crystallization plates with 24 wells (Hampton Research, Aliso Viejo, CA, USA).
12. EasyXtal X-Seal 15-well plates with crystal supports (Qiagen, Valencia, CA, USA).
13. 22 mm siliconized glass circle cover slides.
14. Compressed gas can.
15. Stereomicroscope 12:1 zoom range with polarizing filters (e.g., Olympus SZX12).

3 Methods

3.1 Construct Preparation

3.1.1 Construct Design

Success of structural studies depends on the choice of a riboswitch sequence. As in crystallographic studies with proteins, it is important to choose a representative RNA which forms a stable compact structure and does not contain flexible regions. We typically begin crystallization experiments by choosing 3–5 typical riboswitches from thermophilic species. These riboswitches should have the shortest sequences and contain most of the evolutionarily conserved nucleotides with the least number of bulges and other irregularities. Selected sequences should not adopt a major alternative conformation according to the in silico tests using RNA folding program Mfold [13].

T7 RNA polymerase produces many aberrant transcripts with defects on the termini [14–18]. For instance, T7 RNA polymerase incorporates extra nucleotides in the 3′ end of the transcript. Since heterogeneity at the ends may interfere with formation of crystal contact(s), extended RNA species, sometimes representing over 50 % of the transcribed products, may prevent formation of crystals. To alleviate the problem, we typically add self-cleaving hammerhead or HDV ribozyme sequences immediately downstream of the riboswitch sequence [19]. Schematics of metabolite-sensing domain constructs with adjoined hammerhead [20] and HDV [21] ribozymes are shown in Fig. 1a, b, respectively.

T7 RNA polymerase can transcribe in vitro from both supercoiled and linear DNA templates and is highly effective with double- and single-stranded synthetic templates containing a ~17 nt double-stranded T7 promoter. Therefore, a DNA template can be easily prepared by annealing two chemically synthesized DNA oligonucleotides, which reconstitutes the T7 promoter, followed by a target gene sequence. Alternatively, transcription can be conducted from a DNA template prepared by PCR. We have however noticed that for large scale RNA preparation, it is more practical to clone a riboswitch sequence into a plasmid under the control of a T7 promoter, amplify plasmid DNA in *E. coli*, and conduct transcription from the linearized plasmid DNA template.

To make a double-stranded DNA for cloning, we typically use three strategies. The first strategy (Fig. 1c, left) employs a chemical synthesis of two long complementary DNAs which, after annealing, reconstitute the desired riboswitch sequence followed by a ribozyme and a half of the restriction site, usually *Hind*III. The DNA fragment is then ligated at the *Stu*I and *Hind*III restriction sites of the pUT7 plasmid [22] (Fig. 1c) under the control of the class III T7 promoter. The promoter sequence is followed by two guanosines required for efficient initiation of transcription. These two guanosines remain in the plasmid after cleavage by *Stu*I and therefore will be added to the 5′ end of any RNA transcribed from the plasmid.

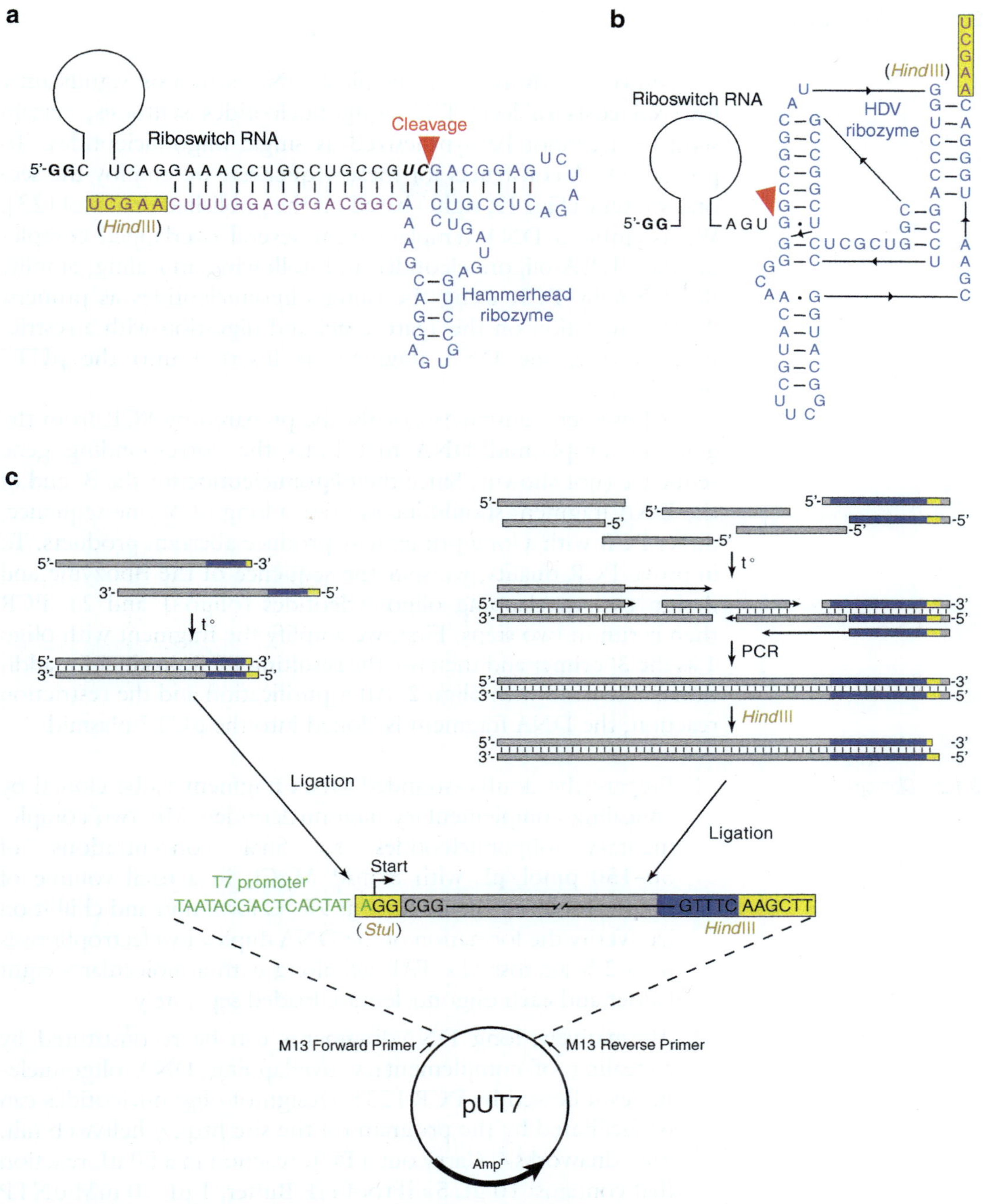

Fig. 1 Design of riboswitch constructs for crystallization. (**a**, **b**) Riboswitch RNAs shown with adjoined hammerhead (**a**) and HDV (**b**) ribozymes. Riboswitch and ribozyme sequences are shown in *black* and *blue colors*, respectively. A pair of the 5′-terminal guanosines encoded in the pUT7 plasmid is in *bold*. A pair of the 3′-terminal nucleotides of the riboswitch sequence required for efficient cleavage of the hammerhead ribozyme is shown in *bold italics*. Cleavage sites of the ribozymes are shown with *red arrows*. Nucleotides encoded by the *Hin*dIII site are shaded by *yellow color*. (**c**) Preparation of template DNA fragments and the cloning strategy. *Top left*, a DNA fragment can be prepared by annealing of two DNA oligonucleotides and ligating into the *Stu*I and *Hin*dIII restriction sites of the pUT7 vector. *Top right*, A DNA template can be also obtained by reconstituting the sequence from multiple overlapping fragments and PCR. After digestion with *Hin*dIII, the fragment is ligated into the pUT vector. *Gray*, *blue* and *yellow boxes* depict riboswitch RNA, ribozyme, and restriction sites. T7 promoter is in *green color*. The DNA insert can be sequenced with M13 forward and reverse primers

Although advances in chemical DNA synthesis significantly lowered costs of long DNA oligonucleotides synthesis, certain sequences cannot be synthesized as single oligonucleotides. To prepare DNA constructs for these sequences, we employ the second strategy (Fig. 1c, right) based on the published protocol [23]. We assemble a DNA template from several overlapped complementary DNA oligonucleotides and, following annealing, amplify the DNA by PCR using the outer oligonucleotides as primers. After purification on the agarose gel and digestion with a restriction enzyme, the DNA fragment is inserted into the pUT7 plasmid.

Riboswitch constructs can also be prepared by PCR from the genomic or plasmid DNA that bears the corresponding gene sequence (not shown). Since the oligonucleotide for the 3′ end of the DNA fragment should incorporate a long ribozyme sequence, direct PCR with a long primer may produce aberrant products. To improve PCR quality, we split the sequence of the ribozyme and design two overlapping oligonucleotides (oligo 1 and 2). PCR then is run in two steps. First, we amplify the fragment with oligo 1 as the 3′ primer and then use the resulting PCR product for additional extension with oligo 2. After purification and the restriction reaction, the DNA fragment is cloned into the pUT7 plasmid.

3.1.2 Cloning

1. Prepare the double-stranded DNA fragment to be cloned by annealing complementary oligonucleotides. Mix two complementary oligonucleotides to final concentrations of 50–150 pmol/μL with 3 mM $MgCl_2$ in a total volume of 10 μL. Incubate the mixture at 95 °C for 3 min and chill it on ice. Verify the formation of the DNA duplex by electrophoresis on a 2 % agarose/1× TAE gel along with a molecular weight ladder and each oligonucleotide loaded separately.
2. Alternatively, long DNA fragments can be reconstituted by annealing of complementary overlapping DNA oligonucleotides followed by PCR [23]. Design of oligonucleotides can be facilitated by the program on the site http://helixweb.nih.gov/dnaworks/. Carry out a PCR reaction in a 50 μL reaction that contains: 10 μL 5× iProof HF Buffer, 1 μL 10 mM dNTP mix, 0.9 μL of both outer primers at a 100 μM stock concentration, 0.45 μL of all inner primers at 10 μM stock concentration, and 0.5 μL of iProof DNA polymerase. PCR typically includes 30 circles of the following steps: 95 °C for 30 s, 40–60 °C for 60 s, 72 °C for 30 s. Purify the PCR product by 1 % agarose gel electrophoresis and gel extraction kit.
3. Prepare the vector DNA by digesting the pUT7 plasmid with *Stu*I and *Hind*III restriction enzymes in NEB 2.1. Purify the vector using 1 % agarose gel electrophoresis and gel extraction kit.
4. Prepare the insert DNA by digesting with *Hind*III restriction enzyme in NEB2.1. Purify the DNA using a DNA clean up kit.

5. Mix 100 ng of the vector with 5–10× molar excess of the insert. Bring the volume of the mixture up to 10 μL with water, add 10 μL of 2× ligation buffer and 1 μL of T4 Quick Ligase. Briefly centrifuge samples and leave at room temperature for 5 min.
6. Transform 5 μL of the ligation mixture into *E. coli* strain DH5α and select the clones on LB/Amp plates.
7. Grow 3 mL bacterial cultures in LB/Amp medium and streak cells from each culture on the LB/Amp plate for future large scale plasmid DNA preparation. Purify plasmid DNA using a Miniprep kit. The inserts can be identified by a comparison of the *Eco*RI- and *Hind*III-digested ligated and control plasmids on a 2 % agarose gel. The clones with inserts of the correct size must be verified by DNA sequencing.

3.2 RNA Preparation

3.2.1 DNA Template Preparation

1. For large-scale plasmid DNA purification, grow bacterial cells of the correct clone in 3 L of LB/Amp medium at 37 °C overnight (*see* **Note 1**) and purify plasmid DNA using a Gigaprep kit. Expect 5–10 mg of plasmid DNA.
2. Prepare the DNA template for in vitro transcription by cleaving 5 mg of the plasmid with 1,000 U of *Hind*III in 5 mL of NEB 2.1 buffer supplemented with 0.25 mL of 1 M Tris–HCl, pH 8.0, at 37 °C for ~4 h. Verify completion of the reaction using 1 % agarose gel electrophoresis. Continue digestion if DNA has not been fully cut.
3. Extract restricted DNA with phenol–chloroform. Dilute the restriction reaction to 10 mL and add 10 mL of phenol–chloroform (1:1) mixture. Vortex the mixture for several min in a 50 mL conical tubes (*see* **Note 3**) and centrifuging at 4,000 × *g* for 5 min. Carefully collect the upper aqueous phase and repeat the extraction with chloroform alone.
4. Precipitate the DNA-containing aqueous phase with ethanol in 50 mL conical tubes. To the DNA solution, add one-tenth volume of 3 M Na-acetate, pH 5.2, and 3 volumes of 100 % ethanol. Incubate the mixture at −20 °C overnight and centrifuge for 30 min at >10,000 × *g*. Discard the supernatant, wash the pellet with ~10 mL of 80 % ethanol, and centrifuge for another 15 min. Discard supernatant in small amounts and repeat centrifugation as needed to ensure none of the pellet is drawn into the pipette. Air dry the pellet and dissolve it in 2–5 mL of water.

3.2.2 In Vitro Transcription

1. Prior to the large scale in vitro transcription, perform small-scale transcription reactions. A typical reaction is done in a 50 μL mixture of 100 mM Tris–HCl buffer, pH 7.9, 30 mM DTT, 2 mM spermidine, 4 mM of each ribonucleotide triphosphate, 2.5 μg of DNA template (50 μg/mL), 15 mM $MgCl_2$, and 2.5 μg of T7 RNA polymerase (*see* **Note 4**).

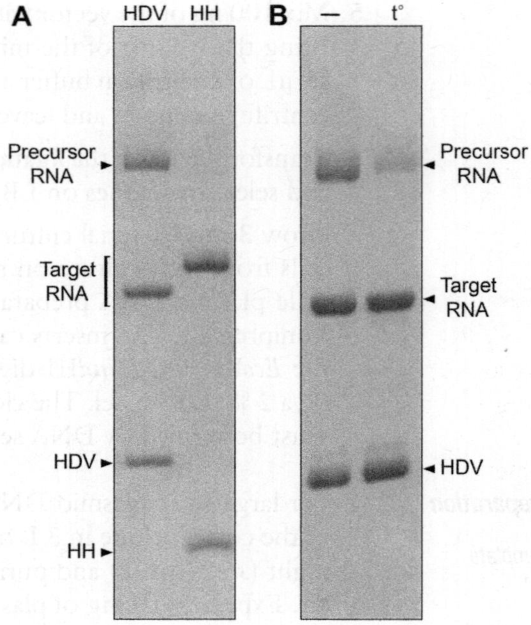

Fig. 2 Transcription and ribozyme cleavage analyzed by PAGE. (**a**) Transcription products for two RNAs followed by the HDV (*left*) and hammerhead (HH, *right*) ribozymes. Note incomplete cleavage of the precursor RNA by the HDV ribozyme. (**b**) Denaturation and renaturation procedure (t°) improves cleavage by the HDV ribozyme

The mixture should be preheated at 37 °C prior to the addition of DNA template and polymerase. Transcription of each DNA template should be routinely optimized by varying concentrations of $MgCl_2$ (10–20 mM), DNA template (40–100 μg/mL), and polymerase (30–100 μg/mL). Incubate the reactions at 37 °C for ~1–3 h. The reactions should get cloudy from the precipitation of magnesium pyrophosphate. Adjust the final concentration of $MgCl_2$ to 50 mM and incubate the reactions for another 30 min to complete ribozyme cleavage from the riboswitch RNA. Verify the transcription yield and the efficiency of the ribozyme cleavage using a 10 % small-scale PAGE (Fig. 2a).

2. Perform large-scale in vitro transcription in a total volume of 25–40 mL under optimized conditions found in analytical transcription reactions for ~3 h at 37 °C. Incubate for another 30 min after adjusting the $MgCl_2$ concentration and store the reactions at −20 °C. Preheat the reagents and water to maximize transcription yield. To avoid polymerase aggregation, add the polymerase drop by drop while swirling the mixture. Analyze the transcription mixture by PAGE.
3. If the ribozyme cleavage is not complete, distribute 1.0 mL aliquots of the transcription reaction to 1.5 mL Eppendorf

tubes and perform five cycles of RNA denaturation and renaturation: 72 °C for 1 min, 68 °C for 5 min, and 37 °C for 10 min (Fig. 2b) [24]. Prior to the large scale renaturation, test a smaller aliquot to make sure that the RNA can withstand several rounds of heating.

4. Thaw the transcription mixture and centrifuge at 4,000 × g for ~20 min. Discard magnesium pyrophosphate pellet and repeat centrifugation one more time. Filter the supernatant through 0.45 and 0.22 μm disposable filtration systems (*see* **Note 5**). Immediately concentrate the mixture to a volume of 1.0–2.5 mL using a 20 mL centrifugation filtration device at 4 °C. Mix the RNA with an equal volume of the loading buffer for loading onto a large polyacrylamide gel.

3.2.3 Large-Scale PAGE for RNA

1. We perform preparative PAGE on a Dual-Adjustable Mega-Gel system (C.B.S. Scientific) modified to hold large (45 cm in length, 50 cm in width) glass plates. Buffer chambers and glass plates for the gels must be scrubbed clean with a detergent, such as RNAse ERASE, rinsed thoroughly with RNase-free water and ethanol.
2. Position three spacers (0.5 cm thick) along the sides and bottom of the glass plate, place another glass plate on top, and clamp the plates together. Prepare 1 L of 10 % PAG/urea solution (*see* **Note 6**) by mixing 0.5 L of 20 % PAG stock solution with 0.5 L of TBE/urea solution. Take 100 mL of PAG and mix with 600 μL of APS solution and 40 μL of TEMED. Pour the gel along the side spacers and allow the gel to polymerize at the bottom with the glass plates standing at about a 30° angle. Make sure that the gel was in contact with the entire length of spacers. Mix the remaining 900 mL of gel with 5.4 mL of APS solution and 360 μL of TEMED and pour on top of the polymerized gel. Insert and clamp the comb (we usually use one or two single-tooth combs) and allow the gel to polymerize for 45 min.
3. Clamp a metal plate against the clean surface of the outer glass plate to ensure the even distribution of heat and mount the plates on the gel stand. Add 1 L of 1× TBE buffer to each buffer chamber and wash the well of the gel. Preheat the gel at 55 W for ~60 min. Load the concentrated transcription mixture and run the gel for ~15 h. One gel can accommodate RNA from 15 to 30 mL of the transcription mixture.
4. After the run, place the gel assembly on the bench, remove the top glass plate, wrap the gel and the remaining glass plate with Saran Wrap, and flip the glass plate and gel onto a clean flat surface. Remove the glass plate and place fluorescent TLC plates under the Saran Wrap. Use an ultraviolet lamp with $\lambda = 365$ nm to locate the dark band of the RNA and cut it out

of the gel with a disposable scalpel (*see* **Note 7**). The gel slices can be stored at −20 °C.

5. Assemble an Elutrap chamber by placing a BT1 trapezoidal impermeable membrane at each end of the chamber between two U-shaped inserts (so that the slanted tip faces the arrow on the chamber) using clean forceps. Place a BT2 rectangular semipermeable membrane between two U-shaped inserts closer to the middle of the chamber, dividing the chamber into two sections: a larger one for gel slices and a smaller one for eluate collection. Tighten both ends of the chambers with a piece of plastic, place the gel slices into the chamber, and place the chamber onto a tray of the Elutrap base. Add 20 mM Tris-acetate buffer to the chambers (enough to cover the gel) and to the base. Close the base with a lid and perform electrophoresis in the cold room at 5 W per Elutrap (20 W for the system loaded with four Elutrap).
6. Collect eluate every 2–3 h for the first 8 h and estimate the RNA concentration using UV spectroscopy at $\lambda = 260$ nm. If necessary, continue electrophoresis overnight at 70 V. Precipitate the plasmid DNA with ethanol, as described in Subheading 3.2.1, **step 4**.

3.2.4 RNA Chromatography

1. Dissolve RNA in ~5 mL of water and filter it through a 0.2 μm 25 mm nitrocellulose syringe filter to remove residual acrylamide particles.
2. To remove urea and other contaminants, purify the RNA using anion-exchange chromatography on a Mono Q column. The column should be cleaned before loading the sample by injecting ~2 mL of 0.5 M NaOH followed by ~2 mL of 10 % acetic acid and a step gradient of 100 % Buffer B. Once the column is clean, load no more than 5–8 mg of RNA per run. After loading the RNA, wash the column with five column volumes (CV) of Buffer A and elute RNA using a 10 CV 0–100 % gradient of Buffer B. Typically, only half of the RNA can be eluted after the first gradient; therefore, repeat elution several times using a step gradient of 100 % Buffer B (*see* **Note 8**). Store the column in 20 % ethanol.
3. Precipitate eluted RNA with ethanol, as described in Subheading 3.2.1. Dry the pellet in the lyophilyzer for 5 min or by air for 30 min, dissolve RNA in ~0.5 mL of water, and determine the concentration by UV spectroscopy.

3.3 Crystallization Screening and Optimization

We typically perform initial crystallization screening on the 96-well double-drop plates in the sitting drop format by the vapor diffusion method using commercial matrix kits. Crystallization trials are set up with a crystallization robot Mosquito at two temperatures (+4 and +20 °C). Setup for optimization of crystallization conditions is done manually by the vapor diffusion method in the hanging drop format using either pre-greased 24-well plates or 15-well EasyXtal X-Seal plates.

1. Distribute 1 mL of each well solution from commercial matrix kits into the Masterblock 96 Deep Well polypropylene block. For the first screening, we typically use Classic Suite, Classic II Suite, and Nucleix kits from Qiagen.
2. To prepare a plate for the Mosquito, dispense 80 μL of each well solution from the Masterblock to wells of the MRC 2 Well Crystallization Plate with a multichannel pipette (*see* **Note 9**). Place AlumiSeal film onto the block and seal the block with a film applicator. Store the block at 4 °C. Seal the crystallization plate with sealing tape. Keep the plates at the desired temperature for no more than 1 day.
3. Prepare riboswitch–ligand complexes by heating 0.2 and 0.5 mM RNA samples with 1.1–5.0× molar excess of the cognate metabolite in 100 μL of Binding buffer for 1 min at 90 °C or for 3 min at 55 °C. Chill samples on ice and incubate at 37 °C for 15 min.
4. Take the crystallization plate to the Mosquito, remove the tape, place the plate onto the moving tray of the robot, and set the proper parameter file (*see* **Note 10**).
5. Dispense 2.4 μL of each riboswitch samples to the wells of the two 8-position micro-reservoir strips. The left and right strips should have the lower and the higher complex concentrations, respectively.
6. Start the Mosquito program to dispense the 0.15 μL volumes of the RNA and crystallization solutions into the wells of the plate. Seal the plate with Clear Seal film. Store the plate at the desired temperature.
7. Allow the drops in the plates to equilibrate for at least 1 day before checking them with a stereomicroscope at a magnification range of ~5–10×. A polarizing filter is highly recommended for locating small crystals (Fig. 3a, b, g, h).
8. Once conditions that produce crystals have been determined, reproduce crystals in 2 μL hanging drops. Prepare 1 mL of reservoir solution and add it to the well of either 24-well or 15-well EasyXtal plates. The procedure below describes how to set up crystallization drops using glass slides and the 24-well pregreased plate. The 15-well EasyXtal plates are supplied with greaseless screw-in crystallization supports which replace glass slides.
9. Take a round glass slide and blow away dust particles from the surface using the compressed gas can.
10. Place a 1 μL drop of the reservoir solution on the glass slide. Add 1 μL of the riboswitch–ligand complex to the reservoir solution drop and mix the drop.

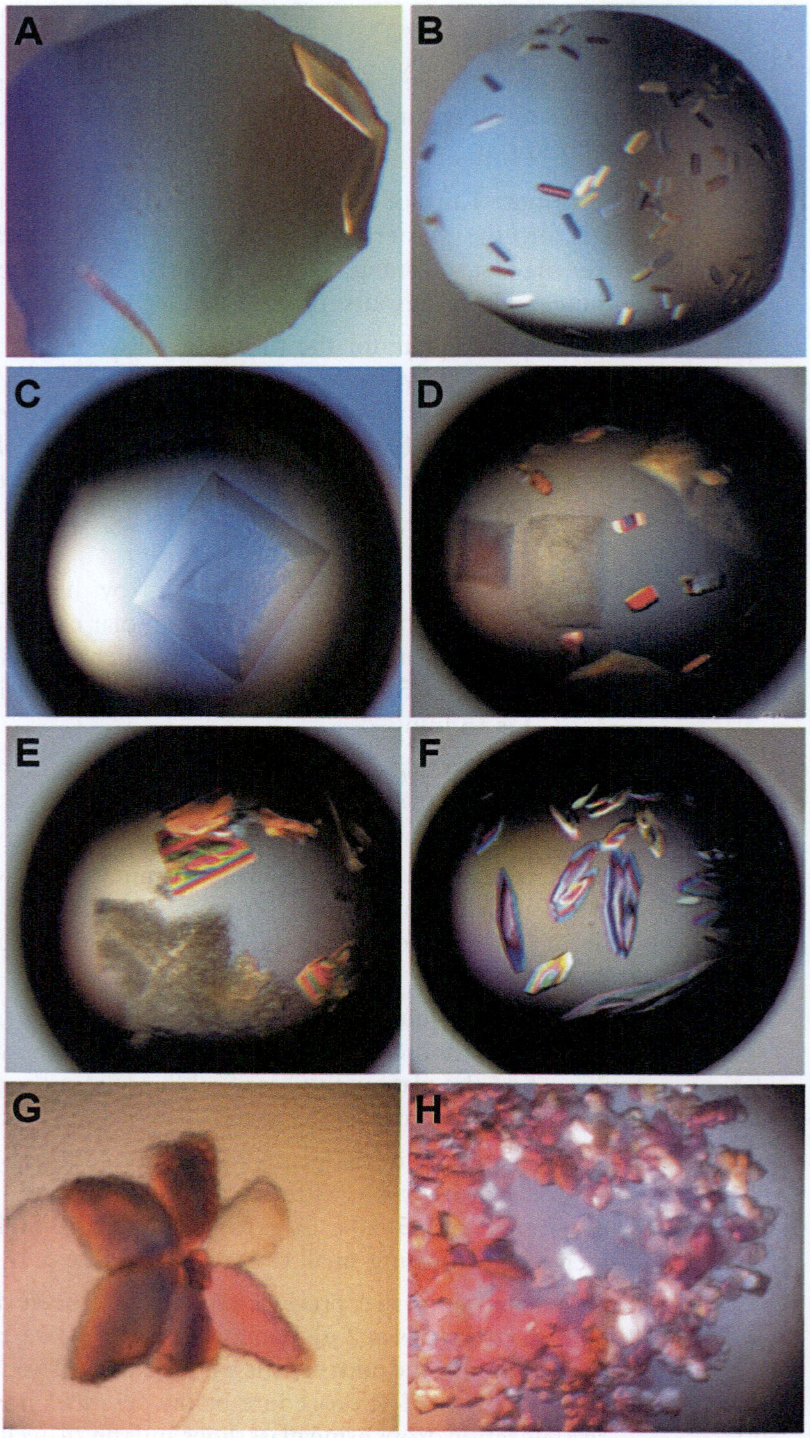

Fig. 3 Crystals of ligand-riboswitch complexes after initial crystallization screening and optimization. (**a**, **b**) Medium (**a**) and small (**b**) size crystals of the fluoride riboswitch obtained after initial screening with Mosquito. (**c**–**f**) Optimization of the fluoride riboswitch crystals. Crystals were grown with addition of 1 mM (**c**), 5 mM (**d**), and 25 mM (**e**) manganese cations and 5 mM cobalt hexamine cations (**f**). Note the change of the crystal shape (**c**–**e**) upon increase of the manganese concentration and the presence of two crystal forms in panel (**d**). (**g**, **h**) Crystal forms of the adenosylcobalamin riboswitch. (**g**) The riboswitch crystals are grown from the phase separation (a drop of liquid in the bottom left part of the crystallization drop) and form a flower-shaped crystalline conglomerate in which individual flat crystals form the petals of the corolla. (**h**) Different forms of crystals growing from phase separation in the same crystallization drop

11. Flip the glass slide and position it over the well. Gently press down on the glass slide to make sure that the edges of the glass slide are fully in contact with the grease.
12. Repeat **steps 9–11** for each additional set of conditions and place the plate in the incubator with the required temperature.
13. If a drop contains crystals or crystal-like material, optimize the crystallization conditions further by, for instance, varying the ratio between the reservoir and sample solutions, the concentration of the precipitant and RNA-metabolite complex, and the pH of the buffer. Crystallization of RNA is also sensitive to the presence of various metal cations, as illustrated in Fig. 3c–f.

4 Notes

1. DEPC inactivates ribonucleases by irreversible modification of catalytic residues. DEPC-treated water and solutions can be prepared as described, for instance, in [23]. We typically add 0.2–1.0 mL of fresh DEPC to 2 L of water, mix well, incubate the solution overnight at room temperature or at 37 °C, and autoclave the following day.
2. Solutions for studies of RNA must be prepared with RNase-free chemicals and DEPC-treated water. All glassware should be heated at 160 °C for at least 1 h to remove RNases. Plasticware must be RNase-free.
3. Make sure that the tubes withstand centrifugation in the presence of the phenol–chloroform mixture.
4. We routinely purify T7 RNA polymerase from *E. coli* using the plasmid described in ref. [25].
5. Ethanol precipitation of the transcription mixture at this step may lead to the loss of RNA due to the formation of partially insoluble precipitate.
6. Use an 8 % gel for RNA over 150 nt and a 12.5 % gel for RNA up to 80 nt.
7. To prevent DNA damage, minimize the exposure time and cover the bands that are not immediately cut with aluminum foil.
8. The column purification works best for well-folded RNAs. Some RNAs, especially long species, can be more efficiently eluted if 0.5 mL of either 10 % acetic acid or 0.5 M NaOH is injected into the column prior to elution with the step gradient. To ensure that the eluted RNA sample has a neutral pH, verify pH using litmus paper; adjust pH with acetic acid, if necessary.
9. Mosquito can set up both hanging and sitting drops.
10. For setting up crystallization at +4 °C, prechill the micro-reservoir block of Mosquito so that sample strips are placed onto the cold surface.

References

1. Mironov AS, Gusarov I, Rafikov R et al (2002) Sensing small molecules by nascent RNA: a mechanism to control transcription in bacteria. Cell 111:747–756
2. Nahvi A, Sudarsan N, Ebert MS et al (2002) Genetic control by a metabolite binding mRNA. Chem Biol 9:1043
3. Winkler W, Nahvi A, Breaker RR (2002) Thiamine derivatives bind messenger RNAs directly to regulate bacterial gene expression. Nature 419:952–956
4. Sudarsan N, Barrick JE, Breaker RR (2003) Metabolite-binding RNA domains are present in the genes of eukaryotes. RNA 9: 644–647
5. Breaker RR (2011) Prospects for riboswitch discovery and analysis. Mol Cell 43:867–879
6. Nudler E, Mironov AS (2004) The riboswitch control of bacterial metabolism. Trends Biochem Sci 29:11–17
7. Winkler WC, Breaker RR (2005) Regulation of bacterial gene expression by riboswitches. Annu Rev Microbiol 59:487–517
8. Serganov A, Nudler E (2013) A decade of riboswitches. Cell 152:17–24
9. Serganov A, Patel DJ (2012) Metabolite recognition principles and molecular mechanisms underlying riboswitch function. Annu Rev Biophys 41:343–370
10. Milligan JF, Groebe DR, Witherell GW et al (1987) Oligoribonucleotide synthesis using T7 RNA polymerase and synthetic DNA templates. Nucleic Acids Res 15:8783–8798
11. Pikovskaya O, Serganov AA, Polonskaia A et al (2009) Preparation and crystallization of riboswitch-ligand complexes. Methods Mol Biol 540:115–128
12. Peselis A, Serganov A (2012) Structural insights into ligand binding and gene expression control by an adenosylcobalamin riboswitch. Nat Struct Mol Biol 19:1182–1184
13. Zuker M (2003) Mfold web server for nucleic acid folding and hybridization prediction. Nucleic Acids Res 31:3406–3415
14. Martin CT, Muller DK, Coleman JE (1988) Processivity in early stages of transcription by T7 RNA polymerase. Biochemistry 27:3966–3974
15. Rong M, Durbin RK, McAllister WT (1998) Template strand switching by T7 RNA polymerase. J Biol Chem 273:10253–10260
16. Pleiss JA, Derrick ML, Uhlenbeck OC (1998) T7 RNA polymerase produces 5′ end heterogeneity during in vitro transcription from certain templates. RNA 4:1313–1317
17. Helm M, Brule H, Giege R et al (1999) More mistakes by T7 RNA polymerase at the 5′ ends of in vitro-transcribed RNAs. RNA 5:618–621
18. Nacheva GA, Berzal-Herranz A (2003) Preventing nondesired RNA-primed RNA extension catalyzed by T7 RNA polymerase. Eur J Biochem 270:1458–1465
19. Price SR, Ito N, Oubridge C et al (1995) Crystallization of RNA-protein complexes. I. Methods for the large-scale preparation of RNA suitable for crystallographic studies. J Mol Biol 249:398–408
20. Martick M, Scott WG (2006) Tertiary contacts distant from the active site prime a ribozyme for catalysis. Cell 126:309–320
21. Ferre-D'Amare AR, Zhou K, Doudna JA (1998) Crystal structure of a hepatitis delta virus ribozyme. Nature 395:567–574
22. Serganov A, Rak A, Garber M et al (1997) Ribosomal protein S15 from Thermus thermophilus-cloning, sequencing, overexpression of the gene and RNA-binding properties of the protein. Eur J Biochem 246:291–300
23. Xiong AS, Yao QH, Peng RH et al (2006) PCR-based accurate synthesis of long DNA sequences. Nat Protoc 1:791–797
24. Walker SC, Avis JM, Conn GL (2003) General plasmids for producing RNA in vitro transcripts with homogeneous ends. Nucleic Acids Res 31:e82
25. Studier FW, Rosenberg AH, Dunn JJ et al (1990) Use of T7 RNA polymerase to direct expression of cloned genes. Methods Enzymol 185:60–89

Chapter 4

In Vitro/In Vivo Production of tRNA for X-Ray Studies

Clément Dégut, Alexandre Monod, Franck Brachet, Thibaut Crépin, and Carine Tisné

Abstract

tRNAs occupy a central role in the cellular life, and they are involved in a broad range of biological processes that relies on their interaction with proteins and RNA. Crystallization and structure resolution of tRNA or/and tRNA/partner complexes can yield in valuable information on structural organizations of key elements of cellular machinery. However, crystallization of RNA, is often challenging. Here we review two methods to produce and purify tRNA in quantity and quality to perform X-ray studies.

Key words In vitro transcription, In vivo production, tRNA purification, Crystallization

1 Introduction

tRNAs are universally conserved key components of the translation machinery, linking the genetic code with the amino acid sequence of proteins. tRNAs have a length of 73–90 nucleotides (nt) and have a characteristic secondary structure in cloverleaf and a tertiary structure in the so-called "L-shape" (Fig. 1). tRNA represents one of the most abundant cellular transcripts and amount up to 15 % of the total cellular RNAs. They belong to the most stable RNAs in a cell [1] with half-lives of hours to days. Folding and stability of tRNA molecules is further controlled by numerous posttranscriptional modifications that cluster primarily in the anticodon-, the D- and T-arms [2]. Since structural and functional integrity of tRNA molecules is crucial for proper cellular functioning, many control pathways evolved, which recognize and degrade misfolded or hypomodified tRNA molecules rapidly [3].

2014 celebrates respectively the 40th anniversary of the first high resolution tRNA crystal structure [4, 5] and the 25th anniversary of the first crystal structure of a complex between an aminoacyl-tRNA synthetase and its cognate tRNA [6]. Forty years later, they are still matter of research in structural biology since advances in biochemistry, crystallization, collection and treatment of X-ray data

Eric Ennifar (ed.), *Nucleic Acid Crystallography: Methods and Protocols*, Methods in Molecular Biology, vol. 1320,
DOI 10.1007/978-1-4939-2763-0_4, © Springer Science+Business Media New York 2016

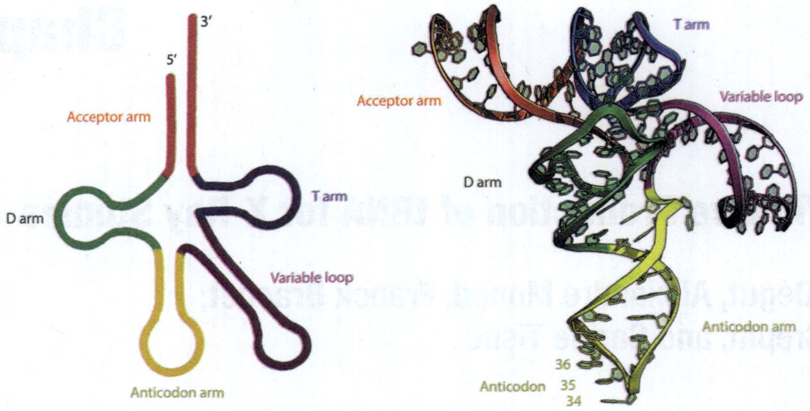

Fig. 1 Secondary and tertiary structures of tRNASec (PDB ID 3W3S) showing the cloverleaf and the "L-shape" organization of tRNA

allow to tackle more complex biological assembly involving, for instance, tRNA in the ribosome [7–10], or in complex with enzymes involved in the processing and/or maturation of tRNA. Moreover, newly discovered functions of tRNA have boosted new structural studies with tRNA molecules. For instance, tRNAs have been shown to act as stress-sensors by regulating numerous metabolic and cellular processes, in both prokaryotes and eukaryotes. For example, in *E. coli*, uncharged tRNAs act as signaling molecules under nutritional stress, activating the stringent response pathway and thereby promoting cell survival [11]. A similar function in gene expression has been described in yeast and human cells [12]. Upon binding to T-box elements located in the 5′ UTR of cistronic bacterial mRNAs, tRNAs are capable of regulating aminoacyl-tRNA synthetase genes or genes involved in amino acid biosynthesis and uptake [13, 14]. Besides, tRNAs can also serve as primers for the replication of viruses, participate in cell wall synthesis, and in the biosynthesis of certain porphyrins and amino acids [15].

These newly discovered functions involve newly discovered partners for tRNA and make that tRNA and/or tRNA co-crystallization is still relevant nowadays. In this chapter, we detail protocols used to produce tRNA by in vivo or in vitro strategies ready for crystallization assay and X-ray studies [16–19]. We will focus on the production of *B. subtilis* $tRNA^{Ser}$ (GGA) that is 92 nucleotides long since it has a large variable loop. This tRNA is under study in one of our group as the substrate of a modifying enzyme.

2 Materials

2.1 In Vitro Production: Preparation of T7 RNA Polymerase

1. pAR1219 plasmid (Sigma-Aldrich, Saint Louis, MO, USA).
2. Electro-competent *Escherichia coli* cells BL21(DE3).
3. 5 ml Ni-NTA HiTrap column (GE Healthcare, Pittsburgh, PA, USA).

4. Äkta purifier (GE Healthcare, Pittsburgh, PA, USA) or similar chromatographic system.
5. 100 W sonicator.
6. LB-agar plates.
7. 2xTY culture media.
8. Luria–Bertani (LB) culture media.
9. Ampicillin.
10. Isopropyl β-D-thiogalactopyranoside (IPTG) 200 mM.
11. Buffer A: 50 mM phosphate buffer pH 8.0, 300 mM NaCl, 10 % glycerol, 0.5 mM DTT, 20 mM imidazole pH 8.0, PMSF at 20 μg/mL.
12. Buffer B: Buffer A supplemented with 500 mM imidazole pH 8.0.
13. Buffer C: 20 mM phosphate buffer pH 7.7, 100 mM NaCl, 50 % glycerol, 1 mM DTT, 1 mM EDTA.
14. Lysis buffer: 20 mM Tris–HCl pH 8.0, 50 mM Na Phosphate pH 8.0, 20 mM NaCl, lysozyme 100 mg/mL, phenylmethanesulfonylfluoride (PMSF) 0.1 M, and optionally: benzamidine 0.1 M, pepstatin 1 μg/mL, leupeptin 2 μg/mL.
15. Dialysis sacks or cassette, MWCO <50 kDa.
16. MiniProtean® protein gel electrophoresis system (Bio-Rad, Hercules, CA, USA).
17. SDS-PAGE gel (10 %) for MiniProtean®.

2.2 In Vitro Production: Preparation of DNA Template for In Vitro Transcription

1. Oligonucleotides for the PCR.
2. PCR tubes.
3. *Pyrococcus furiosus* DNA polymerase Pfu (30 U/μL) in 10× Pfu buffer.
4. 10× Pfu buffer.
5. Orange G loading dye 10×: 0.4 % (w/v) orange G, 30 % (v/v) glycerol.
6. Thermocycler.
7. Agarose.
8. SYBR® Safe (Life Technologies, Carlsbad, CA, USA).
9. 10× TAE Buffer: 48.4 g of Tris-base, 11.4 mL of glacial acetic acid, 3.7 g of EDTA disodium salt, complete to 1 L with water.
10. Agarose gel electrophoresis system.
11. Stock solution containing a mix of each dNTP at 10 mM in water.

2.3 In Vitro Production: T7 Transcription

1. $MgCl_2$ 1 M.
2. Polyethylene glycol (PEG) 4000.
3. Stock solution of rNTP mix at 100 mM.

4. In vitro transcription buffer concentrated ten-times (T10 buffer): 400 mM Tris–HCl pH 8.0, 10 mM spermidine, 50 mM DTT, 1 % Triton X-100.
5. EDTA, pH 8.0 0.5 M.
6. Sodium dodecyl sulfate (SDS) 20 %.
7. Tris–HCl, pH 8.8 1 M.
8. Tris–HCl, pH 6.8 1 M.
9. Blue loading buffer 5×: 300 mM Tris–HCl pH 6.8, 1 % (w/v) SDS, 1 % (w/v) bromophenol blue.

2.4 Gel Purification

1. 10× Tris–Borate–EDTA buffer (TBE): 890 mM Tris base, 890 mM boric acid, 20 mM EDTA pH 8.4.
2. Urea, powder.
3. 10 % (w:v) ammonium persulfate (APS) solution.
4. *N,N,N′,N′*-tetramethylethylenediamine (TEMED), ≥99 %.
5. RNA loading dye: 1× TBE, 8 M urea, 0.01 % xylene cyanol and 0.01 % bromophenol blue.
6. 40 % acrylamide–bis-acrylamide 19:1 solution.
7. High-voltage power supply for gel migration system.
8. DNA sequencing-like gel migration system.
9. DNA sequencing glass plates (420 × 330 × 3.5 mm).
10. Spacers (420 × 20 × 2 mm) for DNA sequencing glass plates.
11. Large comb for DNA sequencing gel (100 × 265 × 2 mm).
12. Large plastic wrap (width larger than 50 cm).
13. Silica plate for UV shadowing.
14. Cutter blade.
15. Rotation instrument (rotating wheel or similar).
16. Syringe with needle.
17. UV lamp.
18. 0.2 % Methylene blue staining solution.
19. Steel spoon.
20. Sodium acetate pH 5.2 3 M.
21. 0.2 μm RNase-free Stericup-GP Express PLUS, polyethersulfone, 150 mL (Millipore, Billerica, MA, USA).
22. RNase-free plastic spoon.
23. DEAE sepharose resin (GE Healthcare, Pittsburgh, PA, USA).
24. Small empty column with tubing.
25. Ethanol ≥99.8 %.
26. Large binder clips.
27. Sealing tape, 48 mm width.

2.5 In Vitro Production: tRNA Concentration and Folding

1. RNA storage buffer: 50 mM Tris–HCl pH 8.0, 100 mM KCl and 10 mM $MgCl_2$.
2. Amicon Ultra-15 concentrators, 10000 MWCO (Millipore, Billerica, MA, USA).
3. Vacuum concentrator.

2.6 In Vivo Production of tRNA

1. Oligonucleotides encompassing the tRNA gene.
2. T4 polynucleotide kinase 3′phosphatase free (10 U/μL).
3. T4 DNA ligase (400 U/μL).
4. PstI-HF, high fidelity restriction endonuclease (20 U/μL).
5. EcoRI-HF, high fidelity restriction endonuclease (20 U/μL).
6. SmaI, restriction endonuclease (20 U/μL).
7. Buffer D: Tris–HCl pH 7.4 1 mM, Mg acetate 10 mM.
8. pBSTNAV plasmid (Addgene ID 45801) (*see* **Note 1**).
9. Electro-competent *E. coli* cells JM101TR strain.
10. Vaccum concentrator.
11. Ethanol, ≥99.5 %.
12. Ethanol, 70 % in H_2O.
13. NEB CutSmart buffer (New England Biolabs, Ipswitch, MA, USA).
14. Phenol.
15. NaCl 1 M.
16. NaCl 5 M.
17. Tris–HCl pH 8.0 2 M.

2.7 Chromatography Purification for In Vivo Production

1. Mono Q HR 10/10 column (GE Healthcare, Pittsburgh, PA, USA).
2. Phenyl Superose HR 10/10 column (GE Healthcare, Pittsburgh, PA, USA).
3. Äkta purifier (GE Healthcare, Pittsburgh, PA, USA) or similar chromatographic system.
4. Tris–HCl pH 8.0 1 M.
5. $MgCl_2$ 1 M.
6. NaCl 5 M.
7. Ammonium sulfate 3 M.
8. Buffer A: 50 mM Tris–HCl pH 8.0, 10 mM $MgCl_2$.
9. Buffer B: Buffer A supplemented with 1 M NaCl.
10. Buffer B2: 50 mM Tris–Cl pH 8.0, 10 mM $MgCl_2$, 1.7 M ammonium sulfate.
11. Hiprep 26/10 desalting column (GE healthcare, Pittsburgh, PA, USA).

12. RNA storage buffer: 50 mM Tris–HCl pH 8.0, 100 mM KCl and 10 mM $MgCl_2$.
13. Amicon Ultra-15 concentrators, 10000 MWCO (Millipore, Billerica, MA, USA).

2.8 NMR Spectroscopy

1. NMR Buffer: 10 mM KPO_4 pH 6.5, 50 mM KCl, 10 mM $MgCl_2$.
2. 3 mm NMR tube.
3. D_2O 99.9 %, 15 μL.
4. NMR spectrometer (600 MHz or equivalent).

2.9 Crystallization

1. Cristalquick, 96-well plates (Greiner Bio-One, Frickenhausen, Germany) or equivalent.
2. Natrix sparse-matrix kit (Hampton Research, Aliso Viejo, CA, USA) or Nucleix sparse-matrix kit (Qiagen, Venlo, The Netherlands).

3 Methods

It is important to always use talc-free gloves during sample preparation and manipulation as human skin is rich of RNases. For materials that cannot be exclusively used for RNA preparation or than are not single use, wash them when possible with 1 M NaOH and rinse extensively with bi-distilled water. When adjusting pH of an RNase-free solution, avoid inserting a pH meter electrode into the solution; instead prefer paper pH-indicator strips. Store RNase-free solution at low temperature or frozen when possible, and keep them sterile. The use of bi-distilled water (18.2 MΩ) is sufficient to assure RNase-free condition, provided that the water purification system is well-maintained.

3.1 In Vitro Production: Preparation of T7 RNA Polymerase

1. Transform by electroporation BL21(DE3) competent cells with the pAR1219 plasmid expressing the T7 RNA polymerase (98 kDa). Add 1 mL of LB medium and incubate the cells at 37 °C for 15 min. Spread 10–100 μL of this solution on a LB-agar plate supplemented with 50 μg/mL of ampicillin. Incubate the plate at 37 °C overnight.
2. The next evening, start a pre-culture (5 mL culture in LB supplemented with 50 μg/mL of ampicillin) inoculated with one isolated colony of bacteria containing the pAR1219 plasmid. Incubate overnight at 37 °C.
3. Start a 1 L culture in 2xTY medium supplemented with ampicillin (50 μg/mL) with the necessary volume of the overnight culture to obtain a starting optical density at 665 nm (OD_{665}) around 0.05. Incubate the remaining culture at 37 °C with

vigorous shaking. When the OD_{665} reached 0.6, add 1 mM (final concentration) of IPTG to induce the expression of T7 RNA polymerase. Let the bacteria grow for 3–4 h.

4. Pellet the bacteria by centrifugation (6,000×*g* for 30 min).
5. Resuspend the pellet in the lysis buffer. Sonicate the cell suspension with ten short burst of 10 s (45 W) followed by intervals of 30 s for cooling. Keep the suspension on ice as much as possible. Remove cell debris by centrifugation at 4 °C, 30 min at 20,000×*g*.
6. Purify the T7 RNA polymerase at 4 °C using a chromatographic system: equilibrate the Ni-NTA column with 5 volumes of buffer A. Inject the supernatant at 2 mL/min, wash the column with buffer A until the baseline is recovered to be sure that all the flow-through is gone. Then, elute the T7 RNA polymerase using a 100 mL gradient (from 0 to 100 % of buffer B). Collect 2 mL fractions. Analyze the fractions on a SDS-PAGE gel (10 %). Pool the fractions that contain the T7 RNA polymerase and dialyze at 4 °C against 2 L of buffer C. The volume in the dialysis chamber will be roughly reduced by a factor 3.
7. Recover the solution and aliquot the T7 RNA polymerase by 1 mL in microtubes for long-term storage at −80 °C. Approximately, 7 mL of T7 RNA polymerase (15 mg/mL) can be obtained from 1 L of culture.

3.2 In Vitro Production: Preparation of DNA Template for In Vitro Transcription

1. DNA template for in vitro transcription of $tRNA^{Ser}$ was designed as described in Fig. 2, using two consecutive PCR (Polymerase Chain Reaction) (*see* **Note 1**). The construct must include the T7 promoter sequence upstream the tRNA gene (in bold in Fig. 2a). The first two oligonucleotides must cover all the sequence of the DNA template and should be designed so that they share complementary sequences between the 3′ of oligonucleotide 1 and the 5′ of oligonucleotide 2 (Fig. 2b). We recommend at least 20 base pairs between oligonucleotides 1 and 2. The two other oligonucleotides are primers for classic PCR that allows the amplification of the DNA template.
2. For the first PCR (Fig. 2b), mix 40 μL of 10× Pfu buffer, 8 μL of the dNTP mix, 1 μL of oligonucleotides 1 and 2 (previously dissolved at 100 μM in water), 8 μL of Pfu DNA polymerase and complete the volume to 400 μL with water. Aliquot the reaction in PCR tubes (50 μL each). On a thermocycler, start with the first step of denaturation by heating to 95 °C for 1.5 min; then repeat two times the following cycle: 35 s of denaturation at 95 °C, 35 s of annealing at 55 °C, and 30 s of elongation at 72 °C. Carry on the reaction by a final step of elongation at 72 °C for 5 min and then hold the samples at 4 °C.

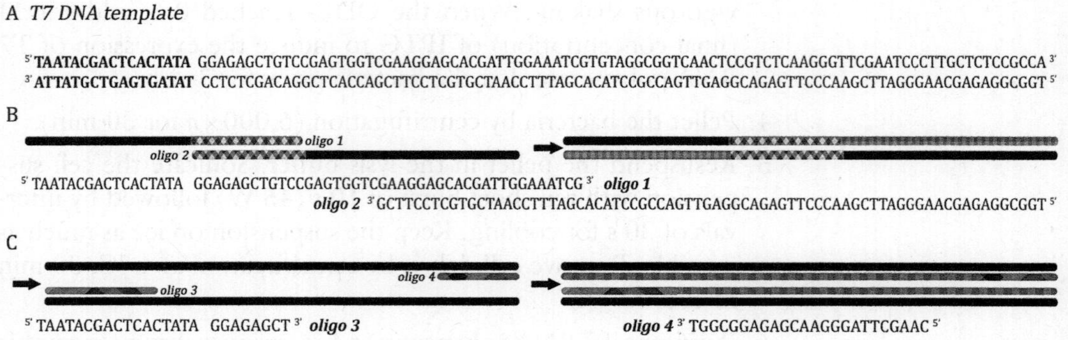

Fig. 2 Design of the DNA template for in vitro T7 transcription. The gene encoding *B. Subtilis* tRNA^{Ser} (GGA) has been assembled by two consecutive PCR. (**a**) DNA template of *B. Subtilis* tRNA^{Ser}. (**b**) Oligonucleotide 1 and 2 are needed during the first PCR to generate the double-stranded DNA template (**c**) Amplification of the DNA template by a second PCR using oligonucleotides 3 and 4

3. For the second PCR (Fig. 2c), mix 120 μL of 10× Pfu buffer, 24 μL of the dNTP mix, 12 μL of oligonucleotides 3 and 4 (previously dissolved at 100 μM in water), 1 μL of the PCR1 product, 24 μL of Pfu DNA polymerase and complete the volume of reaction to 1.2 mL with water. Aliquot the reaction in PCR tubes (50 μL each). Repeat 25 times the PCR cycle described in **step 2**. This PCR product can be used immediately as DNA template for in vitro T7 transcription without any further treatment. For long-term conservation, keep it at −20 °C.
4. Analyze the PCR product on a 2 % agarose gel stained with SYBR® Safe. Weight 0.5 g of agarose, add 2.5 mL of 10× TAE buffer, and complete to 25 mL with water. Melt the agarose powder until complete dissolution in a microwave at 20 % power. Let cool down the solution until you can hold the vessel with your bare hands and add 2 μL of SYBR® Safe. Pour the solution in an agarose gel mold. Put the comb and wait for gel solidification. Take 10 μL of your PCR reaction and add 5 μL of orange G loading dye. Run the gel at 100 V for 15 min in 1× TAE buffer. Orange G dye runs like a 50 base-pair DNA fragment, so the PCR product (109 base pairs) should be slightly above.

3.3 In Vitro Production: T7 Transcription

The optimal in vitro transcription conditions depend on the concentrations of the DNA template, rNTP and/or $MgCl_2$ concentrations (Fig. 3). These concentrations should be adjusted to optimize the yield of the RNA of interest. It is recommended to test small-scale reactions (currently 100 μL) before large-scale production of RNA (5–20 mL). In practice, we usually vary the concentration of $MgCl_2$ in small-scale reactions (Table 1) and analyzed the yield of transcript RNA by SDS-PAGE (*see* **Note 2** and Fig. 3). The conditions described in Table 1 are good starting point. Concentrations

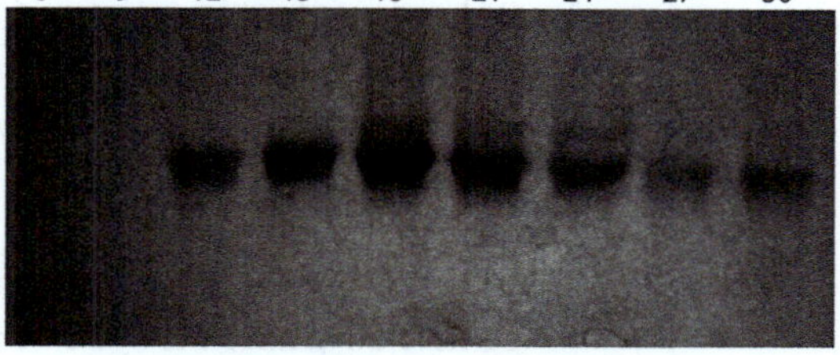

Fig. 3 Sensitivity to the magnesium concentration of the RNA production by in vitro T7 transcription. SDS-PAGE gel at 14 % revealed by UV-shadowing showing the analysis of small-scale assays of Table 1 testing the production of $tRNA^{Ser}$ by T7 transcription

Table 1
Examples of small-scale assays used to define the optimal $MgCl_2$ concentration

Assays	1	2	3	4	5	6	7	8	9
[$MgCl_2$] (mM)	6	9	12	15	18	21	24	27	30
DNA template	10 μL	10 μL	10 μL	10 μL	10 μL	10 μL	10 μL	10 μL	10 μL
T10 buffer	10 μL	10 μL	10 μL	10 μL	10 μL	10 μL	10 μL	10 μL	10 μL
rATP (100 mM)	4 μL	4 μL	4 μL	4 μL	4 μL	4 μL	4 μL	4 μL	4 μL
rUTP (100 mM)	4 μL	4 μL	4 μL	4 μL	4 μL	4 μL	4 μL	4 μL	4 μL
rGTP (100 mM)	4 μL	4 μL	4 μL	4 μL	4 μL	4 μL	4 μL	4 μL	4 μL
rCTP (100 mM)	4 μL	4 μL	4 μL	4 μL	4 μL	4 μL	4 μL	4 μL	4 μL
$MgCl_2$ (300 mM)	2 μL	3 μL	4 μL	5 μL	6 μL	7 μL	8 μL	9 μL	10 μL
H_2O	58 μL	57 μL	56 μL	55 μL	54 μL	53 μL	52 μL	51 μL	52 μL
T7 RNA polymerase (1/10)	4 μL	4 μL	4 μL	4 μL	4 μL	4 μL	4 μL	4 μL	4 μL

of T7 RNA polymerase and DNA template can also be further optimized. For small RNA (<50 nucleotides), the addition of 5 % (w/v) final of PEG 4000 should be tested.

1. To investigate the impact of $MgCl_2$ concentration on the yield of RNA, the mixtures detailed in Table 1 should be prepared in microtubes. The T7 RNA polymerase is used diluted tenfold compared to the stock obtained in Subheading 3.1, **step 7**.
2. Run the test reactions for 3–4 h at 37 °C. The appearance of a white precipitate due to magnesium pyrophosphate is most of the time visible and can be used to monitor the transcription efficiency.
3. Store at −20 °C or analyze immediately by SDS-PAGE. Alternatively, the reaction can be stopped by adding 10 μL of 0.5 M EDTA pH 8.0.

4. To analyze the products of transcription, prepare a SDS-PAGE gel using the same materials as those used for protein SDS-PAGE analysis. The percentage of gel depends on the length of the expected product, i.e., a 14 % gel for a 92 nucleotides tRNA. For 2 gels running back to back, mix 3.5 mL of acrylamide–bis-acrylamide 19:1, 3.75 mL Tris–HCl pH 8.8, 50 μL SDS 20 % and adjust the volume to 10 mL with water. Add 140 μL of APS 10 % and 10 μL of TEMED just before placing the gel between the glass plates. Leave a sufficient space at the top of the glass plates to pour then the stacking gel. For the stacking gel, mix 0.5 mL of acrylamide–bis-acrylamide 19:1, 630 μL of Tris–HCl pH 6.8, 50 μL of SDS 20 % and complete to 5 mL with water. Add just before pouring, 100 μL of APS 10 % and 5 μL of TEMED. Put the comb immediately. The stacking gel should take one third of the total volume of the gel.
5. Take an aliquot of each small-scale reaction (typically 5 μL) and add 5 μL of blue loading buffer. Do not heat the samples before loading them on the gel. Load your samples on the wells and run the gel for 1 h at 24 mA.
6. The visualization is then done by UV-shadowing. Carefully remove one of the glass plate, leaving the gel on the other one. Put a clear plastic wrap over the gel and pull carefully the gel away from the second glass plate. Put the plastic wrap bearing the gel on a piece of silica plate and then illuminate the gel with a portable UV lamp (254 nm). You should see clear shadows that correspond to the RNA in the gel. The gel corresponding to the assays described in Table 1 and revealed by UV-shadowing is shown in Fig. 3. The best yield in our case is obtained for a concentration of $MgCl_2$ around 18 mM. This concentration is thus used for large-scale preparation of this tRNA. The yield of RNA is very sensitive to the concentration of $MgCl_2$. We recommend carrying out these small-scale assays each time that a new preparation of DNA template is used, even if the protocol of preparation is the same.

3.4 *In Vitro* Production: tRNA Purification Using Denaturing Polyacrylamide Gel

For large-scale in vitro transcription reaction (more than 5 mL), sequencing-like gel migration system can be adapted to purify the RNA transcripts. It is often the best method to obtain a homogeneous sample of tRNA ready for crystallization assay. The problem with in vitro transcription is that it generates $n-1$ and/or $n+1$ product in addition to the desired RNA. These species ($n-1$, $n+1$) can be a bottleneck to get diffracting crystals (*see* **Note 3**).

1. One volume of RNA loading buffer is added to one volume of transcription reaction. For really large transcription (up to 25 mL), in order to minimize the loading volume, a 8 M urea concentration can be possibly obtained by adding the

corresponding urea mass as a powder to the transcription reaction and a 50-fold concentrated xylene cyanol and bromophenol blue solution.

2. The DNA sequencing gel glass plates and spacers are washed and assembled using clamps and tape (at the bottom), in RNase-free conditions. To prepare 300 mL of 12 % polyacrylamide and 8 M urea gel, dissolve 144 g urea in 30 mL 10× TBE and 90 mL 19:1 polyacrylamide–bis-acrylamide. Add just before pouring, 5 mL of APS 10 % and 250 μL of TEMED. Put the large comb for DNA sequencing gel immediately before the gel has polymerized. After polymerization, the well is washed using a syringe with a needle and 1× TBE buffer, and the transcription solution containing the RNA loading buffer is loaded on the gel. It is recommended to check the in vitro transcription efficiency on small 8 M urea gel or SDS-PAGE gel before large-scale assays. The appropriate percentage of 19:1 polyacrylamide gel is determined according to the size of the RNA molecule (Table 2). The gel is pre-run for 2 h in 1× TBE running buffer to reach an optimal migration temperature. Set the power to 65 W for one large gel and 80 W for two gels running together. The well is washed with running buffer before loading the transcription reaction. Syringe and needle are used for these two steps. The migration runs for 6–8 h depending on the percentage of polyacrylamide gel.

3. There are two options to visualize the band corresponding to the tRNA molecule:
   (a) Using the UV-shadowing approach. The gel is removed from the system, protected with a thin plastic wrap, placed on a silica plate and exposed to UV (wavelength close to 254 nm) using a short-wave UV lamp. Silica reflects the UV while the RNA bases absorb UV. A shadow corresponding to the tRNA band can thus be observed on the silica.
   (b) Using methylene blue. The gel is removed from the system. A small band is cut along the gel, stained with a 0.2 %

Table 2
Percentage of 19:1 polyacrylamide–bis-acrylamide gel according to the size of the RNA

RNA molecule (nucleotides)	19:1 Acrylamide (%)	10 mL gel (mL)	50 mL gel (mL)	300 mL gel (mL)
Less than 50	15	3.75	18.75	112.5
50	12	3	15	90
100	10	2.5	12.5	75
More than 100	8	2	10	60

For each percentage, the volumes of acrylamide–bis-acrylamide 19:1 have been indicated to prepare a 10 mL, 50 mL, or 300 mL gel solution, respectively methylene blue solution for few seconds and destained with water. In that case, a small quantity of tRNA is lost, compared to the first option.

4. The band corresponding to the tRNA is excised with a blade and placed into a 50 mL Falcon tube. The gel is then crushed with an RNase-free steel spoon. tRNA is eluted by diffusion with 40 mL of 0.3 M sodium acetate pH 5.2 on the rotating wheel at 4 °C for several hours. The time required depends on the size of the RNA, the acrylamide gel concentration and the size of the crushed gel particles. The eluate is filtered on a 0.2 μm RNase-free Stericup-GP Express PLUS. Acrylamide crumbs are recovered with an RNase-free plastic spoon and put back in the same 50 mL Falcon tube for a second elution. This step can be repeated once.
5. The sample (in 0.3 M sodium acetate pH 5.2) is concentrated using DEAE sepharose resin and a small column. A small column apparatus was linked by tubing to a 50 mL syringe that is used as a reservoir. Before use, the homemade system is cleaned with 1 M NaOH and rinsed extensively with water. 1 mL of DEAE sepharose resin is packed into the column. The resin is equilibrated with 0.3 M sodium acetate pH 5.2. The sample is then loaded and the system is washed with 10 mL of 0.3 M sodium acetate pH 5.2. The tRNA is finally eluted with 7 mL of 3 M sodium acetate pH 5.2 and ethanol-precipitated by adding 21 mL of ethanol. The sample is then placed at −20 °C.

3.5 *In Vitro* Production: tRNA Concentration and Folding

1. After ethanol precipitation, the tRNA is centrifuged at 4 °C and 6,000 × *g*. The pellet is then washed three times with 70 % ethanol, dried under vacuum, and dissolved in 1 mL of pure water for 1 mL of transcription reaction. For the tRNA folding, the sample is incubated for 1 min at 85 °C and slowly (i.e., 45 min) cooled to room temperature by switching off the heating of water bath. The tRNA can be further concentrated, and buffer exchanged by using an Amicon centrifugal concentration unit. The sample can be stored in RNA storage buffer. Typically, a concentration of 10 mg/mL is a good stock concentration to conduct crystallization assays with this tRNA. Adding 10 mM $MgCl_2$ in the final buffer can greatly improve tRNA folding and is sometimes necessary to stabilize the tertiary interactions between the D- and T-loops. Keep your sample at −20 °C.
2. The folding of tRNA can be analyzed by NMR spectroscopy (*see* **Note 4** and Fig. 4). Put the tRNA in the NMR buffer at a concentration around 0.2 mM in a 3 mm NMR tube. In this kind of tube, only 150 μL of sample is needed. Put 135 μL of tRNA and 15 μL of D_2O. Record a 1D proton spectrum using a watergate sequence to suppress the water signal [20]. The imino protons are carried by G and U nucleotides and resonate

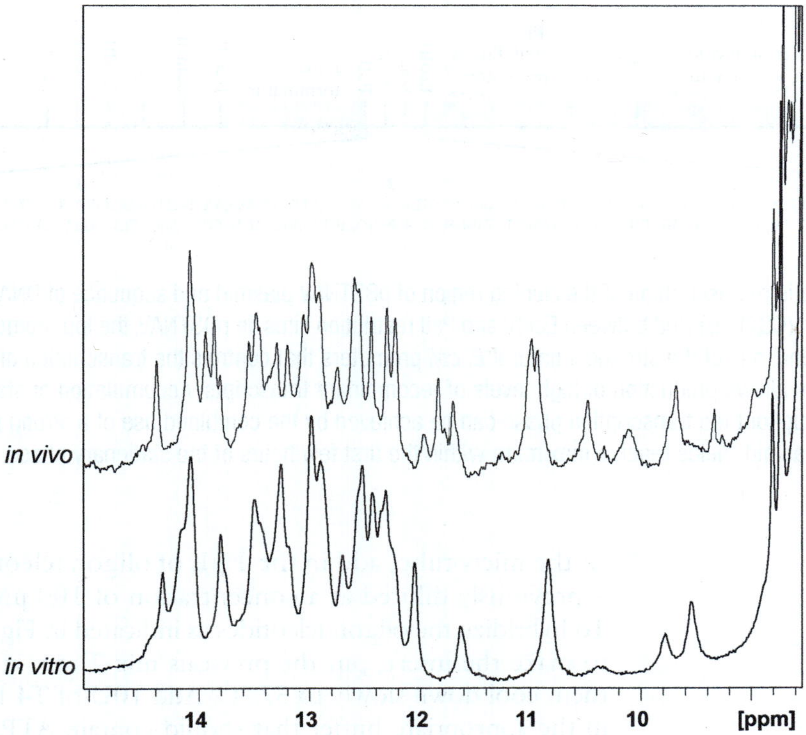

Fig. 4 1D ^{1}H NMR spectra of tRNA^{Ser} produced by in vitro transcription (*lower spectrum*) and in vivo in *E. coli* (*top spectrum*). The spectra were recorded at 35 °C on a Bruker DRX 600 MHz NMR spectrometer. The samples were put in a 10 mM KPO_4 pH 6.5 with 50 mM KCl and 10 mM $MgCl_2$

between 10 and 15 ppm. They constitute the signature of base-pair within a folded RNA. The NMR spectrum of tRNA^{Ser} imino protons produced in vitro is consistent with one molecular species and showed about 25 peaks (uncertainty due to overlap), which is consistent with the number of base-pairs expected in the 3D structure of this tRNA. These data support folding of the tRNA into its correct tertiary structure.

3.6 In Vivo Production: Cloning a tRNA Gene in the pBSTNAV Vector

tRNA can also be easily produced in vivo in *E. coli* using the pBSTNAV vector [16, 17, 19, 21]. The cloning procedure of the tRNA^{ser} gene in this plasmid and the expression and purification of tRNA^{Ser} are described in the following paragraphs.

1. Phosphorylate oligonucleotides 2, 3, 5 (Fig. 5). In a microtube, put 1 μL of each oligonucleotide previously diluted at a concentration of 100 μM in water. Add 10 U of T4 polynucleotide kinase 3′-phosphatase free and 1.5 μL of the buffer recommended by the manufacturer. Complete with water to 15 μL (final volume of the reaction). Incubate at 37 °C for 30 min. Leave 10 min at 65 °C to inactivate the kinase.

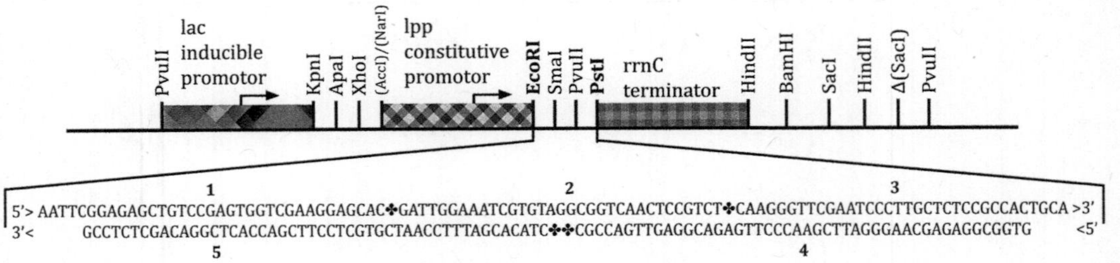

Fig. 5 Schematic representation of the cloning region of pBSTNAV plasmid and sequence of DNA coding for *B. subtilis* tRNA^{Ser} (GGA) inserted between EcoRI and PstI restriction sites. In pBSTNAV, the lpp promoter upstream of the tRNA gene, one of the strongest natural *E. coli* promoters that controls the transcription of the mRNA of the lipoprotein, allows production of high levels of recombinant transcripts. Accumulation of stable recombinant RNA throughout the transcription phase can be achieved by the combined use of a strong promoter and a multicopy plasmid. Yields reach a maximum within the first few hours of the stationary phase

2. In the microtube, add in the 1 μL of oligonucleotides 1 and 4 previously diluted at a concentration of 100 μM in water. To hybridize the oligonucleotides as indicated in Fig. 4 in order to make the insert, put the previous mix 2 min at 95 °C, let them cool down slowly to 37 °C. Add 10U of T4 DNA ligase in the appropriate buffer that should contain ATP. Complete with water to a volume of reaction of 30 μL. Incubate for 1 h and 30 min at 37 °C.
3. Preparation of the pBSTNAV plasmid. To linearize the pBSTNAV plasmid, digest it with PstI and EcoRI restriction enzymes. For 10 μg of plasmid, add 3 μL of PstI-HF and EcoRI-HF, 10 μL of NEB CutSmart buffer and complete the volume of the reaction with water to reach a volume of 100 μL. Incubate 1 h30 at 37 °C. Add 250 μL of ethanol and 10 μL of NaCl 5 M and pellet plasmid by centrifuging the mix for 15 min at 4 °C and at 7,000 × *g*. Remove carefully the ethanol and dry the sample in a vacuum concentrator. Dissolve the linearized plasmid in 30 μL of water.
4. To insert and ligate the oligonucleotide containing the tRNA^{Ser} gene into the linearized pBSTNAV plasmid, mix 2 μL of the linearized plasmid, 15 μL of the insert, 4 μL of the T4 DNA ligase, 3 μL of the manufacturer recommended T4 ligase buffer (X10), and complete the volume to 30 μL. Incubate one night at 14 °C. Put the mix for 2 min at 95 °C to inactivate the ligase.
5. Counter-selection by SmaI (*see* **Note 5**). After the ligation step, add 100 μL ethanol and 3 μL NaCl (5 M) and recover the pellet DNA by centrifuging for 15 min at 4 °C and 7,000 × *g*. Redissolve the plasmid in 10 μL of water. Add 3.5 μL of SmaI and 1.5 μL of NEB CutSmart buffer. Incubate at 20 °C for 1 h 30 min. Then, add 50 μL of ethanol and 1 μL NaCl 5 M, centrifuge for 15 min at 4 °C and 7,000 × *g*. Carefully remove

the supernatant and wash by ethanol 70 % (add 50 μL of cold ethanol 70 % mix and centrifuge for 15 min at 4 °C and 7,000 × *g*). After the wash, remove as much as possible ethanol and dry the sample in a vacuum concentrator. Dissolve DNA in 15 μL of water and store at −20 °C.

3.7 In Vivo Production: Expression in E. coli and Total tRNA Extraction

1. Transform *E. coli* electro-competent cells (JM101TR strain) by electroporation. The bacteria are grown overnight at 37 °C in LB medium supplemented with 50 μg/mL of ampicillin (*see* **Note 6**).
2. Recover the bacteria by centrifugation for 30 min at 6,000 × *g*.
3. Resuspend the bacteria in 10 mL of buffer D. Add 10 mL of phenol saturated in the same buffer.
4. Mix slowly 1 h at room temperature. Centrifuge for 30 min, 8,000 × *g* at 15 °C. Carefully recover the soluble phase.
5. Do a counter-extraction with 2 mL of buffer d. Recover again the soluble phase and add 0.1 volume of NaCl 5 M and 2 volumes of ethanol to the soluble phase that contains total nucleic acids extracted from the bacteria. Centrifuge for 30 min, 8,000 × *g* at 4 °C.
6. Resuspend the pellet in 5 mL of NaCl 1 M (*see* **Note 7**). Centrifuge for 30 min at 4 °C and at 8,000 × *g*. Recover the supernatant and add 2.5 volume of ethanol. Centrifuge for 30 min, 8,000 × *g* at 4 °C. Discard the supernatant and resuspend the pellet in 2 mL of Tris–HCl pH 8.0 2 M and incubate the mix at 37 °C for 2 h to insure a full deacylation of the tRNA molecules (*see* **Note 8**).
7. Precipitate the total extract of tRNA by adding 2 volumes of ethanol. Centrifuge for 30 min, 8,000 × *g* at 4 °C. Redissolve the pellet in RNA storage buffer (or equivalent).

3.8 In Vivo Production: tRNA Purification Using Chromatography Columns

tRNAs produced in vivo in *E. coli* have the advantages to be correctly folded and matured by the enzymes of the host. Therefore, even if the purification by denaturing polyacrylamide gel is still an option, it could be advantageous to optimize a protocol of purification using anion-exchange and hydrophobic chromatography. This protocol is detailed below.

1. First, wash the ÄktaPurifier (or other FPLC purification system) and the columns with 1 M NaOH and rinse them extensively with bi-distilled water to work in RNase-free conditions. After the total tRNA extraction (*see* Subheading 3.7), load the tRNA sample on a Mono Q HR 10/10 column (8 mL) previously equilibrated with buffer A. Wash by several volume of column with buffer A. Then, separate the tRNAs using a gradient from 35 to 55 % of buffer B with a length of 240 mL and a flow of 4 mL/min. Collect the eluate putting 2 mL/fraction.

Noticeably, the gradient has to be optimized according to the size and the sequence of the tRNA to purify.

2. You then need to analyze the content of the fractions to identify the fractions containing the overexpressed tRNA. One simple experiment is to perform gel-shift assay using SDS-PAGE (Fig. 5). First, design a DNA that is complementary to the overexpressed tRNA in a region that is not well-conserved in the other *E. coli* tRNAs. In our case, we select a DNA complementary to the anticodon arm and the large variable loop (Fig. 6). Take 5 μL of each fraction to analyze and mix with 1 μL of the complementary DNA (previously dissolved in water at 100 μM). In a thermocycler, heat the samples at 95 °C for 2 min and cool down to 4 °C, The loading blue is added after this operation, the shift will not appear if it is present when samples are heated.
3. Load the samples on a 14 % SDS-PAGE gel (*see* Subheading 3.3, **step 4**). Visualize the RNA present in each fraction by UV-Shadowing. $tRNA^{Ser}$ can be easily discriminated from *E. coli* tRNAs by a shift of its species in presence of the complementary DNA (arrows in Fig. 5.). Indeed, the band of $tRNA^{Ser}$ has disappeared in the presence of the DNA and a band of higher molecular weight has appeared.
4. Pool the fractions that contain the tRNA of interest. The next step consists in further purifying the tRNA by hydrophobic chromatography. First, put in your tRNA sample the quantity of ammonium sulfate in powder necessary to have your sample in 1.7 M ammonium sulfate before loading it on the Phenyl

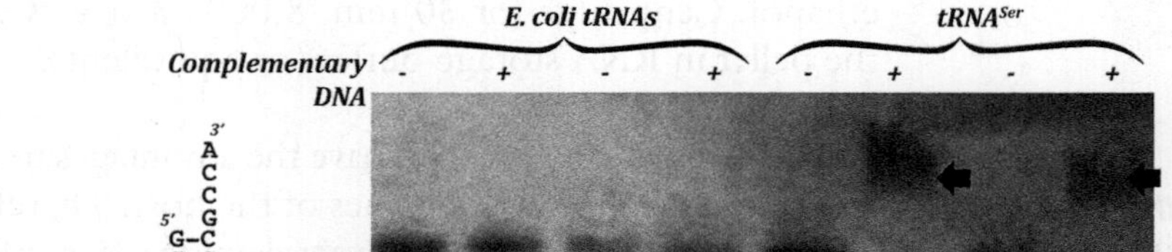

Fig. 6 Gel-shift assays (SDS-PAGE 14 %) realized on fractions obtained after the Mono Q column. This simple experiment allows to discriminate the tRNA of interest from those of *E. coli* using an hybridization test of a complementary DNA to tRNAs present in the different fractions

Superose. Equilibrate the Phenyl Superose with buffer B2 Then, separate the tRNAs using a gradient from 10 to 45 % of buffer A with a length of 240 mL and a flow of 0.7 mL/min. Collect 1 mL/fraction.

5. Analyze the fractions with a 14 % SDS-PAGE gel (*see* Subheading 3.8, **step 3**). The high concentration of ammonium sulfate (>1 M) prevents the good migration of RNA samples in SDS-PAGE. You should take 5 μL of each fraction and dilute them in 20 μL of water to avoid migration problem. Visualize the bands by UV-shadowing. Pool the fractions that contain the tRNA of interest (*see* **Note 9**).
6. The tRNA of interest has then to be desalted to get rid of the large amount of ammonium sulfate. This can be done by extensive dialysis in RNA storage buffer or by using a desalting column (Hiprep 26/10 desalting column, GE healthcare). Concentrate to the desired concentration using an Amicon centrifugal concentrator. Typically, a concentration of 10 mg/mL is a good storage concentration to conduct crystallization assays with this tRNA. Keep your sample at −20 °C.
7. Optionally, you can check the homogeneity of your tRNA sample and its folding by recording a 1D NMR spectrum (Fig. 4). The NMR spectrum of $tRNA^{Ser}$ produced in vivo compared to that produced in vitro is noticeably different.

3.9 tRNA Crystallization Free or in Complex

1. Prepare the tRNA sample as described in previously.
2. Automated initial crystallization screen can be performed using a robot using 96-well microplates. For initial screening, sparse matrix crystallization screens (Nucleix or Natrix) give good results. Several concentrations of tRNA should be tested (typically 3 mg/mL, 5 mg/mL and 10 mg/mL) (*see* **Note 11**). The robot mixes the same volume (0.7 μL in our case) of tRNA sample and of the reservoir solution. Screening of crystallization conditions is carried out with the sitting drop, vapor diffusion technique. Put the crystallization plate in an incubator set at a temperature between 4 and 25 °C. We usually tested 4 and 18 °C.
3. Check for crystals the next day, in 3 days, and then every week. Crystals may appear after several hours, after several months, or never. It is important to check for crystals on a regular basis as they can disappear. If precipitate is observed in most of the crystallization condition, then proceed to new assays but lower the concentration of tRNA.
4. When crystals are observed (Fig. 7), optimize crystallization conditions by varying the ratio between the volumes of the sample and that of the reservoir solution in the drop, the concentration of tRNA in the drop, the concentration of the precipitant or salts in the reservoir solution, the buffer, the pH,

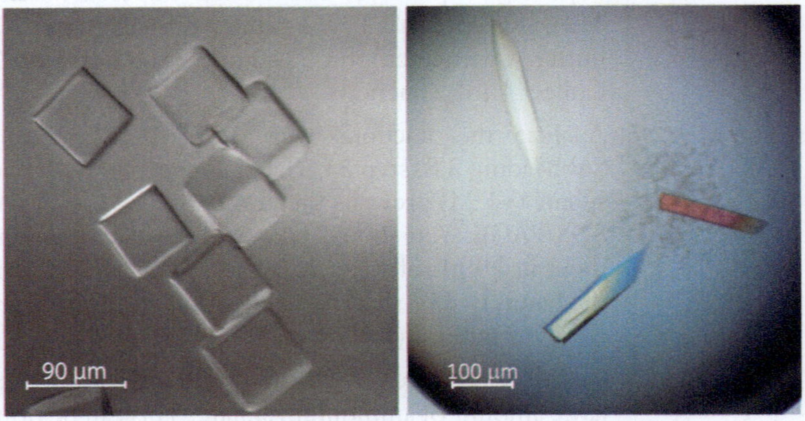

Fig. 7 Examples of crystals obtained for (**a**) tRNA^{Ser} produced in vivo and (**b**) tRNA^{Leu} produced in vitro and co-crystallized with LeuRS (*see* **Note 10**) [18]. The conditions of crystallization are 100 mM HEPES pH 7.5 and 1.5 M sulfate lithium for tRNA^{Ser} and 0.2 M sodium Nitrate, 20 % PEG 3350 for tRNA^{Leu} in complex with LeuRS

various additives, ... For tRNA^{Ser}, the diffraction quality of crystals was improved by modifying the precipitant and its concentration in 2 μL drops.

4 Notes

1. In vitro run-off transcription with T7 RNA polymerase [22] is commonly used to prepare milligram quantities of RNA required for crystallization studies. For efficient transcription, the promoter-region of the DNA template should be doubled-stranded and the first nucleotide in the RNA product should be a guanine. T7 RNA polymerase tends to add one or more random nucleotides (usually adenosine) at the 3′-end of the run-off transcript, leading to problems with 3′ homogeneity. For tRNA run-off transcription, the DNA template can be either a purified PCR product containing the tRNA gene and the T7 promoter or a linearized plasmid. In the case of tRNA^{Ser}, the higher yield of production was obtained for a DNA template that is a PCR product. We thus detail this protocol.
2. For the use of a linearized plasmid, the template plasmid must contain a restriction site incorporated at the 3′-end of the desired RNA sequence. For tRNA, the commonly used restriction enzyme is BstNI that allows to generate a CCA sequence at the 3′-end of the transcript tRNA [23]. The vector is usually a pUC18 or pUC19. When we used a PUC18 vector carrying the tRNA^{Ser} gene linearized with BstNI restriction enzyme, other large species of transcript products were observed. This feature reduced the quantity of tRNA transcript.

3. Although SDS-PAGE is commonly restricted to protein electrophoresis, this technique is well suited for RNA production analysis [16, 17, 19, 21]. Gel migration is much faster, bands are well-defined and the migration is not sensitive to salts present in the samples like in urea PAGE. Furthermore, these gels are not denaturing for RNA and thus hybridized RNA or RNA/DNA complexes can be detected if they are quite stable (i.e., if they involved a large number of base pairs, more than 10). We used this feature to easily detect the fractions containing the overexpressed tRNA during chromatographic steps of purification (Fig. 6).

4. Homogeneous samples are crucial for the success of crystallization. Crystals containing tRNA molecules are very often packed in such a way that two tRNA molecules form a pseudo-continuous helix with their 3′-ends [16, 17, 19, 21]. To facilitate such packing, it is important to ensure that 3′-end of tRNA are homogeneous. Unfortunately, run-off transcription with T7 RNA polymerase usually results in product RNA that are heterogeneous at its 3′-end. The gel purification of the product of in vitro transcription enables us to eliminate unincorporated NTP, products of aborted transcription, template DNA and most importantly serves to produce RNA with homogeneous ends. A nucleotide-level resolution on a preparative gel can be obtained. In case of heterogeneity at the 3′-end of the tRNA, we recommend to avoid overloading of the sample. Several bands can be observed and it could be interesting to cut all major bands to use them in crystallization trials, of course separately!

5. The refolding of the tRNA prior to crystallization assays is very important, especially for large molecules. The ability of magnesium to cut RNA at high temperatures has to be taken into account. Usually, the high temperature denaturation of RNA is performed without magnesium and it is added at a later step when the sample is cooled down at room temperature.

6. The counter selection by SmaI is essential to avoid false positive clones, i.e., clones transformed by the pBSTNAV vector that does not carry the insert sequence encompassing $tRNA^{Ser}$ gene.

7. Since high yields of recombinant tRNA are required, it is quite important that fast growth rates are achieved during expression. Expression in rich medium at 37 °C should thus be favored, as it will support the fastest growth rates. For large-scale expression, use 2xTY medium supplemented with 0.4 % (w/v) glucose in a fermentor equipped with pH and dissolved oxygen (pO2) control. Glucose contributes to increasing the growth rate, but requires real-time neutralization of acids produced by fermentation and hence must be done into a fermentor. Optimal aeration is also a key factor, most importantly at high

cell densities. For medium scale expression, 2xTY cultures (without glucose) in flat culture flasks (Fernbach) are a simpler yet quite efficient alternative. With the strong lpp promoter on a multicopy plasmid, stable recombinant tRNA will accumulate during all their transcription phase. Yield will therefore be maximum in the first few hours of the stationary phase. This leads to a very straightforward protocol: liter-sized culture flasks are typically inoculated at low density with freshly transformed bacteria and allowed to grow overnight (15 h) at 37 °C under vigorous shaking. Putting only 500 mL of 2xTY medium in liter-sized culture flasks allows a better oxygenation of cells and thus provides a simple way to improve yield.

8. Only the smallest nucleic acids like tRNAs and 5S RNA will be soluble in 1 M NaCl. This step is not obligatory but it enables to get rid of the largest nucleic acids like chromosomes for instance.
9. This step enables the nucleophilic deacylation of the 3′-end of tRNA.
10. 10–20 mg of pure tRNA per liter of culture can be routinely obtained. The yield and the purity will depend on the tRNA sequence and notably on the posttranscriptional modifications carried by the overexpressed tRNA. Complex modifications, notably in the anticodon, can be incorporated at a substoichiometric level [16]. This can be a source of heterogeneity preventing crystallization.
11. For the crystallogenesis, a parameter to consider in order to prepare the protein-tRNA sample, is the ratio between the two partners. Most of the time, a little excess of nucleic acids is required. For example, in the case of *E. coli* LeuRS-tRNAleu complex, the best crystals were obtained with a 1:1.2 ratio.

Acknowledgements

C. D. and C. T. thank Marjorie Catala for her help in optimizing in vitro T7 transcription on tRNA.

References

1. Grosjean H, De Henau S, Doi T, Yamane A, Ohtsuka E, Ikehara M, Beauchemin N, Nicoghosian K, Cedergren R (1987) The in vivo stability, maturation and aminoacylation of anticodon-substituted Escherichia coli initiator methionine tRNAs. Eur J Biochem 166: 325–332
2. Motorin Y, Helm M (2010) tRNA stabilization by modified nucleotides. Biochemistry 49:4934–4944
3. Phizicky EM, Hopper AK (2010) tRNA biology charges to the front. Genes Dev 24: 1832–1860
4. Robertus JD, Ladner JE, Finch JT, Rhodes D, Brown RS, Clark BF, Klug A (1974) Structure of yeast phenylalanine tRNA at 3 Å resolution. Nature 250:546–551
5. Suddath FL, Quingley GJ, McPherson A, Sneden D, Kim JJ, Kim SH, Rich A (1974) Three-dimensional structure of yeast

phenylalanine transfer RNA at 3.0 angstroms resolution. Nature 248:20–24

6. Rould MA, Perona JJ, Soll D, Steitz TA (1989) Structure of E. coli glutaminyl-tRNA synthetase complexed with tRNA(Gln) and ATP at 2.8 A resolution. Science 246:1135–1142
7. Selmer M, Dunham CM, Murphy FV 4th, Weixlbaumer A, Perty S, Kelley AC, Weir JR, Ramakrishnan V (2006) Structure of the 70S ribosome complexed with mRNA and tRNA. Science 313:1935–1942
8. Schmeing TM, Voorhees RM, Kelley AC, Gao YG, Murphy FV 4th, Weir JR, Ramakrishnan V (2009) The crystal structure of the ribosome bound to EF-Tu and aminoacyl-tRNA. Science 326:688–694
9. Jenner L, Demeshkina N, Yusupova G, Yusupov M (2010) Structural rearrangements of the ribosome at the tRNA proofreading step. Nat Struct Mol Biol 17:1072–1078
10. Demeshkina N, Jenner L, Yusupova G, Yusupov M (2010) Interactions of the ribosome with mRNA and tRNA. Curr Opin Struct Biol 20:325–332
11. Wegrzyn G, Wegrzyn A (2008) Is tRNA only a translation factor or also a regulator of other processes? J Appl Genet 49:115–122
12. Li Y, Zhou H (2009) tRNAs as regulators in gene expression. Sci China C Life Sci 52: 245–252
13. Green NJ, Grundy FJ, Henkin TM (2010) The T box mechanism: tRNA as a regulatory molecule. FEBS Lett 584:318–324
14. Zhang J, Ferre-D'Amare AR (2013) Co-crystal structure of a T-box riboswitch stem I domain in complex with its cognate tRNA. Nature 500:363–366
15. Giege R (2008) Toward a more complete view of tRNA biology. Nat Struct Mol Biol 15: 1007–1014
16. Tisne C, Rigourd M, Marquet R, Ehresmann C, Dardel F (2000) NMR and biochemical characterization of recombinant human tRNA(Lys)3 expressed in Escherichia coli: identification of posttranscriptional nucleotide modifications required for efficient initiation of HIV-1 reverse transcription. RNA 6: 1403–1412
17. Barraud P, Schmitt E, Mechulam Y, Dardel F, Tisne C (2008) A unique conformation of the anticodon stem-loop is associated with the capacity of tRNAfMet to initiate protein synthesis. Nucleic Acids Res 36:4894–4901
18. Palencia A, Crépin T, Vu MT, Lincecum TL Jr, Martinis SA, Cusack S (2012) Structural dynamics of the aminoacylation and proofreading functional cycle of bacterial leucyl-tRNA synthetase. Nat Struct Mol Biol 19: 677–684
19. Ponchon L, Catala M, Seijo B, El Khouri M, Dardel F, Nonin-Lecomte S, Tisné C (2013) Co-expression of RNA-protein complexes in Escherichia coli and applications to RNA biology. Nucleic Acids Res 41:e150
20. Piotto M, Saudek V, Sklenar V (1992) Gradient-tailored excitation for single-quantum NMR spectroscopy of aqueous solutions. J Biomol NMR 2:661–665
21. Meinnel T, Mechulam Y, Fayat G (1988) Fast purification of a functional elongator tRNAMet expressed from a synthetic gene in vivo. Nucleic Acids Res 16:8095–8096
22. Studier FW, Moffatt BA (1986) Use of bacteriophage T7 RNA polymerase to direct selective high-level expression of cloned genes. J Mol Biol 189:113–130
23. Sampson JR, Uhlenbeck OC (1988) Biochemical and physical characterization of an unmodified yeast phenylalanine transfer RNA transcribed in vitro. Proc Natl Acad Sci U S A 85:1033–1037

Chapter 5

Polyacrylamide Gel Electrophoresis for Purification of Large Amounts of RNA

Mélanie Meyer and Benoît Masquida

Abstract

Polyacrylamide gel electrophoresis (PAGE) constitutes a powerful technique for the efficient purification of RNA molecules dedicated to applications that require high purity levels. PAGE allows for the fractionation of RNA obtained from cell extracts, chemical or enzymatic synthesis, or modification experiments. Native or denaturing conditions can be chosen for analytical or preparative-scale separations and the nucleotide resolution can be tuned by changing the percentage and reticulation of the gel material. In this protocol, we focus on the preparation of milligram-scale amounts of ~200 nucleotides (nt) RNA molecules that were used in subsequent crystallization experiments.

Key words Gel electrophoresis, RNA purification, In vitro transcription, RNA degradation

1 Introduction

In the landscape of techniques used for RNA purification, polyacrylamide gel electrophoresis (PAGE) remains a method of choice. It has been in use for >50 years and has been extensively described, especially in the laboratory handbook from Sambrook and Russell [1]. As stated there, the gel material is obtained by a radical chain reaction leading to the polymerization of vinyl groups from acrylamide monomers. The resulting long chains are simultaneously cross-linked together by adding bisacrylamide moieties. To separate macromolecules, the acrylamide/bisacrylamide ratio, as well as the concentration of polyacrylamide in the gel material (typically between 5 and 20 %), can be varied to modify the gel reticulation. These properties confer to the polyacrylamide gel material a resolving power and a loading capacity that hold it as a fair competitor of other purification process, such as gel filtration or ion-exchange chromatography [2]. Moreover, there is, a priori, no need for RNA engineering unlike other RNA purification strategies that are based on the adjunction of affinity tags in order to purify the RNA of interest under native conditions (*see* [3] for

Eric Ennifar (ed.), *Nucleic Acid Crystallography: Methods and Protocols*, Methods in Molecular Biology, vol. 1320,
DOI 10.1007/978-1-4939-2763-0_5, © Springer Science+Business Media New York 2016

review). In fact, PAGE can also be performed under native or denaturing conditions at different temperatures, which offers enough diversity to optimize purification protocols. Likewise, temperature gradient gel electrophoresis (TGGE [4, 5]) can in principle also be brought into play in order to identify the best conditions for the purification of a given RNA. A handful of different RNAs can be purified simultaneously by one user, which enables their parallel study in subsequent experiments.

In this protocol, we focus on the purification of RNAs produced by large-scale in vitro transcription reactions of up to 10 mL volumes [6, 7], which were further used for crystallization experiments [8, 9]. In this context, several dozens of different RNA constructs usually need to be studied to yield suitable crystals. The systematic use of PAGE allows for both the purification and the conformational study of these RNAs under denaturing or native conditions, respectively. PAGE is thus a powerful method to facilitate the selection of constructs with satisfying refolding properties, a prerequisite to obtain crystals. Moreover, when discrete conformations are resolved, purification under native conditions can also be achieved to overcome the coexistence of alternative conformations in the sample mix, typically resulting from the application of RNA refolding protocols based on heating/cooling steps [10, 11]. This chapter presents the different steps from the assembly of the apparatus to harvesting of the purified RNA.

2 Materials

Prepare all solutions using ultrapure water (optionally DEPC treated). Except when they contain acrylamide, solutions are filtered through a 0.22 μm sterile filter and kept at room temperature, unless otherwise specified. Acrylamide and bisacrylamide solutions are commercially available in order to prepare gel solutions with a user-defined reticulation. However, acrylamide/bisacrylamide reticulation ratios of 19:1, 29:1, or 37.5:1 (available from Carl Roth, Karlsruhe, Germany, or Bio-Rad, Hercules, CA, USA) optimized for macromolecule separation of increasing molecular weight, respectively, are available as premix gel solutions (*see* **Note 1**).

2.1 Gel Apparatus Components

1. Two glass plates, with dimensions 35×40 cm^2 and 35×42 cm^2.
2. 0.2 cm thick gel spacers.
3. Four paper clips for holding the glass plates and spacers together.
4. (Optional) Water-soluble siliconizing fluid (silane- or siloxane-containing solutions from Thermo Fisher Scientific, Waltham, MA, USA, or Sigma-Aldrich, Saint Louis, MI, USA) (*see* **Note 2**).
5. Adhesive tape (5 cm wide roll brown tape).

6. 0.2 cm thick comb forming loading wells of at least 4 mL.
7. PAGE gel apparatus (i.e., Model S2 Sequencing Gel Electrophoresis Apparatus—Life Technologies, Carlsbad, CA, USA) (*see* **Note 3**).
8. Power supply for gel apparatus.

2.2 Denaturing PAGE Components

1. 8 M Urea stock solution.
2. 40 % Acrylamide/bisacrylamide (19:1) stock solution (i.e., Rotiphorese® Gel 40 (19:1)—Carl Roth, Karlsruhe, Germany).
3. Tris–borate–EDTA buffer (TBE 10×): 216 g of Tris base, 55 g of boric acid, 40 mL of 0.5 M EDTA pH 8.0, adjust with water up to 1 L.
4. APS 10 % stock solution: Weigh 1 g of ammonium peroxodisulfate to dissolve in 10 mL of water (*see* **Note 4**). Store at 4 °C.
5. TEMED: *N*,*N*,*N*′,*N*′-tetramethylethylenediamine. Store at 4 °C.

2.3 Sample Preparation and Migration

1. Absolute ethanol. Store at −20 °C.
2. Urea loading buffer 2× with or without tracking dyes: 8 M urea, 0.025 % xylene cyanol, 0.025 % bromophenol blue.

2.4 RNA Recovery

1. Transparent plastic film.
2. Silica plate (several adjoined thin-layer chromatography (TLC) plates).
3. UV lamp 254/302 nm with light emitted from above the bench.
4. Sterile scalpel blades.
5. 10 mL Terumo® syringes.
6. 50 mL BD Falcon™ conical tubes.
7. Roller mixer.
8. Analytical 0.2 μm filter units for 50 mL tubes (Thermo Scientific Nalgene, Rochester, NY, USA).
9. 3 M Ammonium acetate pH 5.2.
10. Ultrapure (>99.5 %) absolute ethanol for analysis. Store at −20 °C.
11. UV spectrophotometer, Nanodrop (Thermo Fisher Scientific, Waltham, MA, USA) or equivalent.

3 Methods

3.1 Preparation, Polymerization, Casting, and Mounting of the Gel

1. Prepare a 25 % acrylamide/bisacrylamide/urea (19:1:8 M) stock gel solution, by dissolving 480 g of urea in 625 mL of acrylamide/bisacrylamide 40 % (19:1). Make up the volume to 1 L by addition of ultrapure water (*see* **Note 5**).

2. Thoroughly clean the glass plates first with water and then with ethanol.
3. (Optional) Siliconize one glass plate to ease taking down the gel at the end of the migration (*see* **Note 6**). Pour 5 mL of siliconizing solution onto the clean plate laid under the hood and spread over the whole plate using a paper towel. Allow the remaining liquid to evaporate thoroughly before cleaning the plate a second time.
4. Assemble glass plates with spacers by aligning the bottom of the glass plates with one extremity of each spacer. Clip the gel mold, and seal its sides and bottom with tape. Make sure that the tape is not punctured in the lower corners of the gel mold to avoid leakage of the acrylamide solution upon casting.
5. Prepare a gel solution (*see* **Note 7**) by mixing a volume of the 25 % stock gel solution from **step 1** required to obtain the target acrylamide/bisacrylamide percentage with the volume of 8 M urea solution required to prepare the amount of solution for the mold from **step 4**. Take into account that the final gel solution should contain 1× or 0.5× final TBE concentration. Typically, to purify RNA of 200 nucleotides we prepare an 8 % polyacrylamide solution.
6. (Optional) Degas the solution under vacuum in a side-arm flask at least fourfold larger than the volume of acrylamide solution to avoid overflow through the side arm.
7. Pour the gel solution in a clean beaker tenfold larger than the volume of the gel solution. Trigger polymerization by adding 528 μL of freshly prepared 10 % APS solution followed by 50 μL of TEMED per 100 mL of gel solution. Mix thoroughly by shaking for few seconds and immediately cast the gel within the glass mold. Slightly tilt the mold to favor the entrance of the gel solution on one side to avoid trapping air bubbles.
8. When the mold is full, lay it over the bench and gently insert the combs.
9. Wait until the gel solution remaining in the beaker is fully polymerized, which indicates that the polymerization within the gel mold is complete. Do not discard non-polymerized gel solutions (*see* **Note 8**).
10. Remove paper clips, tape, and combs.
11. Mount the gel onto the electrophoresis apparatus.
12. Add 1× or 0.5× TBE buffer to the chambers.
13. Wash the urea out of the wells with TBE buffer from the top chamber using a 1 mL pipet.
14. Pre-run the gel for half an hour at 50 W to reach a temperature around 50 °C.

3.2 Sample Preparation and RNA Migration

1. Precipitate the large-scale (5–10 mL) transcription reaction by addition of 2.5 volumes of cold pure ethanol.
2. Spin the tubes containing the crude RNA at 20,000 × *g* for 20 min at 4 °C in 50 mL BD Falcon™ conical tubes (*see* **Note 9**).
3. Discard the supernatant and resuspend the pellet in 1 mL of water (*see* **Note 10**).
4. Add 1 mL of urea loading buffer 2× containing or not (*see* **Note 11**) tracking dyes to the resuspended pellet.
5. After washing out again the urea from the wells, load the sample into a large well and apply power (50 W).

3.3 RNA Recovery

1. After migration, i.e., when the bromophenol dye runs out, take down the gel and wrap it between two transparent plastic sheets. Place the wrapped gel on a silica plate and observe the RNA migration pattern by UV shadowing (*see* **Notes 12** and **13**).
2. Swiftly frame the band with an indelible felt pen to minimize the exposure time to UV radiations [12].
3. Using a clean scalpel blade, cut the RNA band of interest out of the gel.
4. Crush the gel band by pressing the acrylamide through a syringe with the plunger in a 50 mL BD Falcon™ conical tube.
5. To elute passively the RNA overnight, add 11 mL of water to the crushed gel and place the tube at 4 °C on a roller mixer.
6. Filter the sample through a 0.2 μm sterile filtration unit for 50 mL tubes (Nalgene®).
7. Add 10 % of ammonium acetate 3 M pH 5.2 to the filtrate and precipitate with 2.5 volumes of cold pure ethanol.
8. Spin the tubes containing the purified RNA at 27,000 × *g* for 20 min at 4 °C (*see* **Note 9**).
9. Discard the supernatant and resuspend the RNA pellet in a volume of water to obtain an RNA concentration around 20 mg/mL (*see* **Note 14**).
10. Determine the RNA concentration from OD measurements at $\lambda = 260$ nm.
11. Aliquot the sample and store the individual tubes containing the pure RNA at −80 °C.

4 Notes

1. Acrylamide and bisacrylamide powders are highly toxic chemicals. For this reason, ready-made solutions should be obtained from chemical providers. Avoid allowing acrylamide droplets from drying out on laboratory material used to prepare gel

solutions since acrylamide particles may be carried away in the airstream and hazardously be breathed by people.

2. To facilitate gel unmolding, glass plates can be siliconized on their side in contact with the gel.
3. The gel apparatus should possess a metal plate to homogenize the temperature within the gel during the separation run.
4. APS solutions should be as fresh as possible. Prepare stock solution on a 2-weekly basis and store at 4 °C.
5. The dissolution of urea in aqueous solutions is endothermic and thus requires heating. Use a glass beaker on a heated magnetic stirrer (~100 °C). Avoid overheat to prevent urea and acrylamide degradation.
6. Siliconizing fluids containing silane or siloxane derivatives are harmful and should be used with caution under well-ventilated hoods. Avoid skin and eye contact by wearing adequate individual protective equipment. This step is very useful in the case of thin gels with low acrylamide percentage (5–8 %), which is not the case in the protocol presented.
7. Since RNA is mostly single stranded under denaturing PAGE conditions, the choice of the best percentage of polyacrylamide can be guided by knowledge of the expected mobilities of xylene cyanol and bromophenol dyes (Table 7.7 of [1]).
8. Polymerized acrylamide is harmless and can be discarded in a regular trash can. Always neutralize acrylamide solution leftovers by polymerization before discarding.
9. Be aware of using a rotor dedicated to 50 mL conical bottomed tubes. Using a rotor for spherical bottomed tubes will result in destruction of the tube and loss of the sample and may cause imbalance, eventually damaging the centrifuge.
10. The presence of magnesium phosphate in the pellet makes it difficult to resuspend. Pipet up and down until the whole pellet is dissolved, which can take several minutes. The resulting solution will be cloudy.
11. To avoid co-migration of the RNA with tracking dyes, which could cause subsequent purification problems, mix the RNA sample with an equal volume of 8 M urea and load in the main wells, while a sample containing only loading buffer with tracking dyes is loaded in an independent well at the side of the gel.
12. When the RNA concentration is very high, the difference between the refractive index of the gel containing and not containing the RNA sample allows for naked-eye localization of the RNA bands during electrophoresis.
13. Use a handheld UV lamp with a 254/302 nm wavelength. Wear protective goggles, gloves, and lab coat since UV radiation is hazardous to the skin and eye.

14. According to the desired concentration, the volume of water may vary. Nonetheless apply the water directly onto the pellet, hold the tube steady until dissolution, or slightly agitate without dispersing the water drop. The volume can be adjusted after determining the OD of the sample.

References

1. Sambrook J, Russell DW (2001) Molecular cloning: a laboratory manual. Cold Spring Harbor Laboratory Press, Cold Spring Harbor, NY, USA
2. Easton LE, Shibata Y, Lukavsky PJ (2010) Rapid, nondenaturing RNA purification using weak anion-exchange fast performance liquid chromatography. RNA 16:647–653
3. Batey RT (2014) Advances in methods for native expression and purification of RNA for structural studies. Curr Opin Struct Biol 26:1–8
4. Chadalavada DM, Bevilacqua PC (2009) Analyzing RNA and DNA folding using temperature gradient gel electrophoresis (TGGE) with application to in vitro selections. Methods Enzymol 468:389–408
5. Henco K, Harders J, Wiese U, Riesner D (1994) Temperature gradient gel electrophoresis (TGGE) for the detection of polymorphic DNA and RNA. Methods Mol Biol 31:211–228
6. Milligan JF, Uhlenbeck OC (1990) Synthesis of small RNAs using T7 RNA polymerase. Methods Enzymol 180:51–62
7. Beckert B, Masquida B (2011) Synthesis of RNA by in vitro transcription. Methods Mol Biol 703:29–41
8. Meyer M, Nielsen H, Oliéric V, Roblin P, Johansen SD, Westhof E, Masquida B (2014) Speciation of a group I intron into a lariat capping ribozyme. Proc Natl Acad Sci U S A 111:7659–7664
9. Meyer M, Masquida B (2014) cis-Acting 5′ hammerhead ribozyme optimization for in vitro transcription of highly structured RNAs. Methods Mol Biol 1086:21–40
10. Treiber DK, Williamson JR (1999) Exposing the kinetic traps in RNA folding. Curr Opin Struct Biol 9:339–345
11. Wu M, Tinoco I (1998) RNA folding causes secondary structure rearrangement. Proc Natl Acad Sci U S A 95:11555–11560
12. Kladwang W, Hum J, Das R (2012) Ultraviolet shadowing of RNA can cause significant chemical damage in seconds. Sci Rep 2:517

Chapter 6

Use of the U1A Protein to Facilitate Crystallization and Structure Determination of Large RNAs

Adrian R. Ferré-D'Amaré

Abstract

The preparation of well-ordered crystals of RNAs with complex three-dimensional architecture can be facilitated by engineering a binding site for the spliceosomal protein U1A into a functionally and structurally dispensable stem-loop of the RNA of interest. Once suitable crystals are obtained, the U1A protein, of known structure, can be employed to facilitate preparation of heavy atom or anomalously scattering atom derivatives, or as a source of partial model phases for the molecular replacement method. Here, we describe the methods for making U1A preparations suitable for cocrystallization with RNA. As an example, the cocrystallization of the tetracycline aptamer with U1A is also described.

Key words Crystallization, Crystallizability, Spliceosomal protein, RNA-binding domain (RBD), Protein purification

1 Introduction

Determination of the three-dimensional structures of large RNAs with complex architectures by X-ray crystallography requires the preparation of well-ordered crystals. Although the first RNA structure to be successfully determined, that of yeast $tRNA^{Phe}$, employed crystals of an intact molecule isolated from its biological source, the majority of subsequent RNA structure determinations have employed either synthetic or in vitro-transcribed molecules. This makes it possible to engineer for increased "crystallizability" the RNAs whose structures are of interest. Because of the high thermodynamic stability of local structure, which is dominated by the A-form double helix, the periphery of many RNAs can be engineered to alter molecular surface properties without adversely affecting their overall folds or active site structures [1]. One approach that has been particularly successful since its invention in 1998 [2] is to replace a solvent-exposed, functionally dispensable stem-loop of an RNA of interest with the cognate binding site for the spliceosomal protein U1A [3]. The engineered RNA is

Eric Ennifar (ed.), *Nucleic Acid Crystallography: Methods and Protocols*, Methods in Molecular Biology, vol. 1320,
DOI 10.1007/978-1-4939-2763-0_6, © Springer Science+Business Media New York 2016

complexed with the strongly basic RRM-I domain of the protein (hereafter "U1A"), and the RNA:U1A complex is subjected to crystallization. It has been found empirically that cocrystallization of such a complex often results in better ordered crystals than crystallization of the parental form of the RNA in the "naked" state. Moreover, if good crystals are obtained, the U1A protein can be used to facilitate structure determination, either by using it to incorporate heavy or anomalously scattering atoms into the crystals for experimental phasing [for instance [3, 4]] or by providing partial model phases for molecular replacement [for instance [5, 6]]. The use of U1A as a "crystallization module" in several structure determination projects has been reviewed elsewhere [6–12]. Here, we provide detailed methods for the preparation of recombinant U1A protein suitable for use in facilitating RNA crystallization. Because conditions vary from RNA to RNA, it is not possible to provide a general crystallization protocol. Instead, we describe procedures for the in vitro transcription of an engineered tetracycline aptamer RNA and its cocrystallization with U1A [13] as an example of an RNA structure determination that was facilitated by U1A, both for the preparation of well-ordered crystals and for experimental phase determination.

2 Materials

2.1 Experimental Considerations

1. Except where noted, prepare all solutions using deionized, DEPC-treated water (*see* **Note 1**).
2. All solutions used for chromatography need to be filtered through 0.2 μM filters.
3. During purification, protein-containing solutions should be kept on ice or at 4 °C.
4. Purified U1A should be stored at −80 °C, but can be used at room temperature.
5. Personal protective equipment (lab coat, safety goggles, gloves) should be worn where appropriate, and all waste disposal regulations must be followed.

2.2 U1A Protein Expression and Purification Components

1. Lysis buffer: 20 mM HEPES-KOH pH 7.5, 10 % (v/v) glycerol, 100 mM KCl, 0.05–0.1 trypsin inhibitor unit of aprotinin per liter, 5 μg/ml leupeptin, 0.0005 % (v/v) Triton X-100, 0.5 mM EDTA, 0.5 mM phenylmethylsulfonyl fluoride (PMSF, *see* **Note 2**). It is not necessary to use DEPC-treated water for this solution, as it will encounter abundant bacterial nucleases.
2. 5 % (w/v) Neutralized polyethyleneimine (PEI) solution [14].
3. Concentrated HCl solution.
4. Ammonium sulfate, 2 kg.

5. 0.2 μl pore filters.
6. French press, EmulsiFlex, or similar high-pressure cell disruption device.
7. Chromatography system.
8. 30 ml bed-volume SP-Sepharose Fast Flow (GE Healthcare Life Sciences, Pittsburgh, PA, USA).
9. Superdex-75 PG (GE Healthcare Life Sciences, Pittsburgh, PA, USA).
10. Buffer A: 100 mM KCl, 25 mM HEPES-KOH pH 7.5, 0.5 mM EDTA, and 0.5 mM PMSF.
11. Buffer B: 100 mM KCl, 25 mM HEPES-KOH pH 7.5, 0.5 mM EDTA, 0.5 M KCl, and 0.5 mM PMSF.
12. Buffer C: 10 mM potassium phosphate, pH 7.5 and 50 mM KCl.
13. Buffer D: 10 mM potassium phosphate, pH 7.5, 0.5 M ammonium sulfate, and 50 mM KCl.
14. Storage buffer: 10 mM Hepes-KOH pH 7.5, 0.1 mM EDTA.
15. 0.5 M NaOH.
16. 10 ml bed-volume CHT-I hydroxyapatite column (BioRad, Hercules, CA, USA).

2.3 RNA Transcription and Purification Components

1. Neutralized ribonucleoside triphosphates (rNTPs): For economy, lyophilized sodium salts of rNTPs can be purchased.
2. Transcription buffer: Specific conditions need to be optimized for each RNA sequence. As a starting point, transcription reactions contain 30 mM Tris–HCl pH 8.1, 25 mM $MgCl_2$, 0.01 % (v/v) Triton X-100, 2 mM spermidine hydrochloride, 2.5 mM each of the four NTPs, 1 mM DTT, 0.05 g/l T7 RNA polymerase, and 1 unit/ml of inorganic pyrophosphatase (*see* **Note 3**).
3. Tris-borate-EDTA buffer: A 10× stock is prepared by mixing 432 g of Tris base, 220 g of boric acid, 160 ml of 0.5 M sodium EDTA (pH 8.0), and sufficient water to make 4 l. Solubilization is hastened by autoclaving for 30–40 min.
4. Denaturing polyacrylamide gels: These are prepared by dissolving urea with water, 10× TBE buffer, and acrylamide: bisacrylamide solution. We find that a commercially available 40 % total, 29:1 acrylamide:bisacrylamide solution is adequate for most RNA separations. The acrylamide content of the gel needs to be varied depending on the size of the RNAs being purified. The total volume of gel solution needed will depend on the electrophoresis apparatus used. Table 1 indicates the amounts of urea, acrylamide solution, and water needed for different gel compositions.

Table 1
Preparation of gel mixture for polyacrylamide electrophoresis[a]

Gel percentage	Volume of 40 % acrylamide stock solution (ml)	Purified water (ml)
4	10	43
6	15	37
8	20	31
10	25	27
12	30	16
15	38	8
20	50	0

[a]Quantities given are for preparing 100 ml of gel mixture. For this amount, 48 g of solid urea and 10 ml of 10× TBE need to be included in addition to the acrylamide and water volumes tabulated

5. 50 % Aqueous ammonium persulfate: Should be prepared ahead of time, and stored indefinitely in small aliquots at −20 °C until use.
6. Tetramethylethylenediamine (TEMED).
7. Absolute ethanol, ≥99.8 %.
8. Aqueous ethanol, 70 % (v/v).
9. 250 mM EDTA (pH 8.0).
10. Formamide.
11. Vacuum aspirator.
12. Handheld ultraviolet light.
13. Thin-layer chromatography (TLC) silica plate, 20 × 20 cm or larger.
14. Ultrafiltration device, 3 kDa molecular weight cutoff.
15. 1 M KCl.
16. Sequencing gel electrophoresis system.

2.4 U1A-RNA Cocrystallization Components

1. 5× RNA preincubation buffer: 250 mM Tris–HCl pH 7.5, 25 mM $MgCl_2$, 2.5 mM 7-chlortetracycline, and 1.25 mM spermine hydrochloride. Keep at −20 °C in the dark.
2. Crystallization reservoir: 50 mM Hepes-KOH pH 7.0, 20 mM $MgCl_2$, and 12.5–15 % (w/v) PEG 8000.
3. Crystallization plates (Linbro, VDX, or equivalent).
4. Sealing film for crystallization plates.
5. Cryoprotecting solution: 50 mM Hepes-KOH pH 7.0, 20 mM $MgCl_2$, 15 % (w/v) PEG 8000, 0.5 mM spermine, 0.5 mM 7-chlortetracycline, and 30 % (v/v) glycerol.

3 Methods

3.1 U1A Protein Purification

1. Dissolve 100 ml of 50 % (w/v) PEI in 800 ml of water. Adjust the pH to 7.9 by adding concentrated HCl (38–40 ml) with stirring. After diluting to 1 l, the solution is sterilized by filtering through 0.2 μm filters. This solution can be stored for a long term at 4 °C.
2. Prepare saturated ammonium sulfate by dissolving 800 g of solid ammonium sulfate in sufficient water to make a 1 l solution (heating gently, do not exceed 40 °C) and filtering through a 0.2 μM pore filter while warm. Upon cooling to room temperature in a closed bottle, excess ammonium sulfate will crystallize out of the solution. These crystals can be left in the bottle, but care should be taken to only use the supernatant saturated solution.
3. Resuspend *E. coli* from a 9 l U1A-RBD growth and induction (*see* **Note 4**) in 75 ml of ice-cold lysis buffer. Avoid vigorous shaking. Bacterial clumps can be broken down by repeatedly forcing the suspension through a 25 or 50 ml serological pipette. Any residual clumps can be removed using a double thickness of cheesecloth, or a fine metal sieve (such as a repurposed household metal mesh coffee filter).
4. Lyse bacteria using a cell disruption device. Add PMSF immediately prior to lysis. It is important to avoid heating the resuspended bacteria; therefore, sonication is not recommended. If a high-pressure device is not available, several cycles of freeze-thawing complemented with lysozyme can be employed, but protein yield will decrease. Add sufficient PMSF immediately after lysis to a nominal final concentration of 1 mM.
5. Clarify the lysate by centrifugation at 4 °C for 30 min at 75,000 × *g* (*see* **Note 5**).
6. Transfer the supernatant to fresh centrifuge tubes, and add 1/10 volume of neutralized 5 % PEI solution. Mix by gentle inversion and incubate for 10 min on ice.
7. To remove PEI-bound nucleic acids, centrifuge at 4 °C for 30 min at 75,000 × *g*. Discard the pellet.
8. Place supernatant in a suitable container (e.g., an Erlenmeyer flask) on ice. While stirring gently with a magnetic stirrer, add dropwise sufficient saturated ammonium sulfate solution to a final concentration of 35 % saturation (not w/v). Stir for 15 min after addition is complete.
9. To remove contaminant proteins, centrifuge at 4 °C for 30 min at 38,000 × *g*. Discard the pellet.
10. To the supernatant, add sufficient solid ammonium sulfate to achieve a final concentration of 80 % saturation (not w/v).

This requires the addition of 316 g of solid ammonium sulfate per liter of 35 % saturated supernatant. Stir magnetically at room temperature until complete dissolution of the salt.

11. Centrifuge at 4 °C for 30 min at 38,000 × *g*. Discard the supernatant.
12. Gently dissolve the pellet in the smallest possible volume of a solution of the same composition as Buffer A (add PMSF immediately prior to use), but with 50 mM KCl. Dialyze overnight using 3,000 Da nominal molecular weight cutoff membranes against 20× volume of Buffer A.
13. After dialysis, eliminate precipitated proteins by centrifuging at 4 °C for 30 min at 75,000 × *g*. Decant supernatant taking care not to dislodge the pellet. Discard pellet.
14. Equilibrate an SP-Sepharose Fast Flow column in Buffer A. Load dialyzed protein and run a 10-column volume gradient to 100 % Buffer B. U1A elutes at a conductivity approximately equivalent to 300 mM KCl. Concentrate U1A peak by ultrafiltration.
15. Further purify U1A by size-exclusion chromatography on a Superdex-75 PG column equilibrated and run in Buffer A. U1A elutes at a volume consistent with a molecular mass of ~11 kDa.
16. At this point in the procedure, it is advisable to clean the chromatography system to eliminate nucleases. In our laboratory, this is done by running a solution of 0.5 M NaOH through the entire system and then flushing it with DEPC-treated water (*see* **Note 6**). All further solutions employed in the procedure are made with DEPC-treated water and buffer components are of crystallization grade.
17. Equilibrate a CHT-I hydroxyapatite column in Buffer C. To the protein, add enough potassium phosphate (pH 7.5) solution to a final concentration of 10 mM. Load protein on equilibrated column and run a 20-column volume gradient to 100 % Buffer D. The protein elutes at a conductivity approximately equivalent to 250 mM ammonium sulfate (*see* **Note 7**).
18. Dialyze protein against 20 volumes of storage buffer. Concentrate by ultrafiltration using 3,000 Da molecular weight cutoff membranes, quantify, and store at −80 °C (*see* **Note 8**).

3.2 Transcription and Purification of Tetracycline Aptamer

1. rNTPs should be dissolved in water and neutralized with NaOH prior to use (it is sufficient to use pH indicator paper with a legibility of one pH unit).
2. In order to make 100 mM stock solutions of rNTPs, one gram of the salt should be brought up to a final volume of 16.5, 17.1, 16.5, and 17 ml for ATP, CTP, GTP, and UTP, respectively. After filtering through 0.2 μm syringe filters, these solutions can be aliquoted and stored at −20 °C until use.

3. Prepare double-stranded DNA template (*see* **Note 9**) by the polymerase chain reaction (PCR), as described [13].
4. Prepare transcription reactions by adding 1/10 volume of PCR reaction to the transcription reaction. Perform transcription at 37 °C for 2–4 h.
5. Add two volumes of absolute ethanol (chilled to −20 °C) to the reaction. Mix well by inversion and incubate at −20 °C for 30 min.
6. Centrifuge at 7,000 × *g* for 10 min.
7. Decant supernatant. Rinse pellet (without dislodging) with −20 °C 70 % (v/v) aqueous ethanol. Decant ethanol.
8. Remove most of the ethanol from the pellet by placing it under vacuum (a vacuum aspirator or a centrifugal drier—used without the rotor—is sufficient) for 5–10 min.
9. Dissolve pellet in the smallest possible volume of 250 mM sodium EDTA (pH 8.0). Add 0.9 volumes of formamide and 0.1 volume of 10× TBE buffer. Heat to 50 °C for 5 min.
10. Load RNA onto pre-run 8 M urea polyacrylamide gel, and run until separation is achieved (*see* **Note 10**). Tracking dyes (for instance bromophenol blue and xylene cyanol) dissolved in neat formamide can be loaded into a small parallel lane.
11. Visualize band by briefly shadowing the gel with a handheld ultraviolet light while the slab gel is kept over a fluorescent indicator-containing TLC plate (*see* **Note 11**).
12. Excise the portion of the gel containing the RNA with a sterile scalpel or razor blade, transfer to a sterile polypropylene centrifuge tube, and crush using a sterile spatula. Add approximately 25 ml DEPC-treated water per ml of excised gel.
13. Gently mix or tumble overnight at 4 °C.
14. Filter the gel-water suspension through 0.2 μm filter, and discard gel fragments.
15. Using either a centrifugal or a gas-pressure-driven ultrafiltration device equipped with a 3,000 Da nominal molecular weight cut-off membrane, concentrate eluted RNA to a small volume. Dilute at least 25-fold with 1 M KCl, and concentrate again. Then, sequentially, dilute 25-fold or more with DEPC-treated water and concentrate three times, so that the final concentration of KCl is less than 1 mM. Concentrated RNA can be stored at 4 °C until use.

3.3 Cocrystallization of U1A and the Tetracycline Aptamer

1. Incubate RNA and a stoichiometric amount of U1A in preincubation buffer at a final macromolecular complex concentration of 0.25 mM for 20 min at 37 °C.
2. Set up hanging-drop vapor-diffusion experiments by placing 0.5 ml of reservoir solution in the reservoir. Then, mix 1 μl of this solution with 1 μl of preincubated RNA:protein:chlortetracycline mixture to make the hanging

drop (*see* **Note 12**). Seal the system and incubate in the dark at 21 °C for 2–3 weeks. Cube-shaped crystals appear within 4 weeks and grow to maximum dimensions of 0.15 μm³ over two additional weeks.

3. For diffraction experiments, crystals can be cryoprotected (*see* **Note 13**) by exchanging the mother liquor over the course of 10–15 min with a cryoprotecting solution. Crystals are then mounted in nylon loops and flash-frozen by plunging into liquid nitrogen.

4 Notes

1. Although most laboratory-grade purification systems will deliver high-resistivity (18.2 MΩ/cm at 25 °C) deionized water, this may still contain ribonucleases (RNases). Residual RNases can be destroyed by treatment with the alkylating agent diethylpyrocarbonate (DEPC). Unreacted DEPC can in turn be destroyed by autoclaving. Mix 1 ml of DEPC with 4 l of purified water in a bottle with a tightly sealing cap. Shake vigorously for 10 s. Loosen the cap, cover loosely with aluminum foil, and autoclave for at least 30 min. DEPC is highly toxic. It should be handled only in a chemical fume hood and with appropriate personal protective equipment. Unreacted DEPC is converted into ethanol and carbon dioxide through autoclaving.
2. PMSF is a powerful neurotoxin. It is important to avoid any direct contact with PMSF-containing solutions. Because PMSF is slowly destroyed by water, it needs to be dissolved freshly prior to use, or dissolved in an anhydrous solvent and kept free of water. A convenient method to achieve this is to dissolve the PMSF in dry isopropyl alcohol and to store it over 3 Å molecular sieves. Molecular sieves need to be activated by heating to 200 °C for ~15 h (or more) in a vacuum oven, and then allowing them to cool to room temperature in a desiccator prior to use. Isopropyl alcohol itself (which forms an azeotrope with water) can be dried by stirring over activated molecular sieves.
3. Because magnesium pyrophosphate has very low solubility, accumulation of pyrophosphate during transcription leads to sequestration of magnesium ion in a precipitate. Since RNA polymerase requires free magnesium ion for activity, this decreases transcription yield. To overcome this, inorganic pyrophosphatase can be added to the reaction [9].
4. Expression and induction of U1A are described elsewhere [8, 15].

5. Longer centrifugation times will be needed if a preparative centrifuge/rotor combination capable of these centrifugal forces is not available.
6. It will be necessary to first establish if the chromatography system being used tolerates exposure to 0.5 M NaOH. Otherwise, other cleaning methods will be needed. Personal protective equipment must be worn at all times when handling sodium hydroxide solutions.
7. We find that for optimal separation no more than 15 mg of U1A should be loaded on a 10 ml bed-volume CHT-I column. All solutions that come into contact with the CHT-I column need to contain 10 mM phosphate.
8. U1A protein has a molecular mass of 11.3 kDa. A 1 g/l solution of the wild-type and double-mutant [15] proteins produce an absorbance at 280 nM of 0.46 and 0.34, respectively. Although nucleases can prove serendipitous [16], in general it is important that the U1A preparation be scrupulously nuclease free. This can be assayed as described [10].
9. For this example, an RNA with a limited amount of 3' heterogeneity proved sufficient. For other crystallization projects, homogeneous RNA prepared, for instance, through ribozyme cleavage [17, 18], may be necessary.
10. We employ 3 mm thick polyacrylamide slab gels. For a 10 ml volume transcription, a gel that is 32 cm wide and 23 cm high is sufficient. To initiate polymerization, the gel mixture is supplemented with 0.001 % (v/v) of tetramethylethylenediamine (TEMED) and 0.002 % (v/v) of a 50 % (w/v) aqueous solution of ammonium persulfate.
11. UV irradiation can damage RNA [19]. It is imperative to limit exposure of the RNA to a few seconds. It is also essential to wear UV-protective clothing and face and hand protection.
12. Different geometries for crystallization experiments are discussed by McPherson [20].
13. A tabulation of the range of cryoprotectants that have been employed successfully for RNA and RNA:protein complex crystals appears in ref. 12.

Acknowledgements

We thank K. Nagai (Medical Research Council Laboratory of Molecular Biology, Cambridge, UK) for generously sharing U1A expression plasmids. This work was supported by the Intramural Research Program of the NIH, National Heart, Lung and Blood Institute.

References

1. Ferré-D'Amaré AR, Zhou K, Doudna JA (1998) A general module for RNA crystallization. J Mol Biol 279:621–631
2. Ferré-D'Amaré AR, Zhou K, Doudna JA (1998) Crystal structure of a hepatitis delta virus ribozyme. Nature 395:567–574
3. Nagai K, Oubridge C, Jessen TH, Li J, Evans PR (1990) Crystal structure of the RNA-binding domain of the U1 small nuclear ribonucleoprotein A. Nature 348:515–520
4. Rupert PB, Ferré-D'Amaré AR (2001) Crystal structure of a hairpin ribozyme-inhibitor complex with implications for catalysis. Nature 410:780–786
5. Kulshina N, Baird NJ, Ferré-D'Amaré AR (2009) Recognition of the bacterial second messenger cyclic diguanylate by its cognate riboswitch. Nat Struct Mol Biol 16: 1212–1217
6. Xiao H, Murakami H, Suga H, Ferré-D'Amaré AR (2008) Structural basis of specific tRNA aminoacylation by a small in vitro selected ribozyme. Nature 454:358–361
7. Ferré-D'Amaré AR, Doudna JA (2000) Methods to crystallize RNA. In: Beaucage SL, Bergstrom DE, Glick GD, Jones RA (eds) Current protocols in nucleic acid chemistry. Wiley, New York, pp 7.6.1–7.6.10
8. Ferré-D'Amaré AR, Doudna JA (2000) Crystallization and structure determination of a hepatitis delta virus ribozyme: use of the RNA-binding protein U1A as a crystallization module. J Mol Biol 295:541–556
9. Rupert PB, Ferré-D'Amaré AR (2004) Crystallization of the hairpin ribozyme: illustrative protocols. Methods Mol Biol 252: 303–311
10. Ferré-D'Amaré AR (2010) Use of the spliceosomal protein U1A to facilitate crystallization and structure determination of complex RNAs. Methods 52:159–167
11. Kulshina N, Edwards TE, Ferré-D'Amaré AR (2010) Thermodynamic analysis of ligand binding and ligand binding-induced tertiary structure formation by the thiamine pyrophosphate riboswitch. RNA 16:186–196
12. Ferré-D'Amaré AR (2013) Crystallization of RNA for structure determination by X-ray crystallography. In: Klostermeier D, Hammann C (eds) Structure and folding of RNA. deGruyter, Berlin, pp 319–333
13. Xiao H, Edwards TE, Ferré-D'Amaré AR (2008) Structural basis for specific, high-affinity tetracycline binding by an in vitro evolved aptamer and artificial riboswitch. Chem Biol 15:1125–1137
14. Burgess RR (1991) Use of polyethyleneimine in purification of DNA-binding proteins. Methods Enzymol 208:3–10
15. Oubridge C, Ito N, Teo C-H, Fearnley I, Nagai K (1995) Crystallization of RNA-protein complexes II. The application of protein engineering for crystallization of the U1A protein-RNA complex. J Mol Biol 249:409–423
16. Klein DJ, Ferré-D'Amaré AR (2009) Crystallization of the *glmS* ribozyme-riboswitch. Methods Mol Biol 540:129–139
17. Ferré-D'Amaré AR, Doudna JA (1996) Use of *cis*- and *trans*-ribozymes to remove 5′ and 3′ heterogeneities from milligrams of in vitro transcribed RNA. Nucleic Acids Res 24:977–978
18. Price SR, Ito N, Oubridge C, Avis JM, Nagai K (1995) Crystallization of RNA-protein complexes I. Methods for the large-scale preparation of RNA suitable for crystallographic studies. J Mol Biol 249:398–408
19. Kladwang W, Hum J, Das R (2012) Ultraviolet shadowing of RNA can cause significant chemical damage in seconds. Sci Rep 2:1–6
20. McPherson A (1999) Crystallization of biological macromolecules. Cold Spring Harbor Laboratory Press, Cold Spring Harbor

Chapter 7

Fab Chaperone-Assisted RNA Crystallography (Fab CARC)

Eileen Sherman, Jennifer Archer, and Jing-Dong Ye

Abstract

Recent discovery of structured RNAs such as ribozymes and riboswitches shows that there is still much to learn about the structure and function of RNAs. Knowledge learned can be employed in both biochemical research and clinical applications. X-ray crystallography gives unparalleled atomic-level structural detail from which functional inferences can be deduced. However, the difficulty in obtaining high-quality crystals and their phasing information make it a very challenging task. RNA crystallography is particularly arduous due to several factors such as RNA's paucity of surface chemical diversity, lability, repetitive anionic backbone, and flexibility, all of which are counterproductive to crystal packing. Here we describe Fab chaperone assisted RNA crystallography (CARC), a systematic technique to increase RNA crystallography success by facilitating crystal packing as well as expediting phase determination through molecular replacement of conserved Fab domains. Major steps described in this chapter include selection of a synthetic Fab library displayed on M13 phage against a structured RNA crystallization target, ELISA for initial choice of binding Fabs, Fab expression followed by protein A affinity then cation exchange chromatography purification, final choice of Fab by binding specificity and affinity as determined by a dot blot assay, and lastly gel filtration purification of a large quantity of chosen Fabs for crystallization.

Key words RNA crystallography, Chaperone assisted crystallography, Fab (antigen binding fragment), Phage display selection

1 Introduction

Our understanding of RNA's role in life processes has greatly expanded since the initial discovery of structured cleaving ribozymes over three decades ago. Besides the canonical genetic information transfer function, noncoding RNA has many manifestations mediating genetic expression and regulation such as histone modification, DNA methylation, metabolite sensing, post transcriptional gene silencing, intron splicing, trans RNA processing, and ribozymes involved in rolling circle replication [1–3]. Noncoding RNAs have also been implicated in disease processes such as cancer, autism, obesity, Alzheimer's, and atherosclerosis [4–14]. RNA often assumes complex tertiary structures to execute its functions by interacting with proteins, other RNA, DNA, or small molecules;

Eric Ennifar (ed.), *Nucleic Acid Crystallography: Methods and Protocols*, Methods in Molecular Biology, vol. 1320, DOI 10.1007/978-1-4939-2763-0_7, © Springer Science+Business Media New York 2016

therefore, determining high resolution RNA structures will blaze the trail inroads to further understanding RNA mediated genetic regulation and pave the way to structure based drug design. Far more protein structures, however, have been determined than RNA; of the 81,103 X-ray diffraction structures currently in the PDB, less than 2.5 % contain RNA. This reflects the inherent difficulty of RNA crystallography and current lack of a systematic method to accomplish RNA crystallization.

Crystal packing of RNA is difficult because of the large amount of anionic surface area; the repetitive anionic phosphate backbone can create imperfectly replicable packing, i.e., "slippage." RNA's limited chemical diversity relative to proteins is also seen as an impediment to packing [15, 16]. Both chemicals and RNases degrade RNA, making the macromolecular sample more heterogeneous, impeding high order ergo quality crystal formation. RNA structure is also more flexible than a protein of similar mass, perhaps again due to anionic repulsion, which also is counterproductive to a regular crystalline lattice. In addition to crystal formation difficulties, phasing is also problematic, as heavy metal "soak and pray" methods are used to include large atoms to obtain phase information by methods such as MAD/SAD (multi-/single wavelength anomalous dispersion) [16, 17]. Fab CARC provides surface protein chemical diversity for crystal contact formation to promote packing, makes the solvent exposed surface less anionic, and reduces flexibility of the complex [18–20]. In addition, the presence of the large 48 kDa Fab fragment in a crystal allows for molecular replacement determination of phases [18, 19].

Fab CARC successfully crystallized *Tetrahymena* Group I intron P4P6 domain [18] (PBD: 2R8S) and Class I Ligase [19] (3IVK) structures. Fab mediated crystal contacts provided 61 % and 59 % of the total solvent accessible surface area buried in crystal contacts for P4P6 and the Ligase, respectively. Molecular replacement with Fab domains was sufficient to obtain phase information in both of these structures, and crystal structures of the Class I Ligase chaperoned with Fab protein proved much less arduous to phase than the U1A version [19, 21]. Molecular replacement avoids the additional step of making batches of crystals to obtain phase information from (potentially sporadic) heavy metal inclusion. Multiple batch production can introduce additional difficulty in crystal solving and phasing if the crystals are not isomorphic before and after heavy metal soaking. This significance cannot be understated, since as obtaining crystals is a formidable challenge, having to obtain multiples for phasing and "soak and pray" multiplies the difficulty.

Fabs are also beneficial as they can be selected against an unengineered target structure using the phage display technique, requiring little previous knowledge of its functional residues while using the U1A protein crystal chaperone requires mutating in is binding loop [22, 23]. Some other phasing approaches such as metal binding motif engineering also require this knowledge [24]. Binding to a specific location is still an option, however, as the Class I Ligase Fab BL3-6 has a portable loop that can be mutated into an RNA target. Furthermore, this technique can also be combined with macromolecule surface engineering approaches, building upon their strengths. Engineering is a broad topic that cannot be treated thoroughly here, but both RNA and Fab can be engineered to improve crystallizability [17, 20, 25–28]. For example, Fabs can undergo alteration of surface residues to improve crystal packing [20].

Raising Fab against an RNA target is prerequisite for Fab CARC. As such, phage display is important since the traditional approach of injecting an antigen into a host to generate antibodies is not viable for RNA due to chemical lability and RNase activity. Complementarity Determining Regions (CDRs) of Fab libraries can be engineered to improve Fab affinity and specificity for targets [29–32]. Several different phage libraries [18, 19, 31, 32] with different CDRs exist and are used by us in selection, and a binding Fab's CDRs can be even further affinity matured once it has been isolated in selection [19].

As Fab CARC is a voluminous procedure, an overview flow chart is shown below. More common knowledge such as transcribing RNA [33], native and denaturing (urea for nucleic acid, SDS for protein) polyacrylamide gels [20, 34], agarose gels, gamma ^{32}P ATP 5′ radiolabelling [35], making/transforming electrocompetent/chemical competent cells, single stranded DNA (ssDNA) preparation from phage [20], spectrophotometric nucleic acid and protein concentration determination, and restriction enzyme and Kunkel (*dut*$^-$ *ung*$^-$) cloning [20] is not covered in detail in the subsequent methods for brevity. The beginning of each procedure contains a brief description to aid understanding and continuity as well as other practical information.

Clone and transcribe RNA targets → choose libraries and perform phage display selection → phage ELISA to identify individual initial Fabs → miniprep and send clones for sequencing to pick unique and correct Fab clones → make ssDNA → Kunkel (*dut*$^-$ *ung*$^-$ cloning) to add TAG stop → transform into 34B8 cells → colony PCR to identify TAG clones → small scale Fab expression → protein A affinity column purification → High S cation exchange column purification → native gel assay to check complex formation and specificity → radiolabel untagged RNA target and a non-target RNA as control → do dot-blot (filter binding) assay to pick tight and specific Fabs → express large Fab culture → Protein A purification → High S cation exchange column purification twice → Superdex 200 AKTAprime plus gel filtration → transcribe crystallization quantity of RNA → native gel to establish Fab–RNA stoichiometry → make Fab RNA complex and setup crystal screenings with

288 conditions each at 15 and 4 °C in 96 well sitting drop plates → check crystals and optimize conditions with 24 well hanging drop plates → fish crystals → send to synchrotron → solve crystal structures.

2 Materials

Generally, solutions and labware are sterilized by either autoclaving or sterile 0.2 μm filtration. We use ultrapure Milli-Q water ("ddH_2O") from Millipore (18 MΩ cm at 25 °C) with an additional Biopak polisher to remove RNases. RT, room temperature, defined as ~25 °C; O/N, overnight, is 12–16 h.

2.1 Phage Display Selection Reagents

1. Aerosol barrier (filter) tips for pipetting phage and radioactive solutions.
2. 1 ml long reach ART barrier pipette tips (Thermo Fisher Scientific, Waltham, MA, USA).
3. Microcentrifuge tubes, certified RNase free.
4. Sterile RNase free 15 and 50 ml graduated tubes.
5. 50 ml Oakridge tubes.
6. 250 ml baffled flasks.
7. Avanti J-20 XP centrifuge (Beckman Coulter, Fullerton, CA, USA) or equivalent.
8. UV-Visible spectrophotometer.
9. Quartz cuvettes for nucleic acid and protein concentration determination.
10. Disposable plastic cuvettes to monitor bacterial cell growth by turbidity.
11. 10× PBS: 1.37 M NaCl, 30 mM KCl, 80 mM Na_2HPO_4, 15 mM KH_2PO_4, pH 7.2, autoclave. All solutions are pH adjusted with saturated NaOH solution or concentrated HCl. "PBS" in protocol refers to 1× PBS.
12. 10 % Tween 20; Tween 20 cannot be autoclaved.
13. 1 M $MgCl_2$.
14. 4 M NaCl.
15. 10 mM D (+) biotin in DMSO, do not filter DMSO (solvates filter membrane).
16. 5 % RNase free BSA (50 mg/ml), DNase and protease free.
17. 10 mg/ml RNase free streptavidin (SAV).
18. 100 % Glycerol.
19. PEG/NaCl: 20 %(w/v) Polyethylene Glycol 8000 + 2.5 M NaCl.

20. 2YT/Amp/Kan: 2YT media, 100 μg/ml ampicillin, 25 μg/ml kanamycin (we make all antibiotic stock solutions 1,000× concentrated, 0.2 μm syringe filtered, and stored aliquoted in −20 °C).
21. 2YT/Tet: 2YT + 10 μg/ml tetracycline.
22. 10 mg/ml tetracycline: dissolve in 50 % ddH_2O and 50 % ethanol, syringe filter.
23. Solution A: 80 μl 10× PBS, 100 μl 5 % BSA, 5 μl 10 % Tween 20, 5 μl 0.5 M EDTA pH 8.0, 12.5 μl 1 M $MgCl_2$, saturating amount of any target ligand, ddH_20 to total volume of 750 μl.
24. Solution B: 20.2 μl 10× PBS, 1.01 μl 10 % Tween 20, 1.01 μl 0.5 M EDTA pH 8.0, 2.52 μl 1 M $MgCl_2$, 21.47 μl 5 % BSA, saturating amount of any target ligand, ddH_2O to total volume of 190 μl.
25. Binding buffer (PTEM): 1× PBS, 0.05 % Tween 20, 2.5 mM EDTA pH 8.0 (x+2.5) mM $MgCl_2$, where x = $MgCl_2$ required for a particular target's folding in mM. If RNA target folding requires a ligand, add an appropriate amount.
26. PTEM/gly: PTEM, 5 % glycerol.
27. Anneal RNAtag/biotin complex (50 μM): 2.5 nmol RNAtag (need 0.5 nmol for first round, less for later rounds), 2.75 nmol (1.1 eq.) biotinylated DNA oligo, 0.5 μl 1 M Tris–HCl pH 7.5, 2.5 μl 4 M NaCl, ddH_2O to 50 μl.

    Place at 90 °C for 1 min, then 10 min at room temperature, and freeze.

    When diluting RNA-tag/biotin complex always keep NaCl concentration at 200 mM.
28. Bleach, detergent, i.e., Alconox, sodium dodecyl sulfate (SDS), plastic squirt bottles, other labware for phage cleaning.
29. Streptavidin MagneSphere® Paramagnetic Particles (Promega, Madison, MI, USA).
30. MagneticSphere® Magnetic Separation Strand (Promega, Madison, MI, USA).
31. Agarose (agarose gels), ethidium bromide.
32. *E. coli* XL1-Blue cells, sterile culture tubes.
33. Stop RNase Inhibitor RX (5 Prime, Hilden, Germany).
34. SUPERase RNase inhibitor (Life Technologies, Carlsbad, CA, USA).
35. LB/Amp petri dishes, 37 °C incubator for growth (300 rpm): 12.5 g LB Miller + 7.5 g Agar (not agarose!) in 500 ml H_2O, autoclave, cool to 50 °C, add proper antibiotics.
    - ~25 ml in each sterile petri dish.
    - Cool on bench for 24 h to two overnights to solidify and dry, store at 4 °C.

36. Sterile 96 well plates.
37. 50 °C water bath.
38. tRNA mix from *E. coli*.
39. Biotinylated RNase A: 10 mg DNase/Protease free RNase A (Thermo Fisher Scientific, Waltham, MA, USA), EZ-Link Sulfo-NHS-LC-Biotin (Thermo Fisher Scientific, Waltham, MA, USA); follow manufacturer's instructions to biotinylate.
40. KingFisher Flex Magnetic Particle Processor with microplate of strips for selection and corresponding magnetic tip combs (VWR, Radnor, PA, USA).
41. KingFisher programs. Postelution capture removes biotinylated RNase A.

    RNA_2nd: (a) Mix and incubate well A for 15 min. (b) Collect well B beads. (c) Release beads into well A and incubate with mixing for 15 min. (d) Wash beads in well C for 1 min. (5) Beads transferred to well D and washed for 30 s, and same for wells E, F, and G. (f) Elute phage by mixing beads in H for 2 min.; beads disposed in well F.

    Postelution_capture: (a) Collect beads from well B. (b) Wash beads in well C for 20 s. (c) Release beads into well A and incubate for 10 min., followed by disposal of beads in well C. (d) Wash comb tip in well D for 20 s. (e) Collect beads from well E. (6) Release and wash beads for 20 s. in well F. (7) Release and incubate beads with mixing in well A for 10 min. (8) Dispose of beads in well F.

2.2 Phage ELISA Reagents

Do not autoclave binding buffer solutions (Tween 20), instead, use 0.22 μm vacuum filter.

1. Autoclaved 96-well deep-well plate (cover in aluminum foil to autoclave).
2. Nunc Immuno 96 Microwell Maxisorp Plate Clear (Thermo Fisher Scientific, Waltham, MA, USA).
3. Sterile 96 well ELISA plate (denoted "normal" in protocols).
4. Gas permeable adhesive film to cover deep well plate.
5. Adhesive film to cover plates, non permeable.
6. 1 mg/ml NeutrAvidin in 1× PBS; make 1 ml by first solvating with water, then add PBS and 10 % glycerol to prevent precipitation, store at 4 °C.
7. 10× coating buffer: 1 M $NaHCO_3$, pH 9.6, 50 ml in sterile grad tube, syringe filter, store at room temperature.
8. 10 ml 2 μg/ml NeutrAvidin in coating buffer: 1 ml 10× coating buffer, 20 μl 1 mg/ml NeutrAvidin, 9 ml ddH_2O (made during procedure).

9. 20 ml 1× PBS + 1 % g/ml RNase free BSA (amount is for one plate): 2 ml 10× PBS, 0.2 g BSA, 18 ml ddH_2O; store at 4 °C.
10. 500 ml 1× PBS + 0.05 % Tween 20 in 500 ml squirt bottle: 50 ml 10× PBS, 2.5 ml 10 % Tween 20, 447.5 ml ddH_2O. Make another aliquot in reserve, filter, and store in glass bottle.
11. 500 ml PTEM RNA binding buffer in 500 ml squirt bottle: 1× PBS, 0.05 % Tween 20, 2.5 mM EDTA pH 8.0, 12.5 mM $MgCl_2$ or Mg as determined using the formula in selection, saturating amount of any target ligand. Make another aliquot in reserve, filter, and store in glass bottle.
12. 100 ml PEM RNA binding buffer in 500 ml squirt bottle: 1× PBS, 2.5 mM EDTA pH 8.0, 12.5 ml $MgCl_2$ (or Mg as determined using the formula in selection), saturating amount of ligand.
13. 1 M Phosphoric acid (85 % *o*-phosphoric acid = 15.2 M): 32.9 ml 85 % *o*-phosphoric acid, 467.1 ml ddH_2O.
14. LB/Amp petri dishes.
15. Sterile multichannel pipette 25 ml basins and multichannel pipettes (we have 8 channel 0.5–10 μl, 8 channel 5–50 μl,12 channel 50–300 μl).
16. 2YT/Amp media.
17. M13KO7 helper phage (New England Biolabs, Ipswich, MA, USA).
18. 37 °C shaker incubator.
19. 37 °C petri dish incubator.
20. Orbital titer plate shaker.
21. RNAtag/biotin (same as used in selection).
22. 50 °C water bath.
23. Sheared salmon sperm DNA (Life Technologies, Carlsbad, CA, USA).
24. 1 mg/ml Heparin sodium.
25. 1.5 mM Ribonucleoside Vanadyl Complex (New England Biolabs, Ipswich, MA, USA).
26. Stop RNase Inhibitor RX (5 Prime, Hilden, Germany).
27. Allegra X-12R (or equivalent) centrifuge for deep well plate (Beckman Coulter, Fullerton, CA, USA).
28. Bio-Rad iMark plate reader for microtiter plate at 450 nm.
29. Anti-M13 antibody conjugated with horseradish peroxidase (HRP) (Thermo Fisher Scientific, Waltham, MA, USA).
30. Pierce 1-Step Ultra 3,3′,5,5′-tetramethylbenzidine (TMB), 250 ml (Thermo Fisher Scientific, Waltham, MA, USA).

2.3 Fab Expression and Protein A Purification

1. 2YT/Amp.
2. 250 ml baffled flasks.
3. Incubator shaker for 2.8 l baffled flasks.
4. 2.8 l baffled beaded neck flasks.
5. *E. coli* 34B8 or 55244 cells (ATCC, Manassas, VA, USA).
6. CRAP Fab expression media: 1.785 g $(NH_4)_2SO_4$, 0.355 g NaCitrate·$2H_2O$, 0.535 g KCl, 2.68 g Yeast Extract, 2.68 g HyCase SF Casein. After autoclaving and cooling, add the following filtration-sterilized solutions: 55 ml 1 M MOPS, pH 7.3, 5.5 ml 50 % (g/ml) glucose, 3.5 ml 1 M $MgSO_4$, 0.5 ml 1,000× Ampicillin (100 mg/ml). We make this fresh within at most a few days before expression, and if not used the next day, it is stored at 4 °C. For every half liter needed add the following into water, dissolve, and bring to 0.5 l and pH 7.3. Pour into a 2.8 l baffled flask, cover with aluminum foil, and autoclave.
7. UV–Vis spectrophotometer with disposable plastic cuvettes to check cell turbidity at OD_{600nm}, quartz cuvettes.
8. Serological pipettes (5, 10, 25 ml) and electric pipettor for sterile pipetting of cells, media components, etc.
9. Avanti J-20 XP centrifuge, or equivalent (Beckman Coulter, Fullerton, CA, USA).
10. 500 ml centrifuge bottles.
11. Per half liter cell culture, 30 ml lysis buffer (50 mM Tris–HCl pH 8.0, 5 mM $MgCl_2$) is supplemented on the day of lysis with: 9 mg egg-white lysozyme, 2.5 mg bovine pancreas DNase I, and 150 μl 100 mM PMSF ethanol solution.
12. Stir plates and stir bars to mix lysing cell pellets in 500 ml Oakridge bottles.
13. 4 M NaCl.
14. 50 ml Oakridge tubes.
15. 1 and 5 ml HiTrap protein A columns (GE Healthcare, Pittsburgh, PA, USA), multichannel peristaltic pump, tubing, column fittings, peak tubing, ring stand and clamps to suspend columns, sterile 50 ml graduate tubes for buffers for each column, small graduated cylinder and timer to check flow rate, flow collection beakers.
16. TBS: 10 mM Tris–HCl pH 7.5, 150 mM NaCl.
17. Wash buffer: 10 mM Tris–HCl pH 7.5, 500 mM NaCl.
18. Elution buffer: 0.1 M citric acid pH 3.0.
19. 1 M Tris–HCl pH 8.0.
20. Squirt bottle to wash lines between buffers.

21. Regeneration buffer 0.1 M H_3PO_4.
22. Glass fraction collection tubes, plastic wrap or Parafilm.
23. 15 ml 30 kDa MWCO Amicon concentrator (Millipore, Billerica, MA, USA).
24. Centrifuge for Amicon concentrators.
25. Low salt NaOAc buffer: 10 mM NaOAc pH 5.0, 50 mM NaCl.
26. 3,500 MWCO dialysis tubing, 2 and 4 l beakers for dialysis, dialysis clips.

2.4 Fab High S Purification

1. Ring stands, ring stand clamps, tubing, Econo-columns, and Macro-Prep High S Support resin (Bio-Rad, Hercules, CA, USA).
2. Glass fraction collection tubes, flow collection beakers.
3. Low salt NaOAc buffer: 10 mM NaOAc pH 5.0, 50 mM NaCl, filter sterile.
4. High salt NaOAc buffer: 10 mM NaOAc pH 5.0, 550 mM NaCl, filter sterile.
5. Sterile graduated tubes for buffer mixing and loading.
6. UV–Vis spectrophotometer with quartz cuvettes.
7. 15 ml, 50 ml 30 kDa MWCO Amicon concentrator (Millipore, Billerica, MA, USA).
8. Fab storage buffer: 10 mM Tris–HCl pH 7.5, 50 mM NaCl, filter sterile.

2.5 Fab Dot Blot (Filter Binding) Assay

1. PEMdb binding buffer: 1× PBS, 0.1 mM EDTA, 10 mM $MgCl_2$ or Mg as determined in the formula in selection, any ligand required for RNA target, store at 4 °C.
2. Solutions of all specific Fabs being tested.
3. Radiolabelled cognate and non cognate (control) RNA, acrylic blast shields, Geiger counter.
4. Whatman Protran Nitrocellulose membrane.
5. Hybond N+ membranes (GE Healthcare, Pittsburgh, PA, USA).
6. Bio-dot (filter binding or dot blot) binding apparatus (Bio-Rad, Hercules, CA, USA).
7. House vacuum or other small vacuum pump with a trap (e.g., filter flask) for radioactive solution.
8. Tweezers.
9. Filter paper P8.
10. Sterile basins.
11. Multichannel pipettes.

12. Sterile 15 ml graduated tubes.
13. Sterile 96-well plates.
14. Nonpermeable adhesive seals.
15. 50 °C water bath.
16. Stop RNase inhibitor RX (5Prime, Hilden, Germany).
17. 40 mg/ml heparin.
18. Titer plate shaker.
19. Saran wrap, masking tape.
20. Phosphorimager, screens, screen eraser, and ImageQuant visualization software (Molecular Dynamics, Sunnyvale, CA, USA).
21. Graphing software (KaleidaGraph or equivalent).

2.6 Fab Gel Filtration on AKTA Prime Plus

1. Column running buffer: 10 mM Tris–HCl pH 7.5, 150 mM NaCl to reduce nonspecific Fab interaction with column, degassed every day before running the column (sonication or ~10 min house vacuum).
2. Superdex 200 10 × 300 mm column (GE Healthcare, Pittsburgh, PA, USA).
3. AKTA prime plus chromatography system (GE Healthcare, Pittsburgh, PA, USA).
4. Microcentrifuge tubes.
5. UV Visible spectrophotometer.
6. 15 ml 30 kDa MWCO Amicon concentrators (Millipore, Billerica, MA, USA).
7. Centrifuge for Amicon concentrator.
8. Small dewar for $N_{2(l)}$ to flash freeze aliquoted Fabs.
9. Microcentrifuge.
10. 3 and 0.5 ml sterile RNase free syringes, blunt end injection needles.

3 Methods

3.1 Phage Display Selection of Fabs

Before beginning selection, in addition to purchasing equipment/making reagents for the procedure, the RNA target and the RNAtag selection target must already be transcribed, and phage libraries and M13KO7 helper phage must be available. Libraries should be reamplified every 0.5–1 year due to possible protease Fab degradation. A 29 nucleotides tag is attached to the end of a target RNA to make the RNAtag which anneals to the biotinylated DNA oligo (*see* Fig. 1). The tag may interfere with RNA folding, but this is usually not a problem when it is annealed to its complement. If a tag is not needed in a procedure, use the untagged

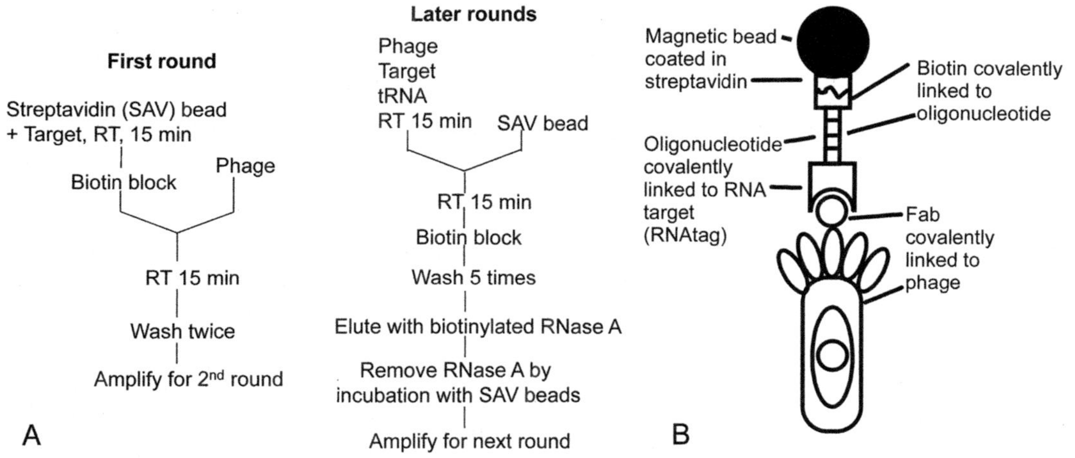

Fig. 1 (**a**) Flow chart of selection rounds. (**b**) Cartoon of macromolecule interactions that occur in selection

target. The theory of selection is to start with a large library (~10^{10-11} diversity) of Fab-phage fusions and with each round of selection enrich the pool of Fabs to contain only those that bind the target specifically and with high affinity. The first step is to do a bead binding assay followed by round one which is different than subsequent rounds. The output from round one is amplified and used as the input in round two. Round two and all other rounds of selection are titered and amplified, and the amplification is used for another round until sufficient enrichment is obtained. It is possible to purify an amplification, do a round of selection, titering, and start the next amplification procedure all in one albeit long day. For a new worker, however, procedures will take longer; the suggestion is to do a round of selection and start the amplification process, followed by amplification purification and the titer the next day.

Figure 1a shows flow charts of selection rounds. Figure 1b is a cartoon of the macromolecule interactions that occur in selection. Streptavidin and biotin bind, an oligonucleotide connected to biotin anneals with the tag of the RNAtag, and the RNAtag is bound by Fab which is fused to M13 phage coat protein III. In later steps, biotinylated RNase A destroys the target to selectively elute the phage and is removed from solution with streptavidin beads to avoid contaminating subsequent rounds.

3.1.1 Phage Handling Precautions

Phage, like RNases, are very stable and easily contaminate reagents, apparatuses, benches, and the entire lab, so handle phage very carefully. Do not spread phage solution on gloves, bench, or floor.

- Use barrier filter tips (pipette tips with aerosol-blocking filter) for handling phage solutions and long reach barrier tips for Oakridge tubes to prevent pipette barrel contamination.

- Place solid disposable labware such as gloves, basins, microcentrifuge tubes, and pipette tips in the biohazardous waste to ensure adequate containment, removal, and disposal.
- Wipe spilled phage solution as soon as possible with a paper towel wetted with 0.1 % SDS solution, as well as bench and pipettes after each experiment.

Treat waste solutions and labware as follows—do not simply discard concentrated phage in the sink or leave phage contaminated labware unattended on the bench or around the sink:

1. Add detergent (e.g., 1 % Alconox final conc.) to phage containing media and reagent waste and keep at RT O/N to kill phage before discarding.

   Resuspend *E. coli* cell pellet from phage amplification with water, add bleach and then add Alconox to approximately 1 %. Keep at RT for O/N.
2. After O/N treatment, pour waste carefully down the drain of the lab sink. Avoid splashing.
3. Before washing labware, soak in a large container with a lid (we have a large plastic storage bin) containing detergent solution (1 % Alconox) overnight. Let all the air out and submerge labware completely.
4. The next day wash the tubes and flasks thoroughly, fill tubes and flasks with water, and wet cycle autoclave. After autoclaving, drain and dry as usual, then dry cycle autoclave for further use.

3.1.2 Bead Binding

First a bead binding experiment must be done to ensure enough SAV beads are used in selection to bind the RNA. Also, the day before any selection round, make a 37 °C, 300 rpm O/N XL1 Blue culture for phage amplification and titer after each round of selection: 2 ml 2YT with Tet in a sterile culture tube.

1. Wash 250 μl of streptavidin beads (make sure to resuspend the beads evenly before pipetting from the bottle) twice with PBS using magnetic holder and resuspend in 250 μl PBS. In the various places in the protocol where it says to wash beads, shake the bead bottle to suspend the beads. Pipet the required volume of beads into a microcentrifuge tube and the place the tube in the magnetic holder. In ~1 min., the beads will separate from the solution, so the storage supernatant can be pipetted from the beads. Add an equal volume to the volume initially pipetted of wash solution (usually PBS), vortex, again magnetically separate out beads, remove solution. Wash a second time, remove solution. The third time keep the solution in the tube to keep the beads (SAV) hydrated until time for use.

2. Aliquot the bead suspension into tubes with following volumes: 3, 6, 15, 30, 60, 120 μl; remove PBS.
3. Dilute 4 μl of the 50 μM RNA-tag/biotin with 76 μl of PTEM.
4. Add 10 μl of RNA-tag/biotin from dilution in **step 3** (25 pmol) to each **step 2** bead tube and shake for 15 min at RT to complex with beads.
5. Use 8 μl of the supernatant from each tube to run a (2–3 %, depending upon target size) agarose gel. Make sure no beads are in the supernatant. Control lane is 8 μl of diluted **step 3** solution.
6. Visually examine the agarose image, and from this, calculate the amount of beads needed to bind 0.5 nmol RNA (it was roughly 1 ml SAV beads for 0.5 nmol P4P6 target).

3.1.3 High Protein First Selection Round

1. Fold RNA-tag/biotin.
   - Add 250 μl PTEM to 10 μl of 50 μM RNA-tag/biotin.
   - 50 °C 20 min, then RT 5 min.
   - Add 2 μl 4 U/μl 5Prime RNase Inhibitor.
2. Prepare phage.
   (a) Pipet 2.7 ml glycerol stock of the phage library into an Oakridge tube. Add 25 ml PBS and 5.4 ml PEG/NaCl, invert to mix.
   (b) Incubate on ice for 5 min, centrifuge 15 min at 27,216 × *g* (15,000 rpm) in JA 25.5 rotor.
   (c) Decant supernatant into phage waste carefully to not disturb pellet. Phage pellets are usually not very large, a white viscous residue on the opposite side of the tube from the centrifuge drive shaft.
   (d) Centrifuge briefly to collect the residual solution and pipet it out with long reach 1 ml barrier tips.
   (e) Resuspend the phage pellet in 750 μl Solution A and transfer into a microcentrifuge tube.
   (f) Add the following to resuspended phage; *see* **Note 1**.
      - 10 μl 20 U/μl Superase Inhibitor.
      - 40 μl 4 U/μl 5Prime RNase Inhibitor.
      - 200 μl 5 mg/ml SAV (removes SAV binders).
3. Wash 1 ml of streptavidin beads (or appropriate amount) with PBS. Never let beads dry. Only remove the last wash immediately before resuspending the beads in a new solution. Start **step 10** for time efficiency.
4. Add RNAtag/biotin to above washed beads and shake for 15 min at RT.

5. Add 5 μl 10 mM free biotin and shake at RT 10 min to block SAV binding.
6. Wash three times with 500 μl PTEM.
7. Add the 1 ml phage solution from **step 2** to the washed beads from **step 6** and shake for 15 min at RT.
8. Wash twice with 1 ml PTEM containing 2 μl 4 U/μl 5Prime RNase Inhibitor.
9. Resuspend in 0.5 ml PBS. If not amplifying same day, store at 4 °C overnight. Do not wait for more than 3 days.
10. To prepare log phase XL1 cells: add 1 ml overnight XL1-blue cells to 20 ml 2YT with Tet; shake at 37 °C for 1–1.5 h; check that OD_{600nm} is between 0.6 and 0.8 A (*see* **Note 2**).
11. To amplify phage add 0.5 ml beads with phage from **step 9** to 5 ml of log phase XL1 cells in sterile culture tube; shake at 37 °C for 20 min.
12. Add M13KO7 helper phage to 10^{10} cfu/ml final concentration; shake at 37 °C for 45 min.
13. Add culture to 30 ml 2YT/Amp/Kan. Shake at 37 °C for 19–20 h.

3.1.4 Phage Purification (Done After Every Selection Round Except the Last)

1. Centrifuge 30 ml 19 h O/N culture which will be input phage for the second selection round, carefully transferring ~30 ml to prevent Oakridge tube overflow (for later rounds, directly pour everything from the 250 ml flask), at 17,418 × *g* (12,000 rpm) in JA25.5 rotor for 10 min. at 4 °C. Chill 11 ml PBS in ice.
2. Transfer the supernatant to a new Oakridge tube for cleaner, more RNase free phage and centrifuge again at 27,216 × *g* (15,000 rpm) in JA25.5 rotor for 10 min at 4 °C.
3. Transfer the supernatant to a new Oakridge tube and add 1/5 culture volume (6 ml) of PEG/NaCl to precipitate phage. Mix by inversion.
4. Incubate for 5 min at 4 °C (on ice), centrifuge at 27,216 × *g* (15,000 rpm) in JA25.5 rotor for 10 min; decant the supernatant into phage waste container.
5. Centrifuge briefly to collect the residual solution and pipet it out into phage waste using long reach barrier tips to prevent phage contamination.
6. Second precipitation further cleans the phage to remove RNase. Resuspend the phage pellet in 10 ml ice-cold PBS, and remove insoluble matter by centrifuging for 5 min at 27,216 × *g* (15,000 rpm) in JA25.5 rotor at 4 °C.
7. Transfer the supernatant containing the phage particles to a new Oakridge tube and add 1/5 volume (2 ml) of PEG/NaCl. Mix by inversion.

8. Incubate for 5 min at 4 °C on ice, centrifuge at 27,216×*g* (15,000 rpm) in JA25.5 rotor for 10 min, and decant the supernatant into phage waste container.
9. Centrifuge briefly to collect the residual solution and pipet it out using long reach barrier tips to prevent phage contamination.
10. Resuspend the phage pellet in 400 μl ice cold PBS and transfer to a new microcentrifuge tube. Set on ice.
11. To save for future use, add 80 % glycerol to bring to 50 % in phage solution (666 μl for every 400 μl phage); mix well and store at −20 °C.

3.1.5 Selection Second Round

1. Pre-fold RNA: 1 μl of 10 μM RNAtag/biotin + 11 μl PTEM, 20 min at 50 °C, 10 min at RT. *For stringent conditions, use 1 μM RNAtag/biotin instead of 10 μM.*
2. PEG precipitate 80 μl of the phage glycerol stock from round 1 **step 11**: add 720 μl 1× PBS and 160 μl of PEG/NaCl, incubate for 5 min at 4 °C on ice, and centrifuge at 18,407×*g* (14,000 rpm) in a tabletop microcentrifuge for 10 min at 4 °C. Pipet the supernatant into phage waste and resuspend the pellet in 100 μl of Solution B.
3. Add above 100 μl phage solution to 60 μl washed SAV beads (beads only, washing buffer removed).
4. Shake at RT for 30 min.
5. Place microcentrifuge tube in magnetic holder to pull beads to side. Transfer supernatant to new tube. Wash beads with 90 μl Solution B, separate beads again, and combine wash solution with supernatant (total volume ~190 μl).
6. Fill wells of KingFisher strips as described in table (**RNA_2nd** program); save the leftover input at 4 °C for titer. *For stringent conditions, use 0.5 μl 100 μg/μl tRNA competitor (competitor to improve Fab binding specificity) instead 10 μg/μl.*
7. Run **RNA_2nd**. This takes about 40 min. At the end of the program, begin subculturing the XL1 by following **step 10**.
8. Transfer contents of well H of **RNA_2nd** strips to well A of new KingFisher strips, and set up the rest of the strips according to the table (**Postelution_capture** program).
9. Run **Postelution_capture**, takes about 25 min. Save well A phage solution in new microcentrifuge tubes and set on ice.
10. To prepare log phase XL1 cells: add 1 ml overnight XL1-blue cells to 20 ml 2YT with

    Tet; shake at 37 °C for 1–1.5 h; check that OD_{600nm} is ~0.6–0.8 A.

11. Reserve 3 μl of eluted output phage for titering in a microcentrifuge tube in ice. The eluted control is also need for titering but does not need to be amplified.
12. If not amplifying on the same day, save remaining output phage at 4 °C; do not wait for more than 3 days. To amplify: add the rest of the eluted phage to 1 ml of log phase XL1 cells in 2YT/Tet; shake 20 min at 37 °C.
13. Add M13KO7 helper phage to 10^{10} cfu/ml, and shake at 37 °C for 45 min.
14. Transfer culture to 30 ml of 2YT with Amp and Kan. Grow at 37 °C for 19–20 h. Next day, purify amplification for use in next round.

3.1.6 Phage Titer (Done After Second and Later Rounds)

1. Grow XL1 in 2 ml of 2YT + Tet at 37 °C O/N the previous day for use in this procedure.
2. To prepare log phase XL1 cells: Add 1 ml overnight XL1-blue cells to 20 ml 2YT with Tet; shake at 37 °C for 1–1.5 h; check that OD_{600nm} is ~0.6–0.8 A.
3. While waiting for cells to grow, make serial dilutions of phage solution with 2YT diluent. The amplified input phage is from **step 6** of second round protocol and is diluted 10^{-5}–10^{-9} ml^{-1} for plating. Eluted output and eluted control phage are both diluted 10^{-1}–10^{-5} ml^{-1} and come from **step 11**.
   - Pipet 90 μl of 2YT in five consecutive wells of a sterile 96 well plate. Three such rows are made, one for amplified input, eluted output, and eluted control.
   - Predilute amplified input to 10^{-4} in microcentrifuge tubes with 2YT.
   - Pipet 10 μl of phage solution into the first of the five wells, one solution for each row. Mix by pipetting.
   - Once phage solution has been added to the first wells, a multichannel pipette can be used to pipet 10 μl of diluted phage solution into the second well. Mix again by pipetting and repeat this dilution procedure for the last three wells.
4. Pipet 45 μl of *E. coli* from **step 2** into wells to make three more rows of five. Add 15 μl of each well of diluted phage from **step 3** into these rows, mix by pipetting, and incubate for 15 min at RT.
5. Mark the bottom of three LB/Amp plates to divide them into five areas in a radial pattern.

   Spot 20 μl *E. coli* cells without phage from **step 2** at the center of the plates to check if the cells are contaminated with phage. Spot 40 μl of each phage-cell mixture from **step 4** onto each of the five areas and spread, being careful not to mix the solutions.

6. Let the plates dry, and incubate the plates at 37 °C O/N.
7. Inspect the plate the next morning; nothing should grow in the middle spot. If it does, the titer plate likely cannot be used; discard it in biohazardous waste and titer again. The most common cause of this is contamination in the cells which retitering often remedies. If not, there may be more widespread contamination. Mixing of the plated solutions also causes this, which would give valid data for unmixed sections. Count the number of colonies on each area and estimate the colony-forming units (cfu) of phage. Use dilutions that are approximately between 20 and 100 colonies for accuracy. For example, say 33 colonies were found in the 10^{-9} dilution factor of the input. Forty microliter were spotted, only 10 μl of which was the diluted phage (15 μl phage/60 μl total = 0.25). The total volume of this that went into selection was 88.5 μl (RNA second program, well A, phage in Solution B). 33 cfu/10 μl × 10^9 dilution factor × 88.5 μl = 2.92 × 10^{11} cfu in original solution for the input. Similarly, exemplar for the 54.13 μl of well H for eluted output and eluted control: 43 cfu/10 μl × 10^3 × 54.13 μ l = 2.33 × 10^5.

   Calculate reduction rate and enrichment values; cfu for input should be on the order of 10^{10}–10^{11} for second and later rounds. Reduction rate should be about 10^7 ideally, though we have observed as low as 10^4.

   – Enrichment = output cfu/control cfu.
   – Reduction rate = input cfu/control cfu.

Second round (and up) **RNA_2nd** program for KingFisher robot: *see* Table 1.

Postelution_capture program for KingFisher robot, *see* Table 2; this program washes the beads, so beads may b used directly from the bottle (wells B and E have a total final volume of 50 μl).

3.1.7 Third (and Subsequent) Rounds

Third and subsequent rounds are the same as the second round, except that more stringent conditions are employed as underlined in steps one and six of the round two protocol. The idea behind selection is to iteratively select for better and better binders, and using more stringent conditions should accelerate this process. If enrichment is negligible with stringent conditions in the third round, then go back and execute round three without stringent conditions followed by a fourth round with stringent conditions, as the initial input was not yet enriched enough in target binders. Our current rule of thumb is that 100-fold enrichment is a good point from which to use the eluted output sans amplification for phage ELISA. If the next round gives 1,000-fold level enrichment, the rule of thumb is to use the previous round's output, as this high enrichment may lead to overenrichment of a few binding Fabs, high degeneracy.

Table 1
RNA_2nd program for KingFisher robot

Well #	Target RNA	Control
A	88.5 µl phage in Solution B 5 µl 4U/µl 5Prime RNase IN 0.5 µl 10 µg/µl tRNA (final 50 µg/ml)[a] RNA/biotin oligo folded in 6 µl PTEM	88.5 µl phage in Solution B 5 µl 4U/µl 5Prime RNase IN 0.5 µl 10 µg/µl tRNA (final 50 µg/ml)[a] 6 µl PTEM
B	30 µl washed beads in 100 µl PTEM	30 µl washed beads in 100 µl PTEM
C	100 µl PTEM/gly 1 µl 0.5 mM free biotin	100 µl PTEM/gly 1 µl 0.5 mM free biotin
D	100 µl PTEM/gly	100 µl PTEM/gly
E	100 µl PTEM/gly	100 µl PTEM/gly
F	100 µl PTEM/gly	100 µl PTEM/gly
G	100 µl PTEM/gly	100 µl PTEM/gly
H[b]	50 µl PBS 3.13 µl 80 % glycerol 1 µl 110 ng/µl biotinylated RNase A	50 µl PBS 3.13 µl 80 % glycerol 1 µl 110 ng/µl biotinylated RNase A

[a]For stringent conditions, use 0.5 µl 100 µg/µl competitor tRNA instead 10 µg/µl
[b]For better pipetting as viscosity causes issues here, make stock solution in microcentrifuge tube with one extra 54.13 µl aliquot; pipet solutions into well H. Use filter tip for RNase A and change gloves afterwards

Table 2
Postelution_capture program for KingFisher robot

Well #	Target RNA	Control
A	Well H of **RNA_2nd**	Well H of **RNA_2nd**
B	20 µl unwashed beads 30 µl PBS	20 µl unwashed beads 30 µl PBS
C	50 µl PBS	50 µl PBS
D	100 µl PBS	100 µl PBS
E	20 µl unwashed beads 30 µl PBS	20 µl unwashed beads 30 µl PBS
F	50 µl PBS	50 µl PBS
G	Empty	Empty
H	Empty	Empty

3.2 Phage ELISA to Choose RNA Binding Fab Clones

Once sufficient enrichment has been obtained, the next step is Phage ELISA, the first screening tool to choose Fab clones based on their target specificity. Figure 2 shows a diagram of the macromolecule connections in a well of a Maxisorp plate in Phage ELISA

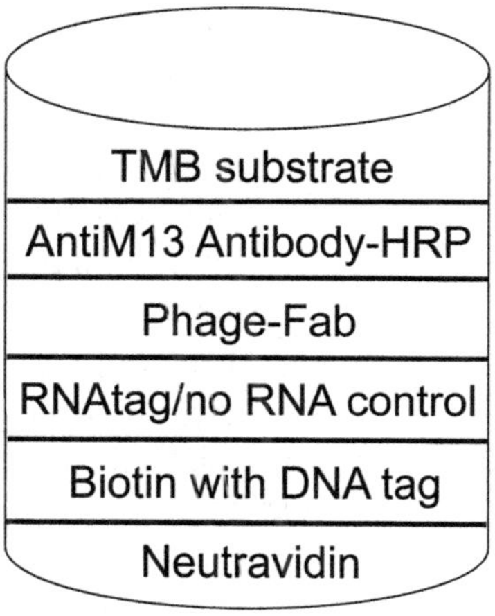

Fig. 2 Macromolecule interactions in a well of a Maxisorp plate in Phage ELISA against an RNA target

against an RNA target, which is similar to that in selection with the addition of an anti-M13 phage antibody conjugated to HRP (horseradish peroxidase). TMB, tetramethylbenzidine, is a substrate for HRP, and the color change is observed in a plate reader. ELISA requires fresh colonies; if a fresh O/N titer plate is not available, the eluted output is regrown as in titering and plated at whatever dilution is needed to obtain 48 individual colonies (48/one ELISA plate). ELISA is a several day process: first fresh colonies must be grown overnight if a fresh titer plate is not available, the next day is inoculation from the fresh plate into a deep well plate for growth overnight (**step 1** below) and NeutrAvidin coating of the Maxisorp plate (**step 2**), and the last day is the ELISA procedure.

1. Grow 48 colonies from a phage titer plate; these colonies have to be from a fresh plate, which means on the same day the plate is obtained from the incubator in the morning, use the colonies in the afternoon. Each colony will be both spotted and inoculated.
   (a) Draw a (48) numbered grid on the bottom of an LB/Amp plate to allow easy retrieval of desired spotted clones later.
   (b) Fill 48 wells of a previously autoclaved 96 well deep well plate with 400 μl 2YT + Amp + 10^{10} ml^{-1} M13KO7 helper phage (this is facilitated by use of multichannel pipettes and a sterile basin, which is discarded in the biohazardous waste).
   (c) We use a 10 μl pipette tip to poke a titer plate colony, then gently spot into one grid square of the petri dish, then inoculate (pipet solution several times to inoculate bacteria) into a well in the deep well plate.

(d) Repeat for all colonies, and cover the deep well plate with breathable (gas permeable) adhesive film. Incubate the deep-well plate at 37 °C, 300 rpm shaking (~20 h, so start growing about 2 p.m. until 10 a.m. next day; we have gone as far as ~24 h to obtain growth since bacteria grow slower with phage). Incubate the petri dish at 37 °C O/N, then save at 4 °C.

2. Fill all 96 wells of the Maxisorp plate with 100 μl 2 μg/ml NeutrAvidin in coating buffer (0.1 M $NaHCO_3$, pH 9.6), cover plate with adhesive film, and store at 4 °C for use the next day.
3. Next day about an hour before end of deep well plate growth, remove coating buffer from Maxisorp plate. To do this quickly and easily in this and subsequent steps, lay out a layer of paper towels on the chemtop in an area that is not a regular bench-work area (to prevent contamination), such as next to a sink. Invert the plate to discard most of the buffer into the sink, then vigorously knock the inverted plate against the paper towels to remove the rest. Block each well with 200 μl PBS, 1 % BSA for 1 h at RT with shaking on a small titer plate orbital shaker; reuse the film to seal the plate while shaking.
4. During this hour, fold RNAtag/biotin:
   - 2.5 μl 50 μM RNAtag/biotin + 97.5 μl RNA binding buffer, 50 °C 20 min; RT 10 min.
   - Add to 4.9 ml RNA binding buffer at RT in a new sterile 15 ml graduated tube and mix.
5. At end of RNA incubation and hour of Maxisorp plate blocking, remove blocking buffer and wash Maxisorp plate four times with 1× PBS/0.05 % Tween 20. To wash, add buffer using 500 ml squirt bottle, empty plate into sink, and knock on paper towels to remove the rest. Repeat four times.
6. Incubate wells in odd numbered columns with 100 μl/well 25 nM RNAtag/biotin in RNA binding buffer, 30 min at room temperature. Even numbered columns receive RNA binding buffer (no RNA control). Sterile basins and a multichannel pipette expedite this.
7. Make 7 ml phage dilution solution (1.5× RNA binding buffer, 0.75 % BSA, 150 μg/ml sheared salmon DNA, 1 mg/ml heparin, 1.5 mM vanadyl ribocomplexes): 1,050 μl 10× PBS, 52.5 μl 10 % Tween 20, 52.5 μl 500 mM EDTA pH 8.0, 131 μl 1 M $MgCl_2$, saturating amount of any ligand, 52.5 mg BSA, 105 μl 10 mg/ml sheared salmon DNA, 175 μl 40 mg/ml heparin, 52.5 μl 200 mM vanadyl ribocomplex, 3 μl 4 U/μl RNase Inhibitor, ddH_2O to 7 ml, in sterile 15 ml graduated tube; do not filter.

8. Pour solution in a sterile basin and use multichannel pipette to transfer 140 μl/well phage dilution solution to columns 1–6 of a normal sterile 96 well ELISA plate.
9. Centrifuge deep well culture plate for 15 min, 2,851 × *g* (3,500 rpm) in a swing-bucket rotor with Beckman Coulter Allegra X-12R. If centrifuging is done and sitting while other steps are being completed, give the plate another brief centrifugation immediately before use to ensure cells are pelleted.
10. Wash the Maxisorp plate in **step (6)** six times with RNA binding buffer in a squirt bottle, invert into sink, and knock on paper towels to remove remaining solution.
11. Align washed and emptied Maxisorp plate, centrifuged deep well plate, and phage dilution solution filled Normal ELISA plate from left to right in the same orientation. Using filter tips with multichannel pipette, transfer 70 μl phage supernatant from deep well plate (be careful to not pipet cell pellet) column 1 to normal ELISA plate column 7 for short term back up storage for an assay repeat, then transfer another 70 μl from deep well plate column 1 to normal ELISA plate column 1. Mix normal ELISA plate column 1 by pipetting, then adjust pipette to 100 μl and transfer 100 μl to Maxisorp plate column 2 (no RNA control), then another 100 μl to Maxisorp plate column 1 (RNA column), eject tips. With a fresh set of eight tips each time, repeat this procedure for the rest of the columns to fill the entire Maxisorp plate. Seal and shake the Maxisorp plate for 30 min. With orbital titer shaker, seal and save the normal ELISA plate at 4 °C, and discard the deep well plate in biological waste.
12. Make diluted anti-M13 antibody/HRP conjugate solution (10 ml, diluted 5,000× in RNA binding buffer, 0.5 % BSA, 1 mM vanadyl ribocomplexes): 10 ml RNA binding buffer, 50 mg BSA, 50 μl 200 mM vanadyl ribocomplex; dissolve, then add 2 μl anti-M13 antibody/HRP conjugate. Place mixture into sterile basin for pipetting.
13. Wash Maxisorp plate six times with RNA binding buffer; add 100 μl diluted Anti-M13 antibody/HRP conjugate to each well. Shake for 30 min at room temperature.
14. Wash six times with RNA binding buffer, then twice with RNA binding buffer without Tween 20.
15. Pour ~10 ml TMB substrate into a sterile basin and add 100 μl across the rows of 12 with a multichannel pipette in the Maxisorp plate to start the reactions on both RNA containing reaction (i.e., A1) and its control (i.e., A2) simultaneously. Wrap in aluminum foil to block light, and incubate 5–10 min at room temperature. Positive wells usually turn blue within a few minutes (*see* **Note 3**).

16. Quench with 100 μl 1 M phosphoric acid (positive wells turn yellow immediately), again adding across the rows with a multichannel pipette to the Maxisorp plate to quench both RNA containing reaction and its control simultaneously.
17. Measure absorbance at 450 nm in a plate reader. The ratio of a well with RNA to its no RNA control well will be used to choose phage-Fab clones, so if an RNA-containing odd well seems particularly colorless (no phage binding), it will not be a viable Fab clone. Either well in such a pair can be rinsed out thrice with ddH_2O by pipetting away solution, then water added for use as a blank. Care must be taken with very high intensity wells. Intensity may bleed into neighboring wells, or they may be out of the range of the instrument. Dilute the well in question and remeasure it until the results are consistent to ensure it is in range. Remove an overly intense solution from its well and remeasure the adjacent wells to ensure their correct reading. Once obtaining accurate readings, those with high ratios (RNA/no RNA control) are potential specific binders. A ration of one (or lower) is nonspecific binding. Several (10–20) higher ratio clones can be chosen for sequencing.

Clones chosen from the previous data are identified by numbering on the 48 grid LB/Amp petri dish. These clones are inoculated and minipreped to harvest a small amount of Fab phagemid to send for sequencing. Returned sequences are checked to see if they are error free Fabs. It is not uncommon to receive degenerate sequences of several clones; only one of each unique clone is carried forward to the next step. For each unique clone, phage are used to isolate single stranded DNA for Kunkel (*dut*$^-$ *ung*$^-$) cloning to incorporate a TAG stop codon, breaking the Fab-phage fusion for expression. Colonies from the Kunkel plate of transformants are inoculated (small amount works better than larger) as template into a colony PCR reaction to check which contain TAG and can therefore express isolated Fab. The glycerol stock of colonies with PCR products is used for expression.

3.3 Fab Expression and Purification with Protein A Column

Once glycerol stocks for each Fab clone are made protein expression and purification can begin. The first steps are expression of Fab in *E. coli* 34B8 (or 55244) followed by protein A affinity purification. Every effort should be made to execute the protein A procedure that begins in **step 6** consecutively with High S (next protocol) the next day to prevent protease degradation of Fabs. If not, freeze the fractions after protein A, but the yield may be poor. Protein A binds to framework residues of the Fab heavy chain variable region; protein A chromatography is done at 4 °C to preserve the Fab. The first day is growth of an O/N starter culture (**step 1**), and this is a convenient time to make the CRAP media (**step 2**) required for the next day's inoculation for 24 h growth (**step 3**).

On day 3 after the 24 h expression, the cell pellet is frozen for 2 h (**step 5**). Protein A chromatography could be done after freezing (starts with **step 6**), but it is advisable to perform the procedure the next day due to time constraints. Initially a small half liter test CRAP culture is expressed since not all Fabs express, the MOPS in CRAP media is somewhat expensive, and Fabs must first be checked for affinity and specificity (later procedures) to determine their viability as crystal chaperones. Once a particular Fab's yield, affinity, and specificity are determined, we pellet a large volume of cell culture for large scale protein A chromatography (eventual use in crystal trays). The procedure below includes both 1 ml column volume (CV) (small test expression) and large scale 5 ml CV protein A purification. Yields are in the range of 1–4 mg Fab/L CRAP. The most that we have done is four 1 ml columns for four unique Fabs' test expressions or two 5 ml columns of the same Fab in 1 day with the UV–Vis spectrophotometer and four channel peristaltic pump set up (automated protein purification can increase efficiency). When doing multiple columns, it is suggested to complete them simultaneously through washing (**step 12**), then release the pump lines and elute them individually.

1. Inoculate 25 ml 2YT/Amp in a 250 ml baffled flask with the Fab glycerol stock (or single colony from a fresh transformation), one flask for each Fab clone being expressed. Incubate starter culture at 30 °C, 300 rpm in large cell shaker O/N.
2. Make CRAP media (*see* Subheading 2). For each unique Fab, start with 0.5 l test culture (*see* **Note 4**).
3. Next day, check OD_{600nm} of 25 ml starter culture (100 μl culture + 900 μl LB/Amp) in disposable plastic cuvette (media blank). Dilution should be 0.6–1 A, so culture 6–10 A. Inoculate 0.5 l CRAP/Amp in 2.8 l baffled flask with 1/100 volume (5 ml) of this starter culture. Shake culture at 300 rpm and 30 °C for 24 h. Next day again check OD; should be ~6–10 A.
4. Centrifuge CRAP culture at 6,693–11,899 × *g* in JLA 10.5 for 10 min at 4 °C, discard supernatant; only fill bottle about halfway to prevent spilling. Centrifuge rest of culture volume in same large centrifuge bottle, discard supernatant.
5. Freeze pellets at −20 °C for over 2 h (or overnight, stopping place in protocol).
6. Take pellets out to thaw and make lysis buffer (*see* Subheading 2). To avoid occasional protein precipitation after elution in **step 14**, purify at most 3 l culture per one 5 ml protein A column.
7. Suspend previously frozen pellets in the above lysis buffer first; then with stir bar, stir vigorously for 60 min at RT. Before starting 60 min., ensure no chunks are still present for best lysis/Fab yield.

8. After 60 min, while stirring, slowly (~1 min) add 3.7 ml 4 M NaCl to above 30 ml lysate (discount volume expansion from pellet resuspension, final [NaCl] = 0.5 M).
9. Spin down in Oakridge tubes in JA 25.5 rotor at 27,216 × *g* for 45 min. Transfer to new Oakridge tubes and spin down for additional 15 min. Transfer supernatant out to a reused, cleaned graduated tube (discard after this use) immediately for loading on protein A column (*see* **Note 5**).
10. While centrifuging, equilibrate the 1 ml protein A column (when purifying >2 l cell culture, use 5 ml protein A column) with 20 CV of TBS at 1 ml/min (5 ml/min for 5 ml column). 1 ml/min ~1 drop 2.5 s^{-1}.
11. Load supernatant from **step 9** onto 1 ml protein A column in refrigerator (4 °C) at 1 ml/min (3 ml/min for 5 ml protein A column). We have our protein A columns and associated setup in a dedicated double glass door refrigerator at all times (and a second refrigerator for High S).
12. Wash with 50 CV of 10 mM Tris–HCl pH 7.5, 500 mM NaCl at 1 ml/min (5 ml/min for 5 ml column). This is a good pause point in the procedure if a break is needed, as the Fab is washed as clean as it will be in this purification. Prepare for fraction checking by turning on UV.
13. Elute column with 0.1 M citric acid pH 3 at 1 ml/min (3 ml/min for 5 ml column); begin to collect 3 ml per tube in tubes prefilled with 1.5 ml 1 M Tris–HCl pH 8.0 immediately after switching line. Mix immediately after elution. For 1 ml columns, three fractions may suffice, while for a 5 ml column, try six fractions. Do not regenerate until last fraction checked shows no protein (*see* **Note 6**).
14. After all the protein is eluted, check fractions in cuvette with UV spectrometer, blanking with Tris prefilled in tubes above. Fractions with a large quantity of protein will be bubbly and slightly viscous. Discard fractions with a negligible yield of protein (~less than 5 %) or with a large amount of DNA contamination as measured by UV for purity.
15. Regenerate the column with 3 CV of 0.1 M H_3PO_4, followed by equilibration with 20 CV of TBS at 1 ml/min (5 ml/min for 5 ml column). Cap the column when not in use to prevent dehydration. Use plastic wrap to cover glass fraction tubes and store overnight at 4 °C for small scale protein A; tomorrow do **step 16** while equilibrating High S resin. For large crystallization, begin overnight dialysis today described in **step 16** for High S tomorrow.
16. Pool and concentrate small scale protein A Fab fractions in 15 ml 30 kDa MWCO Amicon in centrifuge at 4 °C, 2,851 × *g*.

Do not let the solution go so low that the Fab precipitates; we typically stop at ~200 μl.

17. Buffer exchange into low salt NaOAc buffer with the Amicon concentrator. It is good practice to completely mix the protein concentrate with fresh buffer with a pipette each time more buffer is added, as over centrifugation may cause Fab precipitation.

    For large crystallization samples, dialyze pooled Fab fractions in low salt NaOAc buffer overnight in 4 °C. We use 4 l of dialysis buffer in three aliquots per every 6 l of cell culture.

18. Soak an appropriate length of 3,500 MWCO (keeps out RNases which are ~10 kDa) for dialysis. If precipitate occurs in the dialysis bag, remove the pellet by centrifugation and redialyze. For the same protein, next time dilute 2× with low salt NaOAc buffer before dialysis. Take great care so that gloves, beaker used to soak dialysis tube, etc. is very clean and will not contaminate Fab.

3.4 Fab High S Purification

The next round of purification is gravity flow High S cation exchange chromatography to remove proteases, nucleases, and further make the Fab homogeneous. All the buffers, tubes, econo columns, etc. must be kept very clean to limit RNase contamination. The yield from the High S column will decrease 30–50 % from that obtained from the protein A column, and the $OD_{260nm-320nm}/OD_{280nm-320nm}$ ratio will decrease as the Fab becomes cleaner approaching ideally in the limit 0.53 or slightly below. For initial 0.5 l expressions, High S purification is only done once, after which comes affinity and specificity tests. For large scale purifications for crystallization, two rounds of High S column chromatography are performed for extra purity. After the second purification, the Fab is tested with 5′ radiolabelled noncognate RNA for RNase contamination (no more than ~7 % compared to buffer alone) so that the final gel filtration column is not contaminated.

1. Fab purified from protein A column has been pooled and buffer exchanged into low salt NaOAc buffer.
2. To a cleaned econo column, add 5 ml low salt NaOAc buffer, then add required amount of High S resin suspension (manufacturer's listed maximum capacity of High S resin: 45 mg IgG/ml). For Fab < 2 mg, pipet 1 ml suspension (0.5 ml CV); 2 mg < Fab < 5 mg, pipet 2 ml (1 ml CV); for 5 mg < Fab < 15 mg, 4 ml suspension (2 ml CV). Let resin settle, then drain column completely. Wash the column with 20 CV of low salt NaOAc buffer to remove resin storage solution, drain completely.
3. Add the Fab solution from **step 1** onto the column and collect the flow through.

4. Wash with 5 CV low salt NaOAc buffer (final 50 mM NaCl). Then, begin gradient in increments of 10 %, e.g., 90 % low salt/10 % high salt (final 100 mM NaCl), 80 % low salt/20 % high salt (final 150 mM NaCl), or until no protein comes out as checked by UV at 280 nm. Collect in ~4.5 ml fractions. Let one gradient drip completely/almost through column before adding the next. If no fractions show UV protein absorbance, try a higher salt buffer than 100 % of the high salt buffer here, as occasionally some Fabs require more salt to elute.
5. Discard fractions with negligible amount of protein for purity and homogeneity (we discard 5–10 % of a yield, less on the second round of High S). An SDS PAGE can also be a helpful tool as often smaller fragments other than the ~48 kDa Fab are present; it is prudent to discard a low yield fraction that contains a large quantity of small fragments for purity. Empty resin from column and rinse for next use. For non-crystallization purposes, go to **step 6**; for crystallization purposes, go to **step** 7 (*see* **Note** 7).
6. For non-crystallization purposes, pool and buffer exchange with a new Amicon into Fab storage buffer, check concentration with UV (will need dilution for UV; stock is at least 20 μM for dot blot assay), then aliquot into microcentrifuge tubes and store in −20 °C. With the Fab sequence we obtain a calculated molecular weight and molar absorptivity from web.expasy.org/protparam/ to determine molarity.
7. For crystallization purposes, freeze the fractions and purify all remaining cell pellets through one High S to minimize protease degradation, then combine all fractions and purify again through one High S column. After second High S, use a new Amicon to concentrate fractions for loading onto gel filtration column. We load no more than 5 mg in 250 μl in one run on the column.

3.5 Fab Dot Blot (Filter Binding) Assay

After a test culture has been purified once through High S for each Fab, the next step is to check for specific binding in a native polyacrylamide gel using heparin as an anionic competitor. Fabs that complex with RNA in the presence of heparin as shown by a shift in the RNA band (EMSA, electrophoretic mobility shift assay) are then used in a dot blot (filter binding) assay to determine their binding affinity, K_d. Gamma ^{32}P ATP is used to radiolabel untagged cognate RNA target, and a noncognate radiolabelled RNA is used as a control. Two replicates are done for each RNA/Fab pair to determine an average K_d; one set of membranes does eight assays (rows) of ten data points each. A phosphorimager and plate is used to measure the radioactivity, the data points are plotted, and a curve is fit to determine the K_d.

1. Dilute Fabs. These dilutions must be done accurately, as they determine the K_d. Especially the higher concentrations will be viscous, so pipet carefully. Using a normal sterile 96 well plate, make dilutions using PBS of 50 μl of 20 μM, 10 μM, 5 μM, 2.5 μM, 1.25 μM, 0.625 μM, 0.3125 μM, 0.156 μM, 78 nM, 39 nM in for each Fab. Label each column as one concentration and each row as a unique Fab. Seal with adhesive cover, and save at 4 °C.
2. Equilibrate the membranes. Pour PEMdb binding buffer into two sterile plastic containers such as pipette tip lids. Lay out filter paper on chemtop, use clean (RNase free) tweezers to remove cover from Hybond membrane, grab membrane on corner and lay on filter paper, and write "H" on lower right corner with Sharpie for later reference. Immerse Hybond membrane in binding buffer solution in one container, using tweezers and moving container slightly side to side to completely submerge. Trim nitrocellulose to matching size, mark with N at lower right corner, and repeat above for this membrane in second container. Incubate both for 15 min.
3. For one entire 96 well plate, fold 130 × 36 μl = 4,680 μl cognate RNA sample. (Extra sample volume makes it easier to use multichannel pipette basin.) Mix 2 μl radioactive RNA (~650 kcpm ideal, though can use 140) and 4,546 μl binding buffer in 15 ml graduated tube and incubate for 10 min at 50 °C, then RT for 10 min. Add 2 μl 5Prime RNase inhibitor and 130 μl 40 mg/ml heparin and mix well. (Non cognate sample will need to be folded separately, so volume could be split in two.)
4. Combine Fab and RNA. Behind an acrylic shield, pour the folded RNA sample in a multichannel basin, and transfer 36 μl folded cognate radioactive RNA sample using a multichannel pipette to columns 2 through 11 of a normal sterile 96 well plate. (Radioactive tips must be disposed of in solid radioactive waste.) Take the Fab 96 well plate out of the refrigerator, and take 4 μl of each well in a given row (given Fab) and add to a row of the RNA plate. In this way, one complete row across in the RNA plate will now contain a complete series of Fab dilutions. Do in duplicate for each Fab–RNA pair. So, one Fab will have two replicate rows with its cognate target and two replicate rows with the non cognate control.
5. Cover the Fab–RNA complex well plate with adhesive seal and shake gently on the titer plate shaker for 30 min. at room temperature. Pour excess RNA-heparin solution into liquid radioactive waste container; discard the multichannel basin in solid radioactive waste container. The Fab plate can be resealed and stored at 4 °C for several days if a repeat of the assay is needed.

6. Assemble the dot blot apparatus. Blot excess solution from membranes with filter paper and layer them on the apparatus with nitrocellulose membrane on top of Hybond membrane and the N and H at the bottom right hand corner. Tighten diagonally, then apply vacuum and tighten again. Between the apparatus and vacuum use a filter flask as a trap for radioactive solution.
7. Remove the Fab–RNA plate from the shaker and place it behind a shield near where the dot blot apparatus will be used. Pour binding buffer into a multichannel basin and set on ice. Using a multichannel pipette, transfer 100 μl cold binding buffer to first five wells of a row of the dot-blot apparatus with one multichannel pipette (these are reused). As soon as all five solutions are filtered through the membrane, use another multichannel to transfer 40 μl of the complexed Fab-RNA solution from above to the corresponding wells on the dot-blot apparatus. Discard radioactive tips. Immediately after the complex solution has run through membrane, wash with another 100 μl cold binding buffer to prevent the data points from smearing. Allow the solution to drain, and repeat this for the rest of the data points. Place Fab–RNA complex plate in radioactive solid waste.
8. After all samples are done, apply vacuum for extra 5 min. Tilt and drain any excess solution in bottom of apparatus out the vacuum line so it does not rewet the membranes. Carefully disassemble dot-blot apparatus while vacuum is on, and air dry membranes in hood for 15 min., turning over partway through to dry other side. Rinse apparatus thoroughly and leave in designated radioactive area if it still retains some counts.
9. While the membranes are drying, erase the phosphorimager screen.
10. Wrap membranes in saran wrap securely, and securely tape in phosphorimager cassette, placing the H and N in the upper right hand corner (portrait orientation). When the image is scanned after O/N exposure, the less bound RNA (wells with less Fab) will be at the top. Survey laboratory area where experiment was conducted for radiation, and clean up spills.
11. The next day, scan the membrane with PhosphorImager SI and quantify the image in ImageQuant. Nitrocellulose binds protein (ergo Fab-bound RNA), and Hybond binds (unbound) RNA. Fitting a curve with KaliedaGraph to a given Fab's row of percent binding data points (Bound RNA/(Bound RNA + Unbound RNA)) gives the K_d (*see* **Note 8**).

3.6 Fab Gel Filtration on AKTA Prime Plus

After two rounds of High S purification and verified negligible nuclease contamination, Fab gel filtration is performed with a Superdex 200 10 × 300 mm column for final polishing, removing proteases/RNases and increasing Fab homogeneity. Most of the running information can be obtained from the user's manual of the AKTAprime plus and will not be repeated here. This, like High S and protein A chromatography, is done at 4 °C. For Fab purity, discard peak fronting and tailing (~5–10 %), which gives ~70 % recovery yield at the end of the gel filtration process. Buffers should be degassed before use; be careful not to get air in line when changing buffers. For current targets we have not purified a Fab–RNA unit complex for use in crystal trays, though that is an option.

1. The ~250 μl and ~5 mg Fab sample (for small loss) is centrifuged in a benchtop centrifuge and transferred to a new microcentrifuge tube to remove any precipitate. Flush the 0.5 ml sample loop with 3 ml of running buffer, then inject sample. A small volume of sample is retained in the needle to avoid loading air.
2. The sample is monitored at 254 nm. At a flow rate of 0.4 ml/min, we collect ~0.5 ml fractions that begin to elute ~45 min. and continue to collect after the peak has largely flattened for a few minutes due to tubing length from detector to outlet.
3. Between runs we equilibrate the column with 2 CVs (1 CV = ~25 ml) of buffer.
4. Once fractions have been chosen for crystallization with UV (discard 5–10 %), use a new 15 kDa MWCO Amicon to buffer exchange into Fab storage buffer and concentrate to ~14–24 mg/ml (dilute to check by UV) for use in crystallization.
5. Aliquot in 50 μl fractions, immediately freeze in $N_{2(l)}$, and store in −80 °C.

Once Fabs are available, a crystallization quantity and purity of RNA must be produced. After obtaining RNA, a native polyacrylamide EMSA is done with one equivalent of RNA to 1, 1.1, 1.2, 1.3, and 1.4 equivalents of Fab to establish binding stoichiometry. Once this is done, RNA is folded and complexed with Fab at a complex concentration of 12 mg/ml, which we combine with an equal volume of well solution. We set up Hampton Research Natrix, Crystal Screen, and Index, both 1 and 2 screens (288 wells) at both 4 and 15 °C for each Fab. From these initial screening trays, optimization can be done for eventual crystal fishing and mailing for synchrotron diffraction. Optimization is a multifaceted endeavor and cannot be described thoroughly here, though efforts include optimizing precipitate, salt, pH, and the addition of additive screen reagents, as well as seeding methods. Post crystallization treatments such as dehydration can also on occasion assist in diffraction. The final step is to solve the phases and obtain the structure from the diffraction data, a complicated process not addressed here.

4 Notes

1. The choice of RNase inhibitor continuously evolves as better inhibitors are made. Recently, we have tested and found that Superase Inhibitor is more effective at preventing degradation from remaining RNase contamination in phage stock purified from *E. coli* compared to 5Prime Stop RNase Inhibitor. We continue to use Stop RNase Inhibitor for the bulk of our inhibitor needs (in vitro reactions), as Superase is more expensive. In the past, we have also attempted to use vanadyl ribocomplex as an inhibitor in selection, which gave strange ratios. The reduction rate varied widely, and some rounds would result in no enrichment, while repeating the round would sometimes have some enrichment. We suspect some type of nonspecific phage binding to the complex and/or that the complex may be retained with the beads since vanadium is paramagnetic, retaining noncognate phage. It is still used as an inexpensive inhibitor in ELISA, however. Read the manufacturer's instructions, as it may require heating to solvate prior to usage.
2. If the cells take much longer than this to grow, they may be contaminated with other cells or phage. Discard this overnight stock, grow another, and try again the next day. Also, we have used cells that were ~1.1 A.
3. It is important that the TMB be relatively fresh; after about a year, it may not work as well. The choice of TMB is also important; some TMB we have used gave only a weak color change. The order number we have achieved good results with is listed in Subheading 2.
4. We have found that some optimization of CRAP culture volume (increment between 250 ml to 1.5 l) for best yield may be required when expressing in a new shaker or location; the yield may be very sensitive to oxygenation. Also, in **step 3**, we have tried different volumes of the starter culture, and 1 ml has worked as well as the prescribed 5 ml.
5. The following is various useful information on protein A chromatography.
   - We have found that the protein A columns are fairly robust, but it is never a good idea to let air get in a column. If Fab is on the column when it goes dry, heavy yield losses will be incurred.
   - We reuse columns multiple times, though yield should be monitored, such as with a known Fab (Fab yields are somewhat variable but should not alter more than 50 % as a rule of thumb) to ensure the column is not degrading and Fab yield is being lost.

- Cleaning of slow/clogged columns can be accomplished per the manufacturer's instructions.
- Flow rate should be checked about every 15 min, as it is prone to variability with the peristaltic pump tubing set up.
- Litmus paper can be used to monitor buffer changes.
- We rinse feed lines with a 500 ml squirt bottle of ddH_2O when switching solutions to avoid cross contamination and buildup of RNases and proteases in buffers.
- Inspect feed lines after a long period of disuse and change line if mold or other contaminants are present.

6. Occasionally some Fabs do not express, express very poorly, or are heavily contaminated with DNA (apparently DNA binding). A second expression and purification can be attempted, but when the problems persist, these are discontinued. If Fab is contaminated with DNA $[(A_{260nm} - A_{320nm})/(A_{280nm} - A_{320nm}) \sim \leq 0.56$ for clean Fab], a second purification can be also be attempted on the same batch using 100 CV of 0.5 M phosphate, 0.5 M citric acid, pH 4.0 as a wash buffer. For the occasional very well adhered Fab, we have a lower pH elution buffer: 100 mM H_3PO_4, 50 mM NaH_2PO_4, 140 mM NaCl, pH 2. Neutralize with 0.9 ml (1 M NaH_2PO_4, 140 mM NaCl, pH 8.6) per 3 ml elution buffer.
7. One method of cleaning tubing and econo columns, as well as other labware that is contaminated with RNase, is soaking in 0.1 M nitric acid for 15 min. followed by a thorough rinse. To make ~0.1 M nitric acid, add 10 ml ~70 % w/w nitric acid to 1,490 ml ddH_2O.
8. At the high Fab end of a dot blot assay, we sometimes see the top two or three data points drop below the expected curve. We attribute this to two things: higher RNase contamination with a higher concentration of Fab, and secondly nonspecific aggregation of the concentrated Fab. We have seen both of these phenomena in other assays using similar dilutions of Fab as well. These obviously incorrect data points are discarded from a curve fit. Occasionally a data point on the low end will be lower than the rest of the visually reasonable curve; we attribute this to being out of the linear range of the phosphorimager apparatus and also discard it. Our current rule of thumb is that the K_d should be ≤250 nM for a crystallization Fab. Fab should also be complexed with RNA in an in-line probing or HRFP assay to check that the target's global folding is not altered in comparison with an assay without Fab. This may also reveal the Fab's binding interface on the RNA.

References

1. Sabin LR, Delas MJ, Hannon GJ (2013) Dogma derailed: the many influences of RNA on the genome. Mol Cell 49:783–794
2. Serganov A, Nudler E (2013) A decade of riboswitches. Cell 152:17–24
3. Serganov A, Patel DJ (2007) Ribozymes, riboswitches and beyond: regulation of gene expression without proteins. Nat Rev Genet 8: 776–790
4. Pavon-Eternod M, Gomes S, Geslain R, Dai Q, Rosner MR, Pan T (2009) tRNA overexpression in breast cancer and functional consequences. Nucleic Acids Res 37:7268–7280
5. Gutschner T, Diederichs S (2012) The hallmarks of cancer: a long non-coding RNA point of view. RNA Biol 9:703–719
6. Nakatani J, Tamada K, Hatanaka F, Ise S, Ohta H, Inoue K, Tomonaga S, Watanabe Y, Chung YJ, Banerjee R, Iwamoto K, Kato T, Okazawa M, Yamauchi K, Tanda K, Takao K, Miyakawa T, Bradley A, Takumi T (2009) Abnormal behavior in a chromosome-engineered mouse model for human 15q11-13 duplication seen in autism. Cell 137:1235–1246
7. Ziats MN, Rennert OM (2013) Aberrant expression of long noncoding RNAs in autistic brain. J Mol Neurosci 49:589–593
8. Powell WT, Coulson RL, Crary FK, Wong SS, Ach RA, Tsang P, Alice Yamada N, Yasui DH, Lasalle JM (2013) A Prader-Willi locus lncRNA cloud modulates diurnal genes and energy expenditure. Hum Mol Genet 22:4318–4328
9. Sibbritt T, Patel HR, Preiss T (2013) Mapping and significance of the mRNA methylome. Wiley Interdiscip Rev RNA 4:397–422
10. Hulsmans M, Holvoet P (2013) MicroRNAs as early biomarkers in obesity and related metabolic and cardiovascular diseases. Curr Pharm Des 19:5704–5717
11. Tan L, Yu JT, Hu N, Tan L (2013) Non-coding RNAs in Alzheimer's disease. Mol Neurobiol 47:382–393
12. Schonrock N, Gotz J (2012) Decoding the non-coding RNAs in Alzheimer's disease. Cell Mol Life Sci 69:3543–3559
13. Wei Y, Nazari-Jahantigh M, Neth P, Weber C, Schober A (2013) MicroRNA-126, -145, and -155: a therapeutic triad in atherosclerosis? Arterioscler Thromb Vasc Biol 33:449–454
14. Flowers E, Froelicher ES, Aouizerat BE (2013) MicroRNA regulation of lipid metabolism. Metabolism 62:12–20
15. Ferre-D'Amare AR, Doudna JA (2000) Crystallization and structure determination of a hepatitis delta virus ribozyme: use of the RNA-binding protein U1A as a crystallization module. J Mol Biol 295:541–556
16. Golden BL, Kundrot CE (2003) RNA crystallization. J Struct Biol 142:98–107
17. Ke A, Doudna JA (2004) Crystallization of RNA and RNA-protein complexes. Methods 34:408–414
18. Ye JD, Tereshko V, Frederiksen JK, Koide A, Fellouse FA, Sidhu SS, Koide S, Kossiakoff AA, Piccirilli JA (2008) Synthetic antibodies for specific recognition and crystallization of structured RNA. Proc Natl Acad Sci U S A 105: 82–87
19. Koldobskaya Y, Duguid EM, Shechner DM, Suslov NB, Ye J, Sidhu SS, Bartel DP, Koide S, Kossiakoff AA, Piccirilli JA (2011) A portable RNA sequence whose recognition by a synthetic antibody facilitates structural determination. Nat Struct Mol Biol 18:100–106
20. Ravindran PP, Heroux A, Ye JD (2011) Improvement of the crystallizability and expression of an RNA crystallization chaperone. J Biochem 150:535–543
21. Shechner DM, Grant RA, Bagby SC, Koldobskaya Y, Piccirilli JA, Bartel DP (2009) Crystal structure of the catalytic core of an RNA-polymerase ribozyme. Science 326:1271–1275
22. Ferré-D'Amaré AR (2010) Use of the spliceosomal protein U1A to facilitate crystallization and structure determination of complex RNAs. Methods 52:159–167
23. Ferre-D'Amare AR, Zhou K, Doudna JA (1998) Crystal structure of a hepatitis delta virus ribozyme. Nature 395:567–574
24. Keel AY, Rambo RP, Batey RT, Kieft JS (2007) A general strategy to solve the phase problem in RNA crystallography. Structure 15: 761–772
25. Guo F, Gooding AR, Cech TR (2004) Structure of the Tetrahymena ribozyme: base triple sandwich and metal ion at the active site. Mol Cell 16:351–362
26. Golden BL, Gooding AR, Podell ER, Cech TR (1998) A preorganized active site in the crystal structure of the Tetrahymena ribozyme. Science 282:259–264
27. Adams PL, Stahley MR, Gill ML, Kosek AB, Wang J, Strobel SA (2004) Crystal structure of a group I intron splicing intermediate. RNA 10:1867–1887
28. Ferre-D'Amare AR, Zhou K, Doudna JA (1998) A general module for RNA crystallization. J Mol Biol 279:621–631

29. Lee CV, Liang WC, Dennis MS, Eigenbrot C, Sidhu SS, Fuh G (2004) High-affinity human antibodies from phage-displayed synthetic Fab libraries with a single framework scaffold. J Mol Biol 340:1073–1093
30. Sidhu SS, Fellouse FA (2006) Synthetic therapeutic antibodies. Nat Chem Biol 2:682–688
31. Fellouse FA, Esaki K, Birtalan S, Raptis D, Cancasci VJ, Koide A, Jhurani P, Vasser M, Wiesmann C, Kossiakoff AA, Koide S, Sidhu SS (2007) High-throughput generation of synthetic antibodies from highly functional minimalist phage-displayed libraries. J Mol Biol 373:924–940
32. Sherman EM, Holmes S, Ye JD (2014) Specific RNA-binding antibodies with a four-amino-acid code. J Mol Biol 426:2145–2157
33. Kwon M, Strobel SA (2008) Chemical basis of glycine riboswitch cooperativity. RNA 14: 25–34
34. Sherman EM, Esquiaqui J, Elsayed G, Ye JD (2012) An energetically beneficial leader-linker interaction abolishes ligand-binding cooperativity in glycine riboswitches. RNA 18: 496–507
35. Ryder SP, Strobel SA (1999) Nucleotide analog interference mapping. Methods 18:38–50

Chapter 8

Generating Crystallographic Models of DNA Dodecamers from Structures of RNase H:DNA Complexes

Martin Egli and Pradeep S. Pallan

Abstract

The DNA dodecamer 5′-d(CGCGAATTCGCG)-3′ is arguably the best studied oligonucleotide and crystal structures of duplexes with this sequence account for a considerable portion of the total number of oligo-2′-deoxynucleotide structures determined over the last 30 years. The dodecamer has commonly served as a template to analyze the effects of sequence on DNA conformation, the conformational properties of chemically modified nucleotides, DNA–ligand interactions as well as water structure and DNA–cation binding. Although molecular replacement is the phasing method of choice given the large number of available models of the dodecamer, this strategy often fails as a result of conformational changes caused by chemical modification, mismatch pairs, or differing packing modes. Here, we describe an alternative approach to determine crystal structures of the dodecamer in cases where molecular replacement does not produce a solution or when crystals of the DNA alone cannot be grown. It is based on the discovery that many dodecamers of the above sequence can be readily co-crystallized with *Bacillus halodurans* RNase H, whereby the enzyme is unable to cleave the DNA. Determination of the structure of the complex using the protein portion as the search model yields a structural model of the DNA. Provided crystals of the DNA alone are also available, the DNA model from the complex then enables phasing their structures by molecular replacement.

Key words DNA, Molecular replacement, Phasing, Protein–DNA interactions, Ribonuclease H, RNase H

1 Introduction

The crystal structure of the DNA oligonucleotide 5′-d(CGCGAAT TCGCG)-3′, the so-called Dickerson-Drew Dodecamer (DDD), provided the first detailed view of a B-form duplex [1]. In the years since then some 160 structures based on the DDD have been deposited in the Nucleic Acid Database (NDB; http://ndbserver.rutgers.edu) [2], amounting to ca. 15 % of the total number of 1,040 structures of DNA duplexes in the NDB. The DDD exhibited unusual features that allowed a glimpse at the sequence-dependence of DNA structure. Among them were the extremely narrow minor groove in the central AATT section, subsequently

Eric Ennifar (ed.), *Nucleic Acid Crystallography: Methods and Protocols*, Methods in Molecular Biology, vol. 1320,
DOI 10.1007/978-1-4939-2763-0_8, © Springer Science+Business Media New York 2016

analyzed in depth using variations of the DDD sequence in the context of A-tract geometry and DNA bending [3, 4]. The striking water pattern in the minor groove, termed "spine of hydration" became the subject of numerous studies, and ultimately turned out to be two fused spines dissecting the groove [5]. Improvements in the synthetic preparation of oligonucleotides and purification procedures in combination with intense photon beams at third-generation synchrotrons yielded crystal structures of the DDD at atomic resolution [6]. These afforded intricate details of the water structure in the grooves and around phosphates [7] and the locations of mono- and divalent metal ions [5, 8, 9] as well as insights into the relative importance of crystal packing, base sequence and bound cations on DNA conformation [10]. The DDD also proved to be a fertile testing ground for computational simulations of DNA structure, for example by molecular dynamics simulations [11]. Dozens of structures of the DDD in complex with minor groove binding agents paved the road toward the design of highly specific probes and informed the discovery of drugs [12, 13]. The DDD also constitutes a very useful template sequence for analyzing the conformational properties of chemically modified nucleic acids. It contains all four building blocks, thus allowing incorporation of T, A, C or G analogs, crystals can be grown relatively easily, the duplex can be accommodated in several different space groups, including the most common orthorhombic $P2_12_12_1$ type, and crystals typically diffract to medium or high resolution [14–17]. The fact that the native duplex was studied in considerable detail is a further advantage, as the structures of modified DDDs can be compared to reference structures in order to establish the conformational consequences of a particular chemical modification. The DDD sequence also served as the template for the first crystal structures of all-modified DNA (N3′→P5′ phosphoramidate DNA) [18] and RNA (2′-*O*-(2-methoxyethyl)-RNA) [19].

In our efforts directed at the conformational analysis of chemically modified nucleic acids, we encountered quite a few cases where incorporation of a modified nucleotide precluded crystallization of the DDD alone [20–22]. Alternatively, crystals of the modified fragment could be grown, but the structure resisted phasing by molecular replacement using the canonical DDD duplex as the search model [23]. Both limitations can potentially be overcome by co-crystallizing such modified DDDs with *Bacillus halodurans* RNase H (*Bh*RNase H) [24]. RNase H endonucleolytically cleaves the RNA portion of DNA:RNA hybrids, but the enzyme also binds double-stranded RNA (dsRNA) and DNA (dsDNA; although it is unable to cleave either) and single-stranded nucleic acids, albeit with lower affinity [25]. We pursued the structure determination of non-specific RNase H:dsRNA and RNase H:dsDNA complexes. Although attempts to obtain a high-resolution structure of the former complex have thus far failed [26], we found that *Bh*RNase H can be readily crystallized with the native [27] and chemically modified DDDs [20–23]. The structures of (MR) these complexes can be determined with molecular replacement using the RNase H portion as the search model. In cases where crystals of the modified DDD alone are at hand, the availability of the refined structure of the DDD from the complex then provides a handle to phase the crystal of the duplex alone. If the duplex cannot be crystallized without RNase H, the crystal structure of the complex furnishes a model of the modified DDD. The present protocol describes the overall approach of using RNase H complexes as a means to determine the crystal structures of DDDs alone that cannot be readily phased, most likely due to conformational changes as a result of the chemical modification that preclude the use of the native DDD as a viable search model, or to gain access to a model of a modified DDD for which crystals in the absence of protein cannot be grown.

2 Materials

2.1 Oligonucleotide Synthesis and Purification

1. Chemically synthesized native DDD 5′-d(CGCGAATTCGCG) DNA, 1 μmol scale.
2. The modified DDD 5′-d(CGCGAATFCGCG) (dF = 2′-deoxyribo-2,4-difluorotoluylnucleotide, a dT isostere [21]) was synthesized on a 1 μmol scale on an ABI 381A DNA synthesizer with the dF phosphoramidite (gift from Glen Research, Sterling, VA), by using a slightly prolonged wait time (10 min) for the phosphoramidite coupling.
3. The modified DDD 5′-d(CGCGAtcATTCGCG) (tcdA = [(5′R,6′R)-2′-deoxy-3′,5′-ethano-5′,6′-methano-β-D-ribofuranosyl]adenine) that contains an adenosine analog with a tricyclic sugar was obtained from Prof. Christian Leumann (University of Berne, Switzerland). For the preparation of tcDNA phosphoramidite building blocks and oligonucleotide synthesis, please *see* ref. 28.
4. The modified DDDs 5′-d(CGCGAAUsTCGCG) and 5′-d(CGCGAAUsUsCGCG) (Us = 2′-*S*Me-U) were synthesized following published procedures ([29] and cited refs.).
5. The modified DDDs containing 5-chloro-U (ClU) in place of T or C, i.e. 5′-d(CGCGAAClUTCGCG), 5′-d(CGCGAAT ClUCGCG), 5′-d(CGCGAAClUClUCGCG) and 5′-d (CGCGAATTClUGCG), were synthesized as previously reported [30]. For a general description of equipment and reagent setup for oligonucleotide synthesis please *see* ref. 31.
6. Ion exchange or reverse phase column for nucleic acids.
7. High performance liquid chromatography (HPLC) system.
8. 0.45 μM syringe filters.

2.2 Protein Expression and Purification

1. *Bacillus halodurans* (*B. halodurans*) genomic DNA (American Type Culture Collection, ATCC, Manassas, VA, USA).
2. The C-terminal fragment of *B. halodurans* RNase H (*Bh*-RNase H; Met58 to Lys196) with the Asp132 → Asn mutation was cloned into the PET15b vector with an N-terminal His tag and a thrombin cleavage site and expressed in *E. coli* BL21 cells and purified following published procedures [24, 27]. The protein solution was concentrated to ~20 mg/mL.

2.3 Crystallization

1. Nucleic Acid Mini Screen (Hampton Research, Aliso Viejo, CA).
2. Natrix and Crystal Screen I crystallization kits (Hampton Research, Aliso Viejo, CA).
3. Crystallization plates suitable for hanging drops.
4. Siliconized cover-slips.
5. 2-Methyl-2,4-pentanediol (MPD) 35 % (v/v).

2.4 Data Collection and Structure Solution

1. Nylon loop for crystals.
2. Liquid nitrogen and tools for crystal mounting.
3. Data processing program (HKL2000 or similar).
4. CCP4 software suite.
5. PHENIX software suite.
6. The Coot program for model building and visualization.

3 Methods

3.1 DNA Crystallization

1. Purify oligonucleotides either by ion exchange or reverse phase HPLC.
2. After collecting the HPLC fractions, filter solutions through a 0.45 μM syringe filter prior to setting up crystallization droplets.
3. Initial crystallization experiments should be undertaken with the DDD duplex alone. The concentration of the purified oligonucleotide is adjusted to ca. 2–4 mM in RNase/DNase free water. Alternatively, one can constitute the oligonucleotides in 5 mM Tris (pH 7.0), 50 mM NaCl buffer solution. The crystallization trials will typically be done with concentrations of the oligonucleotide that lie between 1 and 2 mM.
4. Strands are then annealed by heating the stock solution to 70–75 °C for 1–2 min., followed by slow cooling to room temperature.
5. Our first choice in terms of initial crystallization conditions to be tested with oligonucleotides is the hanging-drop vapor

diffusion technique in combination with the commercially available Nucleic Acid Mini Screen that consists of 24 different solutions [32]. Equal volumes of 1 or 2 μL of the oligonucleotide and crystallization solutions are mixed on siliconized cover-slips and the droplets equilibrated against a reservoir of 500 μL of 35 % MPD (v/v) in a 24-well plate.

6. The plates are then incubated at 18 °C and checked at regular intervals over the course of several weeks. In many cases, crystals obtained from this screen are suitable for diffraction experiments and data collection. However, it may be necessary to refine the conditions that resulted in initial growth of crystals in order to optimize their size and/or resolution limit. This is best achieved by varying one parameter at A time, e.g. pH, metal ion concentration, volume and concentration of the MPD reservoir solution as well as temperature. If the above sparse matrix screen does not yield diffraction-quality crystals, other conditions should be screened. In our own laboratory we often rely on the Natrix [33] and Crystal Screen I kits [34] from Hampton Research that are comprised of 48 and 50 individual conditions, respectively.

3.2 Phasing DDD Crystal Structures by Molecular Replacement

1. Once crystals are obtained, they are picked up from solution with a nylon loop, frozen in liquid nitrogen and checked for X-ray diffraction, either on in-house instrumentation or at a synchrotron source.
2. Data are processed and scaled with HKL2000 [35]. If the crystals are isomorphous with native DDD crystals, i.e. space group orthorhombic $P2_12_12_1$ with unit cell constants $a \approx 25$ Å, $b \approx 40$ Å and $c \approx 65$ Å, the structure can in all likelihood be determined by the Molecular Replacement (MR) technique, using a native DDD duplex as the search model.
3. MR is performed with commonly used software such as Molrep [36] in the CCP4 suite [37] (or any MR program of choice) in combination with a suitable model.
4. The obtained MR solution is checked for unfavorable packing contacts, and provided packing and the initial values for R-factor and R-free (i.e. values in the mid or low 30s) are indicative of a correct solution, the next steps consist of restrained refinement of atomic positions and temperature factors with, for example, the program PHENIX [38].
5. Visualization of the duplex and electron density as well as manual model building and water placement is performed, for example, using the program Coot [39].

In our investigations directed at the conformational consequences for DNA as a result of chemical modification, we frequently came across modified DDDs that resisted crystallization despite an extensive number of trials. In other cases crystals could be grown but their structures subsequently not determined by MR. Alternative approaches such as multiple isomorphous replacement (MIR), single- or multi-wavelength anomalous dispersion (SAD or MAD, respectively), or combinations thereof (SIRAS or MIRAS, respectively) with suitable derivatives may then be pursued to phase the structure. However, this can result in a potentially protracted search for heavy atom derivatives or require the synthesis of chemically modified strands for SAD/MAD. By comparison, our approach described below involving co-crystallization of such DDDs with *Bh*RNase H and then solving the structure with MR using the protein as the search model offers a quicker solution to overcome the phasing problem.

3.3 DDDs That Resist Crystallization or DDD Crystal Structures That Resist Phasing by MR

1. Co-crystallization with *Bh*RNase H of chemically modified DDDs offers a rapid route to 3D-structural models of the duplexes in cases of DDDs for which crystals cannot be grown or crystals of DDDs that are non-isomorphous with the native DDD crystal form and resist phasing by MR. Crystals of the native DDD in complex with *Bh*RNase H diffract to high resolution (*see* Table 1, Fig. 1) and the structure can be readily determined by MR using the protein as the search model [27].
2. Unlike in the crystal of the native DDD alone, individual strands of duplexes in the complex crystal are related by two-fold rotational symmetry (*see* **Note 1**). Thus, DDD duplexes bound to RNase H display kinks into the major groove at both ends, whereas native duplexes in DDD crystals without protein are located in a general position and feature an asymmetric kink (Fig. 2).

3.3.1 DDDs That Resist Crystallization

1. If crystallization experiments with DDDs that contain modified nucleotides do not yield crystals or in cases where DDD crystals diffract only poorly, co-crystallization with *Bh*RNase H should be attempted (*see* **Note 2**). For crystallization experiments, the annealed DDD is mixed with the protein in a 2:1 M ratio in the presence of 5 mM $MgCl_2$ such that the final protein concentration is ~10 mg/mL.
2. Screening of crystallization conditions is best done with commercially available kits (*see* Subheading 3.1) and once crystals are obtained phasing with MR using RNase H as the search model will readily deliver a structural model of the modified DDD. One should be aware that not all crystals are of a complex, but that in quite a few cases crystals will contain only *Bh*RNase H (*see* Subheading 3.4). Examples of modified DDDs that were crystallized in complex with *Bh*RNase H but for which crystals of the duplex alone could not be grown are listed in Table 1. The complex between *Bh*RNase H and a DDD duplex containing two difluorotoluene residues in place of T is shown in Fig. 3.

Table 1
Selected crystal data and structure refinement statistics for complexes between *Bh*RNase H and native DDD, F-DDD, ClU-DDD, and tcdA-DDD

Complex	*Bh*RNase H/native d(CGCGAATTCGCG)	*Bh*RNase H/F = DFT d(CGCGAATFCGCG)	*Bh*RNase H/U* = 5-ClU d(CGCGAATU*CGCG)[a]	*Bh*RNase H/A* = tcdA d(CGCGAA*TTCGCG)
Crystal data				
Space group	Monoclinic, *C*2	Monoclinic, *C*2	Orthorhombic, $P2_12_12_1$	Tricilinic, *P*1
Unit cell constants *a*, *b*, *c* [Å] *α*, *β*, *γ* [°]	98.40, 66.66, 76.93 90, 122.3, 90	96.18, 66.71, 77.57 90, 121.0, 90	64.08, 64.76, 116.47 90, 90, 90	42.39, 47.50, 55.22 100.9, 101.8, 89.8
Wavelength [Å]	0.9785	1.0000	0.9787	1.0000
Resolution (outer shell) [Å]	1.80 (1.86–1.80)	1.61 (1.64–1.61)	1.49 (1.52–1.49)	1.54 (1.65–1.54)
No. of unique refls. (outer shell)	37,768 (3,643)	52,821 (2,202)	77,514 (3,752)	56,542 (9,820)
Completeness (outer shell) [%]	97.3 (94.7)	98.1 (82.0)	96.9 (95.1)	92.5 (86.6)
R-merge (outer shell)	0.061 (0.248)	0.053 (0.560)	0.071 (0.824)	0.047 (0.480)
I/σ (*I*) (outer shell)	42.7 (4.6)	39.4 (1.7)	27.3 (2.5)	19.5 (4.4)
Refinement parameters				
No. of protein molecules/ DNA strands per asymmetric unit	2 RNase H/2 single strands	2 RNase H/2 single strands	2 RNase H/2 duplexes	2 RNase H/2 single strands
R-work/R-free [%]	0.215/0.241	0.197/0.238	0.189/0.218	0.163/0.217
No. of protein/ DNA atoms	2,169/486	2,154/488	2,244/900	2,154/554
No. of waters/ ions/ligands	167/1 Na^+/–	232/–/4 glycerol	476/2 Mg^{2+}/4 glycer., 1 EGOH	200/–/2
R.m.s.d. bond lengths [Å]/ ang. [°]	0.016/1.7	0.030/2.5	0.009/1.4	0.021/2.0
Avg. B-factor, protein/DNA [Å²]	30.9/29.4	16.9/13.0	22.1/48.5	32.3/34.7

(continued)

Table 1 (continued)

Complex	*Bh*RNase H/native d(CGCGAATTCGCG)	*Bh*RNase H/F = DFT d(CGCGAATFCGCG)	*Bh*RNase H/U* = 5-ClU d(CGCGAATU*CGCG)[a]	*Bh*RNase H/A* = tcdA d(CGCGAA*TTCGCG)
Avg. B-factor, H_2O/ions/ small molecules [Å^2]	37.3/53.2/–	24.4/–/42.5	36.5/13.7/29.0	40.7/–/47.2
PDB entry code	3D0P	3I8D	4HTU	4OPJ

[a]We determined crystal structures of *Bh*RNase H in complex with four different DDDs with incorporated ClU residues. Only one of them is listed here

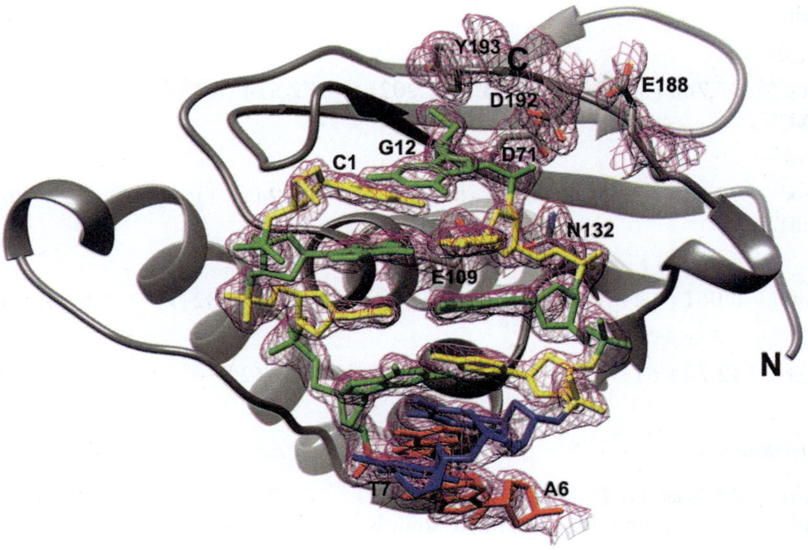

Fig. 1 Overall structure of the native DDD in complex with *Bh*RNase H and quality of the electron density. The DNA duplex is located on crystallographic dyad and only 6 bp are shown. The protein is depicted in a ribbon cartoon colored in gray and the DDD is viewed into the major groove with G, C, A and T nucleotides colored in *green*, *yellow*, *red*, and *blue*, respectively. Protein chain termini and terminal nucleotides are labeled. The Fourier 2Fo-Fc sum electron density around the 6 bp and amino acids at the active site (E109 and N132) or interacting with the terminal base pair is contoured at the 1σ level. Figures were generated with UCSF Chimera [43]

3.3.2 DDD Crystal Structures That Resist Phasing by MR

1. Many DDDs with incorporated chemically modified residues or bound to small molecules crystallize in the same orthorhombic crystal system as the native DDD and with cell constants that deviate only minimally from those of native crystals. However, there are numerous cases where chemical modification causes conformational changes that affect crystal packing, thus resulting in a variety of new crystals forms. Often MR with the native DDD as the search model will then fail

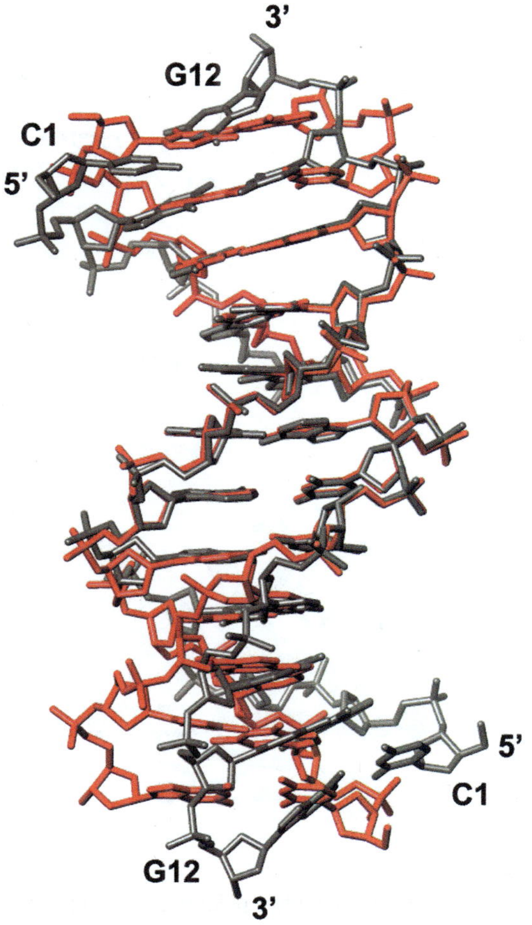

Fig. 2 Superimposition of the DDDs in the crystal structure of the complex with *Bh*RNase H (*gray*, Table 1) and crystallized alone (*red*; PDB ID code 436D [6]). Only phosphorus atoms in the central A-tract region were used for generating the overlay

because altered lattice interactions and/or chemical modification cause the DDD to adopt a conformation that is significantly different from that of the native duplex.

2. Instead of embarking on a potentially time-consuming search for alternative phasing approaches involving heavy atom soaks (the options with DNA are rather limited compared to protein crystals in this respect) or de novo synthesis of oligonucleotides with brominated pyrimidines (*see* **Note 2**), one should first attempt to co-crystallize such DDDs with *Bh*RNase H. MR with RNase H as the search model will then produce a model of the DDD duplex. For example, the 2′-*S*Me-U modification (Us) inserted into the DDD in place of T causes the narrow minor groove in the central section to open up significantly.

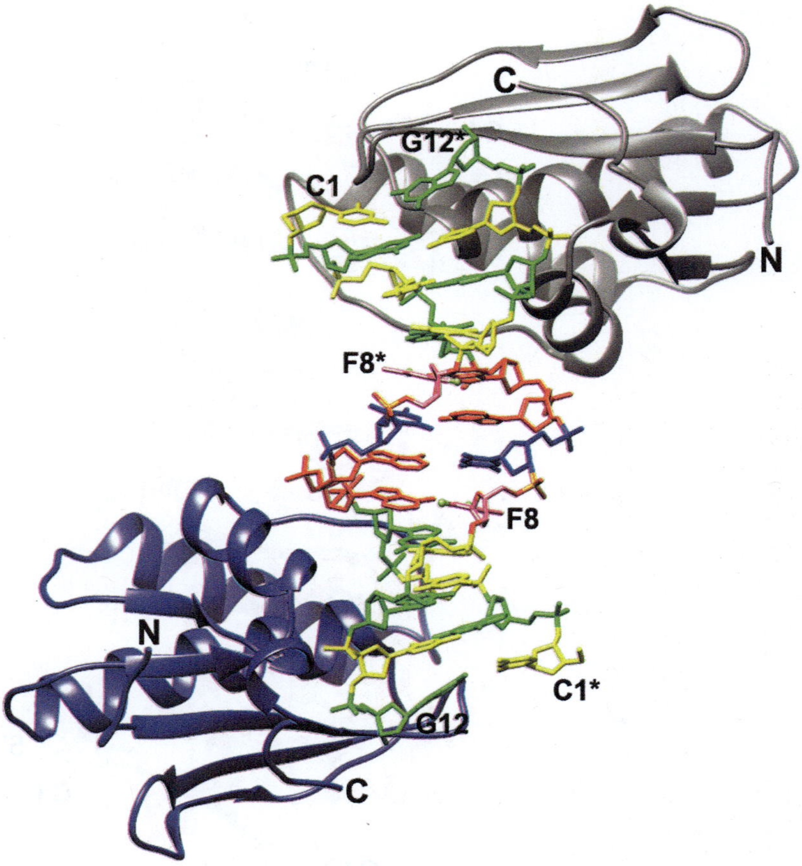

Fig. 3 Structure of a DDD containing two difluorotoluene residues (dF) in complex with *Bh*RNase H. The color code for G, C, A, and T nucleotides is the same as in Fig. 1, dF residues are *pink* with fluorine atoms highlighted as *green balls*, and *asterisks* indicate that the two strands of the DDD are symmetry-related in the crystal structure of the complex

Not surprisingly, the structures of DDDs with either one T or both Ts per strand replaced by Us could not be determined by MR [23]. By comparison, the crystal structures of their complexes with *Bh*RNase H yielded readily to MR with the RNase H search model (*see* Table 2). The modified DDD duplexes from the refined structures of the complexes were subsequently used for phasing by MR of the crystals of the duplexes alone.

3. The success of this approach is of course dependent on rather similar conformations of the duplexes in complex with *Bh*RNase H and in the free form (Fig. 4). Although the number of examples of modifications tested and complexes of DDDs with *Bh*RNase H determined thus far is still somewhat limited, the crystal data summarized in Tables 1 and 2 indicate a considerable variety of packing arrangements between complexes. We take this as evidence in support of the versatility of the co-crystallization approach for phasing of DDD structures (*see* **Notes 3–5**).

Table 2
Selected crystal data and structure refinement statistics for 2′-*S*Me-DDD:*Bh*RNase H complexes

Complex	*Bh*RNase H/U* = 2′-SMe-U d(CGCGAAU*U*CGCG)	*Bh*RNase H/U* = 2′-SMe-U d(CGCGAAU*TCGCG)
Space group	Monoclinic, C2	Monoclinic, C2
Unit cell constants *a*, *b*, *c* [Å] *β* [°]	81.93, 66.62, 38.77 103.3	98.14, 66.61, 77.90 122.1
Wavelength [Å]	0.9785	1.0000
Resolution (outer shell) [Å]	1.60 (1.66–1.60)	1.54 (1.57–1.54)
No. of unique reflections	26,499 (–)	59,362 (2,885)
Completeness (outer shell) [%]	99.4 (96.9)	94.2 (91.8)
R-merge (outer shell)	0.077 (–)	0.047 (0.675)
I/σ (I) (outer shell)	39.4 (1.7)	28.3 (1.3)
Refinement parameters		
No. of protein molecules/DNA strands per asymmetric unit	1 RNase H/1 single strand	2 RNase H/2 single strands
R-work/R-free [%]	0.193/0.261	0.192/0.259
No. of protein/DNA atoms	1,114/245	2,178/498
No. of waters/small molecules	102/3 glycerol	167/3 glycerol
R.m.s.d. bond lengths [Å]/angles [°]	0.030/2.5	0.023/2.3
Avg. B-factor, protein/DNA atoms [Å^2]	59.2/13.0	42.1/32.9
Avg. B-factor, H_2O/small mols [Å^2]	45.2/59.7	41.0/41.2
PDB entry code	3EY1	4OPK

3.4 Crystals of apo BhRNase H

1. Crystals obtained from mixtures of *Bh*RNase H and DDDs in some cases only contain the protein. In addition to the reported crystal form of *Bh*RNase H [24], we have so far identified two additional crystal forms (*see* Table 3).
2. If the unit cell dimensions and space groups of a "complex" crystal match one of these, it is more likely that the crystal obtained contains only RNase H (*see* **Note 6**). However, in order to make sure that the crystal is definitely not of a complex, it is advisable to collect a full data set, conduct a MR search, and inspect the generated Fourier 2Fo-Fc sum and Fo-Fc difference density maps.

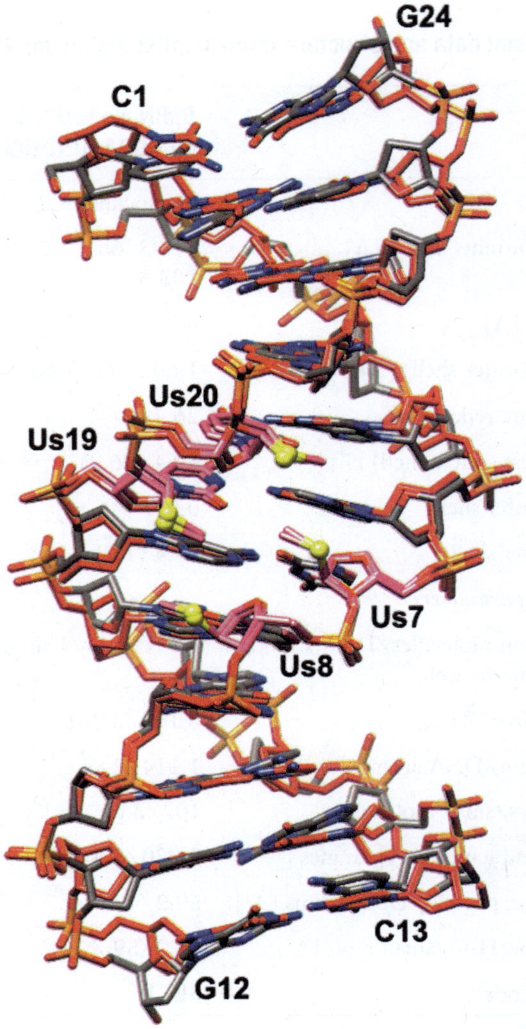

Fig. 4 Superimposition of the structures of the 2′-*S*MeU (Us)-modified DDD bound to *Bh*RNase H (*gray* carbon atoms; Table 2) and the Us-modified DDD crystallized alone (*red* carbon atoms). The central, modified section of the duplex is viewed into the minor groove and sulfur atoms in Us residues (*pink* carbon atoms) are highlighted as *yellow balls*

Table 3
Crystal forms of apo *Bh*RNase H

Space group	Trigonal, $P3_121$ [24]	Monoclinic, $P2_1$	Orthorhombic, $P2_12_12_1$
Unit cell constants a, b, c [Å]	66.8, 66.8, 58.7	34.8, 63.3, 49.7	35.2, 63.8, 80.6
α, β, γ [°]	90, 90, 120	90, 94.0, 90	90, 90, 90

4 Notes

1. With most protein–DNA complexes, successful growth of diffraction-quality crystals is critically dependent on both oligonucleotide length and sequence. In many cases, the duplex mediates packing contacts, for example by end-to-end stacking between base pairs from adjacent duplexes. Crystals of *Bh*RNase H:DDD complexes exhibit various symmetries and associated lattice interactions that are dominated by protein–protein contacts (*see*, for example, refs. 22, 27). The central A-tract section of the minor groove in the DDD is very narrow and RNase H therefore contacts the minor groove in the outer G:C sections, where it is expanded and more similar in width to that in RNA:DNA hybrids. The AATT tetramer therefore serves as a spacer and one can imagine that its length could be varied somewhat without fundamentally altering the interactions between RNase H and the outer G:C sections. Because the proteins would remain bound to the terminal regions of duplexes, the former may mediate packing interactions.
2. Co-crystallizing a DDD with RNase H is a convenient approach to determine the structure of the nucleic acid duplex in cases where it cannot be crystallized alone or when the structure of the duplex alone cannot be determined by molecular replacement. A good case in point is the DDD CGCGAAUsUsCGCG with two 2′-*S*Me-U residues (Us) in place of T. Crystals of this modified DDD can be grown readily from multiple conditions. However, molecular replacement searches with the native DDD as the search model consistently failed. Halogenated pyrimidines such as Br^5U or Br^5C in combination with single wavelength anomalous dispersion (SAD) commonly offer a tested approach to phase a crystal structure that resists molecular replacement (although the use of Br^5U here would have required the synthesis of a special building block because of the 2′-*S*Me-U modification in the above DDD). However, this DDD also appeared to be an almost ideal example for replacement of either one or the other or both of the Us residues by 2′-*Se*Me-U for "SeMe" SAD phasing [40, 41]. We produced all three selenoated DDDs but despite exhaustive attempts none of them gave diffraction-quality crystals. The structure of the complex between CGCGAAUsUsCGCG and *Bh*RNase H then yielded a model of the duplex that was used to determine the structure of the Us-modified DDD alone [23] (*see* Table 2, Fig. 4).
3. Although the co-crystallization approach with *Bh*RNase H is especially useful for structure determination of chemically modified DDDs that resist crystallization alone, there are limits as to the extent (number of modified nucleotides) and

nature of the modification. For example, we were unsuccessful in our attempts to crystallize *Bh*RNase H with the all-peptide nucleic acid (PNA) DDD CGCGAATTCGCG-K that carries a C-terminal lysine.

4. We determined a crystal structure of *Bh*RNase H in complex with a DDD featuring 5-chlorouridine:G (ClU:G) mismatch pairs (*see* Table 1) [22]. This indicates that crystallization of complexes with DDDs exhibiting altered geometry in the outer regions is possible. In the case of the ClU nucleobase analog, the chlorine substituent is directed into the major groove and does not interfere with the protein binding from the minor groove side. However, as with the T:G mismatch, guanine paired to ClU is shifted into the minor groove, but the more exposed 2-amino moiety appears not to interfere with RNase H binding.
5. RNase H binds double-stranded RNAs (dsRNAs) [25] but is unable to cleave RNA opposite RNA (unlike RNA paired opposite DNA). We initially conducted co-crystallization experiments with RNAs in complex with the enzyme from *E. coli*, but were not successful in identifying a crystal with an ordered RNA duplex bound to RNase H [26]. Subsequently, we screened RNA duplexes with lengths of between 8 and 16 nucleotides in co-crystallization experiments with *Bh*RNase H. Although crystals were obtained in many cases, they all contained the enzyme alone and in only one case, the RNA duplex alone. Thus, a crystal structure of an RNase H:dsRNA complex has not been determined to date.
6. Hits obtained from crystallization experiments with complexes between *Bh*RNase H and DDDs may be the desired complex, or alternatively, may contain only the enzyme (*see* Table 3; we have not observed a case where a DDD alone crystallized from the complex mixture). To distinguish between crystals of complexes on the one hand and those of the apo form of RNase H on the other, it may be helpful to use color-labeled oligonucleotides [42]. Many of these are commercially available, e.g. the 5′-Cy5- or 5′-Cy3-labeled oligo-2′-deoxynucleotides from Integrated DNA Technologies, Coralville, Iowa, USA (http://www.idtdna.com).

Acknowledgments

This work was supported by the US National Institutes of Health grant R01 GM055237.

References

1. Wing R, Drew H, Takano T, Broka C, Tanaka S, Itakura K, Dickerson RE (1980) Crystal structure analysis of a complete turn of B-DNA. Nature 287:755–758
2. Berman HM, Olson WK, Beveridge DL, Westbrook J, Gelbin A, Demeny T, Hsieh SH, Srinivasan AR, Schneider B (1992) The nucleic acid database: A comprehensive relational database of three-dimensional structures of nucleic acids. Biophys J 63:751–759
3. DiGabriele AD, Steitz TA (1993) A DNA dodecamer containing an adenine tract crystallizes in a unique lattice and exhibits a new bend. J Mol Biol 231:1024–1039
4. Allemann RK, Egli M (1997) DNA bending and recognition. Chem Biol 4:643–650
5. Minasov G, Tereshko V, Egli M (1999) Atomic-resolution crystal structures of B-DNA reveal specific influences of divalent metal ions on conformation and packing. J Mol Biol 291:83–99
6. Tereshko V, Minasov G, Egli M (1999) The Dickerson-Drew B-DNA dodecamer revisited – at atomic resolution. J Am Chem Soc 121:470–471
7. Egli M, Tereshko V, Teplova M, Minasov G, Joachimiak A, Sanishvili R, Weeks CM, Miller R, Maier MA, An H, Cook PD, Manoharan M (2000) X-ray crystallographic analysis of the hydration of A- and B-form DNA at atomic resolution. Biopolymers 48:234–252
8. Tereshko V, Minasov G, Egli M (1999) A "hydrat-ion" spine in a B-DNA minor groove. J Am Chem Soc 121:3590–3595
9. Egli M (2002) DNA-cation interactions: quo vadis? Chem Biol 9:277–286
10. Egli M, Tereshko V (2004) Lattice- and sequence-dependent binding of Mg^{2+} in the crystal structure of a B-DNA dodecamer. In: Stellwagen N, Mohanty U (eds) Curvature and deformation of nucleic acids: recent advances, new paradigms, ACS symposium, vol 884. Oxford University Press, New York, pp 87–109
11. Young MA, Jayaram B, Beveridge DL (1997) Intrusion of counterions into the spine of hydration in the minor groove of B-DNA: fractional occupancy of electronegative pockets. J Am Chem Soc 119:59–69
12. Neidle S (2001) DNA minor-groove recognition by small molecules. Nat Prod Rep 18:291–309
13. Nanjunda R, Wilson WD (2012) Binding to the DNA minor groove by heterocyclic dications: from AT-specific monomers to GC recognition with dimers. Curr Protoc Nucleic Acid Chem 51:8.8.1–8.8.20
14. Egli M (1996) Structural aspects of nucleic acid analogs and antisense oligonucleotides. Angew Chem Int Ed Engl 35:1894–1909
15. Egli M (1998) Towards the structure-based design of nucleic acid therapeutics. In: Weber G (ed) Advances in enzyme regulation, vol 38. Elsevier, Oxford, pp 181–203
16. Egli M, Pallan PS (2007) Insights from crystallographic studies into the structural and pairing properties of nucleic acid analogs and chemically modified DNA and RNA oligonucleotides. Annu Rev Biophys Biomol Struct 36:281–305
17. Egli M, Pallan PS (2010) Crystallographic studies of chemically modified nucleic acids: a backward glance. Chem Biodivers 7:60–89
18. Tereshko V, Gryaznov S, Egli M (1998) Consequences of replacing the DNA 3′-oxygen by an amino group: high-resolution crystal structure of a fully modified N3′→P5′ phosphoramidate DNA dodecamer duplex. J Am Chem Soc 120:269–283
19. Teplova M, Minasov G, Tereshko V, Inamati G, Cook PD, Manoharan M, Egli M (1999) Crystal structure and improved antisense properties of 2′-*O*-(2-methoxyethyl)-RNA. Nat Struct Biol 6:535–539
20. Pallan PS, Ittig D, Heroux A, Wawrzak Z, Leumann CJ, Egli M (2008) Crystal structure of tricyclo-DNA: an unusual compensatory change of two adjacent backbone torsion angles. Chem Commun 21:883–885
21. Pallan PS, Egli M (2009) The pairing geometry of the hydrophobic thymine analog 2,4-difluorotoluene in duplex DNA as analyzed by X-ray crystallography. J Am Chem Soc 131:12548–12549
22. Patra A, Harp J, Pallan PS, Zhao L, Abramov M, Herdewijn P, Egli M (2013) Structure, stability and function of 5-chlorouracil modified A:U and G:U base pairs. Nucleic Acids Res 41:2689–2697
23. Pallan PS, Prakash TP, Li F, Eoff RL, Manoharan M, Egli M (2009) A conformational transition in the structure of a 2′-thiomethyl-modified DNA visualized at high resolution. Chem Commun 21:2017–2019
24. Nowotny M, Gaidamakov SA, Crouch RJ, Yang W (2005) Crystal structures of RNase H bound to an RNA/DNA hybrid: substrate specificity and metal-dependent catalysis. Cell 121:1005–1016
25. Lima WF, Crooke ST (1997) Binding affinity and specificity of Escherichia coli RNaseH1:

impact on the kinetics of catalysis of antisense oligonucleotide-RNA hybrids. Biochemistry 36:390–398

26. Loukachevitch LV, Egli M (2007) Crystallization and preliminary X-ray analysis of Escherichia coli RNase HI-dsRNA complexes. Acta Crystallogr F 63:84–88
27. Pallan PS, Egli M (2008) Insights into RNA/DNA hybrid recognition and processing by RNase H from the crystal structure of a non-specific enzyme-dsDNA complex. Cell Cycle 7:2562–2569
28. Steffens R, Leumann C (1997) Preparation of [(5′R,6′R)-2′-Deoxy-3′,6′-ethano-5′,6′-methano-β-D-ribofuranosyl]thymine and -adenine, and the corresponding phosphoramidites for oligonucleotide synthesis. Helv Chim Acta 80:2426–2439
29. Lima WF, Nichols JG, Wu H, Prakash TP, Migawa MT, Wyrzykiewicz TK, Bhat B, Crooke S (2004) Structural requirements at the active site of the heteroduplex substrate for human RNase H1 catalysis. J Biol Chem 279:36317–36326
30. Theruvathu JA, Kim CH, Rogstad DK, Neidigh JW, Sowers LC (2009) Base-pairing configuration and stability of an oligonucleotide duplex containing a 5-chlorouracil-adenine base pair. Biochemistry 48:7539–7546
31. Pallan PS, Egli M (2007) Selenium modification of nucleic acids. Preparation of phosphoroselenoate derivatives for crystallographic phasing of nucleic acid structures. Nat Protoc 2:640–646
32. Berger I, Kang C-H, Sinha N, Wolters M, Rich A (1996) A highly effective 24 condition matrix for the crystallization of nucleic acid fragments. Acta Crystallogr D 52:465–468
33. Scott WG, Finch JT, Grenfell R, Fogg J, Smith T, Gait MJ, Klug A (1995) Rapid crystallization of chemically synthesized hammerhead RNAs using a double screening procedure. J Mol Biol 250:327–332
34. Jancarik J, Kim SH (1991) Sparse matrix sampling: a screening method for crystallization of proteins. J Appl Crystallogr 24:409–411
35. Otwinowski Z, Minor W (1997) Processing of X-ray diffraction data collected in oscillation mode. Methods Enzymol 276:307–326
36. Vagin A, Teplyakov A (2010) Molecular replacement with MOLREP. Acta Crystallogr D 66:22–25
37. CCP4 (1994) The CCP4 suite: programs for protein crystallography. Acta Crystallogr D 50:760–763
38. Adams PD, Afonine PV, Bunkoczi G, Chen VB, Davis IW, Echols N, Headd JJ, Hung LW, Kapral GJ, Grosse-Kunstleve RW, McCoy AJ, Moriarty NW, Oeffner R, Read RJ, Richardson DC, Richardson JS, Terwilliger TC, Zwart PH (2010) PHENIX: a comprehensive python-based system for macromolecular structure solution. Acta Crystallogr D 66:213–221
39. Emsley P, Cowtan K (2004) Coot: model-building tools for molecular graphics. Acta Crystallogr D 60:2126–2132
40. Teplova M, Wilds CJ, Wawrzak Z, Tereshko V, Du Q, Carrasco N, Huang Z, Egli M (2002) Covalent incorporation of selenium into oligonucleotides for X-ray crystal structure determination via MAD: proof of principle. Biochimie 84:849–858
41. Pallan PS, Egli M (2007) Selenium modification of nucleic acids. Preparation of oligonucleotides with incorporated 2′-SeMe-uridine for crystallographic phasing of nucleic acid structures. Nat Protoc 2:647–651
42. Jiang X, Egli M (2011) Use of chromophoric ligands to visually screen co-crystals of putative protein-nucleic acid complexes. Curr Protoc Nucleic Acid Chem 46:7.15.1–7.15.8
43. Pettersen EF, Goddard TD, Huang CC, Couch GS, Greenblatt DM, Meng EC, Ferrin TE (2004) UCSF Chimera – a visualization system for exploratory research and analysis. J Comput Chem 25:1605–1612

Chapter 9

Isothermal Titration Calorimetry: Assisted Crystallization of RNA–Ligand Complexes

Cyrielle Da Veiga, Joelle Mezher, Philippe Dumas, and Eric Ennifar

Abstract

The success rate of nucleic acids/ligands co-crystallization can be significantly improved by performing preliminary biophysical analyses. Among suitable biophysical approaches, isothermal titration calorimetry (ITC) is certainly a method of choice. ITC can be used in a wide range of experimental conditions to monitor in real time the formation of the RNA– or DNA–ligand complex, with the advantage of providing in addition the complete binding profile of the interaction. Following the ITC experiment, the complex is ready to be concentrated for crystallization trials. This chapter describes a detailed experimental protocol for using ITC as a tool for monitoring RNA/small molecule binding, followed by co-crystallization.

Key words Microcalorimetry, Crystallization, RNA–ligand complex, ITC, Thermodynamics

1 Introduction

Co-crystallization of nucleic acid/ligand complexes is frequently problematic, even on established biological systems. Much time, efforts, and material can be spared by using preliminary biophysical methods to assess a correct formation of the complex in conditions that are optimal for co-crystallization. Among these approaches, ITC is well-suited for analysis of nucleic acid interactions [1–4] and numerous examples include binding of small ligands [5–10], proteins [11–17] or nucleic acid–nucleic acid interactions [18–21]. A major advantage of ITC over other equivalent biophysical approaches (such as microscale thermophoresis, surface plasmon resonance, nuclear magnetic resonance, or steady-state fluorescence) is that ITC directly measures the heat that is released or absorbed upon a binding interaction [22–25]. Consequently, no labeling of the sample is needed and this sample can be directly used for crystallization experiments. In addition, ITC allows the simultaneous determination of all binding parameters in one single experiment: the affinity constant (K_a), the binding enthalpy (ΔH), the entropy change (ΔS), the Gibbs free energy change (ΔG), and

Eric Ennifar (ed.), *Nucleic Acid Crystallography: Methods and Protocols*, Methods in Molecular Biology, vol. 1320,
DOI 10.1007/978-1-4939-2763-0_9, © Springer Science+Business Media New York 2016

the stoichiometry (*n*). Unlike other techniques, ITC has a very wide range of applications: it is not subjected to any upper or lower molecular weight limits and can be used virtually in any buffers, provided the RNA/DNA buffer is rigorously identical to the ligand buffer. A real limitation of ITC is the requirement of large amount of material; however, this is obviously not an issue if the sample is dedicated for crystallization experiments since crystallization and ITC requirements are quite comparable. For most situations, the complex formed following the ITC titration can thus be used for crystallization by concentrating the sample about ten times. Finally, when complex RNA molecules are used, ITC also provides the opportunity to assess proper folding of the nucleic acids by monitoring the stoichiometry of the interaction: it is indeed commonly observed during large-scale preparation of RNA that a significant portion is misfolded and does not allow ligand binding (the same holds true for proteins). A suboptimal stoichiometry is a strong indicator of such a frequent problem.

In this chapter, we provide a detailed protocol for implementing ITC as a guide to improve co-crystallization of nucleic acid–ligand complexes. An example is presented with the titration of the aminoglycoside antibiotic lividomycin into the homodimeric HIV-1 genomic RNA dimerization initiation site (DIS) loop–loop complex (Figs. 1 and 2) [26–29], followed by co-crystallization experiments. This RNA/ligand interaction also provides a good example of salt-dependent binding, as frequently observed with nucleic acids. Indeed, as many nucleic acid binders, aminoglycosides are positively charged and unspecific interactions due to electrostatic interactions can be observed at low salt concentration, whereas excessively high salt concentration can even prevent the specific ligand binding. It might therefore be advisable to perform preliminary ITC experiments in various salt conditions in order to exclude (or at least reduce) unspecific binding while maintaining specific interactions. For a study of the binding of the HIV DIS RNA to lividomycin, ITC experiments performed in a low-salt buffer (25 mM of potassium chloride) clearly showed two binding events [28] and cannot be fitted with a single site model (Fig. 3). However, in higher-salt conditions (200 mM potassium chloride), only the specific binding was observed (Fig. 3). Such experimental conditions leading to a homogenous RNA/ligand complex should therefore be preferred in the search for crystallization conditions. In addition, such an experiment provides the opportunity to stop the ITC experiment close to the expected stoichiometry (close to 1.0 in this case, depicted by the black arrow in Fig. 3) in order to avoid the use of an excess of ligand. Crystallization experiments can then be performed in the most favorable conditions. In the following example, a more general situation is described with a significant nonspecific binding of a positively charged small molecule ligand and without any prior knowledge of the preferable

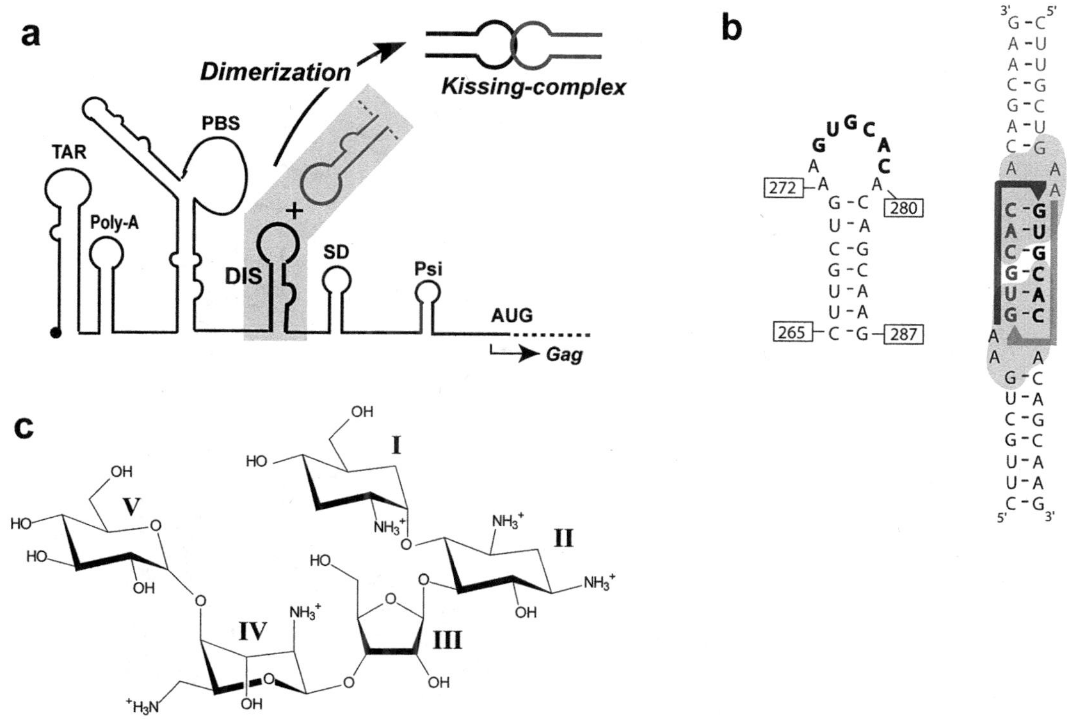

Fig. 1 HIV-1 DIS RNA and aminoglycosides. (**a**) Localization of the DIS in the 5′-untranslated region of the HIV-1 viral genomic RNA and dimerization mechanism through formation of a loop–loop "kissing" complex. (**b**) Sequence and secondary structure of the HIV-1 subtype F stem-loop used in this study. The self-complementary sequence is *bolded*. The loop–loop complex is depicted on the *right*, with the "A-site motif" circled in *light gray*. (**c**) Chemical structure of the aminoglycoside lividomycin

salt concentration to use. A brominated RNA (on the third uridine) is used in order to perform a MAD or SAD phasing following co-crystallization.

2 Materials

2.1 Instrumentation

1. iTC$_{200}$ (Microcal-Malvern, Malvern, UK) or equivalent Isothermal Titration Calorimeter.
2. Crystallization robot Mosquito Crystal (TPP LabTech, Melbourn, UK) or equivalent.
3. NanoDrop ND-1000 Spectrophotometer (Thermo Fisher Scientific, Waltham, MA, USA) or equivalent.
4. Refrigerated Centrifuge.

2.2 Instrument Accessories

1. Hamilton loading syringes for iTC$_{200}$ (Hamilton Company, Reno, NV, USA).
2. BD Luer-Lok 1 ml syringe.

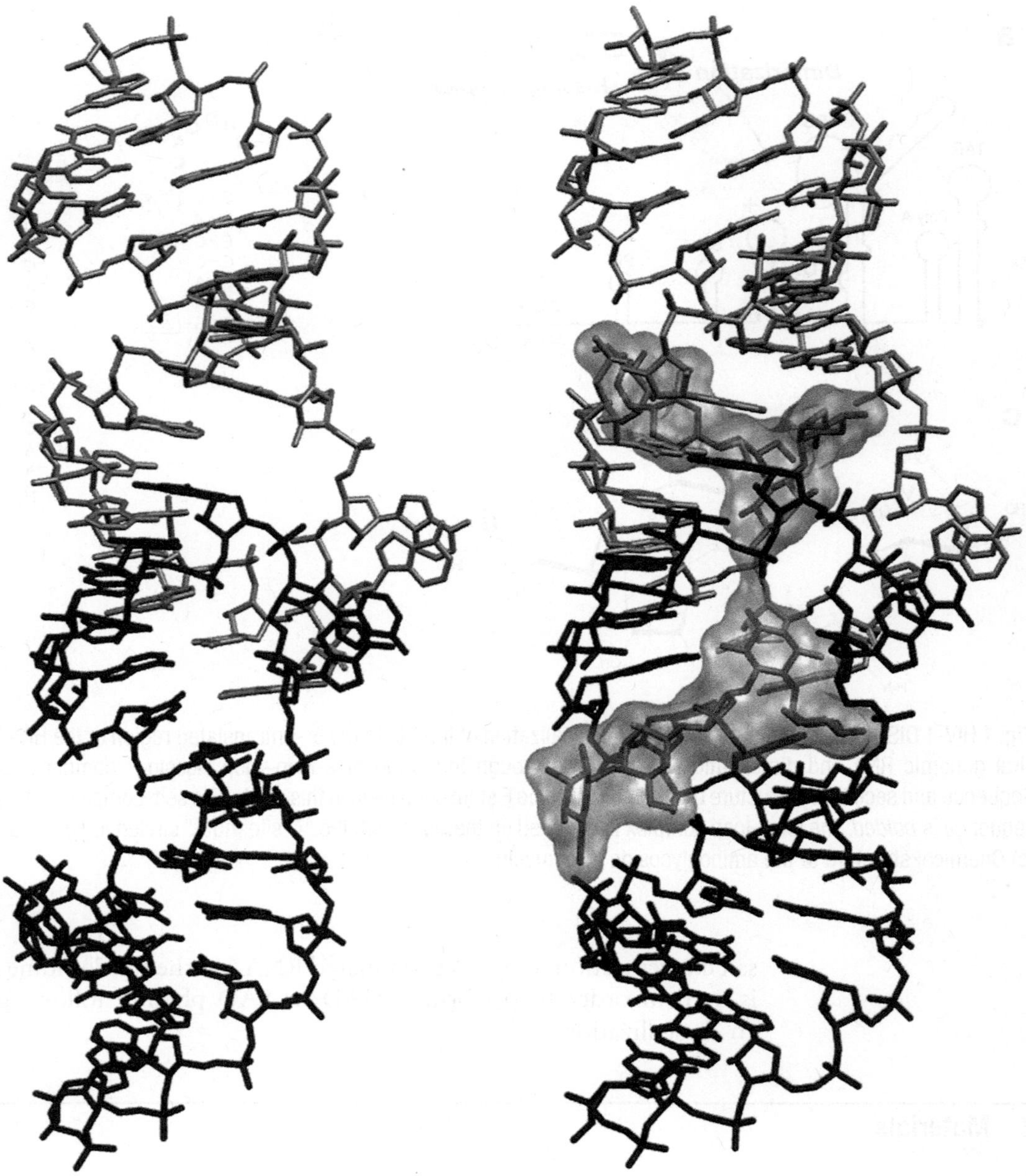

Fig. 2 Crystal structure of the free HIV-1 DIS RNA loop–loop complex (*left*) and bound to two lividomycin aminoglycosides (*right*). The two RNA stem loops are shown in *black* and *gray*, respectively

2.3 RNA Preparation

1. Chemically synthesized (purity >98 %) HIV-1 DIS 23-nucleotides RNA (Br-U3 modified): 5′-CU[BrU]GCUGAAGUGCACACAGCAAG-3′.
2. Dry bath capable of maintaining 90 °C.
3. KCl, powder.
4. 1 M $MgCl_2$. Store at 4 °C.

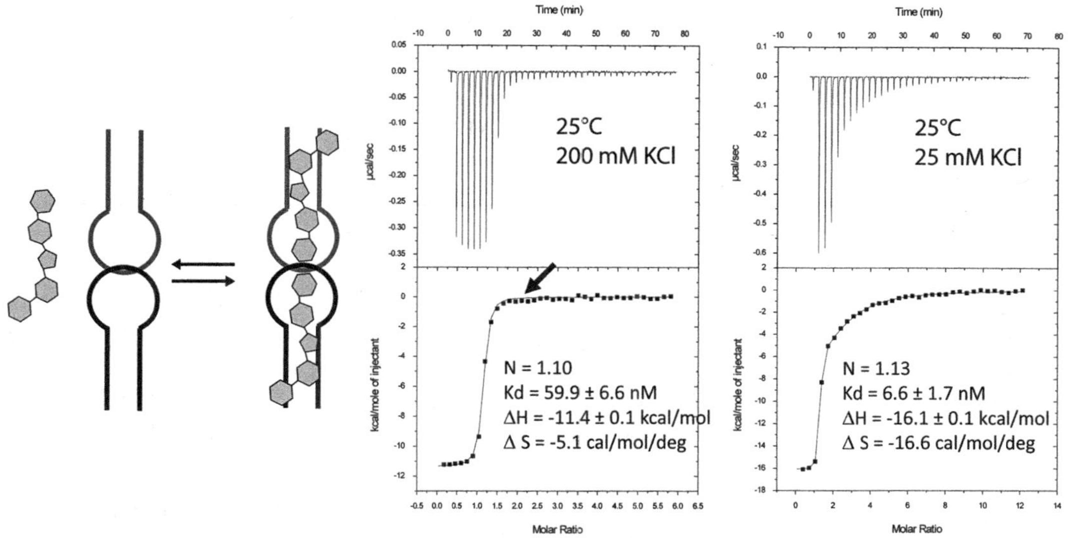

Fig. 3 Examples of titration of DIS RNA with lividomycin at 25 °C. On the *left*, the buffer contains 200 mM KCl and the experiment was performed with 12 μM RNA (in strands) and 350 μM aminoglycoside. On the *right*, the buffer contains 25 mM KCl and the experiment was performed with 6 μM RNA and 400 μM aminoglycoside. The *black arrow* depicts the saturation point where the ITC experiment might be stopped in order to perform crystallization trials in optimized RNA/ligand ratios

5. 1 M Sodium cacodylate, pH 7.0. Store at 4 °C.
6. 10× ITC Buffer: 1,500 mM KCl, 50 mM $MgCl_2$, 200 mM Sodium cacodylate, pH 7.0. Store at room temperature. Prepare 10 ml before use.
7. Amicon Ultra-4 centrifugal filter unit 10,000 MWCO (Millipore, Billerica, MA, USA).

2.4 Aminoglycosides Preparation

1. Lividomycin A Sulfate, powder. Store at 4 °C.
2. 1× ITC buffer: 150 mM KCl, 5 mM $MgCl_2$, 20 mM Sodium cacodylate, pH 7.0.

2.5 ITC Experiment

1. 1× ITC buffer: 150 mM KCl, 5 mM $MgCl_2$, 20 mM Sodium cacodylate, pH 7.0. Prepare 10 ml to wash the sample cell.
2. ITC EXPERT, the software controlling the iTC_{200} calorimeter.
3. Microcal Origin software for ITC data analysis (or equivalent software, such as NITPIC [30] or AFFINImeter: http://www.affinimeter.com/, *see* **Note 1**).

2.6 Crystallization of RNA–Aminoglycosides Complexes

1. Natrix HT crystallization screening kit (Hampton research, Aliso Viejo, CA, USA). Store at 4 °C.
2. 96-well CrystalEX™ microplate, conical flat bottom (Corning, Corning, NY, USA) or equivalent.

3. Crystal Clear Sealing Films (Hampton Research) or equivalent.
4. Amicon Ultra 0.5 ml centrifugal filter unit 10,000 MWCO (Millipore, Billerica, MA, USA).

3 Methods

3.1 RNA and Aminoglycosides Preparation for ITC Experiment

1. In order to prepare 300 μl of RNA at 120 μM for ITC experiment, RNA is refolded in water. To prevent the formation of duplex dimers and favor the hairpin fold, RNA concentration at this step must not exceed 3 μM. RNA is aliquoted into 12 Eppendorf tubes, each tube containing 1 ml of RNA at 3 μM in water.
2. Heat the RNA sample for 5 min at 90 °C and snap cool in water at 0 °C.
3. Add 111 μl of 10× ITC buffer in each Eppendorf tube. The RNA will then be in 1× ITC buffer.
4. Concentrate the RNA solution to about 300 μl with an Amicon Ultra-4 concentrator: centrifuge 10 min at 7500×*g* at 10–12 °C. Higher temperature should be avoided since it might promote the hairpin to duplex conversion during the RNA concentration [31].
5. Read the absorbance at 260 nm using a NanoDrop spectrophotometer. An RNA concentration of 120 μM corresponds to ca. 20 OD.
6. Prepare 1 mL of lividomycin 2,4 mM in 1×ITC buffer.

3.2 Designing the ITC Experiment

3.2.1 Experimental Considerations

In an ITC experiment, the calorimeter directly measures the heat evolved in liquid samples as a result of mixing precise amount of reactants. Heat is produced (exothermic interaction) or absorbed (endothermic interaction) as a result of the interaction between the ligand and the sample molecule (*see* **Note 2**).

ITC monitors these heat changes by measuring the differential power applied to the cell heaters required to maintain a small temperature difference between the reference and sample cells as the binding partners are mixed. This differential power compensation per ligand injection is the raw signal recorded by the experiment. Each ligand injection results in a heat pulse that is integrated with time and normalized for concentration. This will produce a titration curve of kcal/mol vs. molar ratio (ligand/sample). The resulting isotherm is fitted to a binding model to obtain the affinity constant ($K_a = 1/K_d$), stoichiometry (N), and enthalpy of interaction (ΔH). The Gibbs free energy change (ΔG) and the entropy change (ΔS) are obtained via the thermodynamic relationships:

$$\Delta G = -\mathrm{R\,T}\ln K_a = \mathrm{R\,T}\ln K_d$$

$$\Delta G = \Delta H - T\Delta S$$

The appropriate concentration of the component placed in the sample cell (M_T) depends on the binding affinity, the stoichiometry, and heat of binding ΔH [32]. In order to derive accurate binding constant and ΔH, the Wiseman coefficient $c = N \cdot K_a \cdot M_T$ should be larger than 1 but smaller than 1,000 (preferably between 10 and 500) [33, 24].

In practice, required concentrations of sample in the cell for a "traditional" ITC experiment are ranging from 5 to 100 μM. The molar concentration of the ligand in the injection syringe is typically 10–20 times higher (and even higher in the case of low *c* value experiments [34]). This will ensure that the cell material will become saturated by the end of the titration experiment. As a consequence, sample requirement in order to perform an ITC experiment in ideal conditions ($c \sim 100$) might be significant for biological system with a low affinity. This should however not be an obstacle for ITC experiments performed prior crystallization.

The key to successful ITC experiments, one never insists too much on that, is to minimize the heat signal arising from mismatches between the solutions that are mixed (*see* **Note 3**).

3.2.2 Instrument Parameters

On the iTC_{200} microcalorimeter, the titration syringe holds up to 40 μl. Typical injections must not exceed 2 μl and must not be lower than 0.2 μl. In an ITC experiment performed prior crystallization, sample concentrations are large and rather small injections (~1 μl) should be sufficient to ensure a satisfactory signal/noise ratio and a correct sampling of the nucleic acid/ligand saturation. It is recommended to perform a first small injection (0.3–0.5 μl) that may be disregarded in data analysis (*see* **Note 4**). The number of injections per experiment depends on the volume of the injections and should be set so that the sample in the cell will be saturated at the end of the titration. Usually, the complete volume of the syringe is injected into the cell. However, it might be appropriate to stop the titration when the nucleic acid is saturated with the ligand in order to perform crystallization trials with the minimum excess of ligand (Fig. 3). The injection time should be set to twice the injection volume in μl, thus corresponding to a 0.5 μl/s injection speed (i.e. 2 s for 1 μl injections). The spacing between the injections is generally set to 120 s by default, which is often adequate to ensure a correct return to the baseline after each injection. However, the equilibration time is strongly dependent on the rate constants k_{on} and k_{off} for each system [35] and it is not uncommon to have to seriously increase this spacing value to 240 s (even much higher for very slow systems [15]). The filter value corresponds to the integration time that will be used to produce a single data point. Usually set to 5 by default, the filter period might be set to 3 for a better sampling of each peak in view of performing a kinetic analysis of the thermodynamic data [35].

Regarding the calorimeter setup, the feedback mode should be set to "high gain," which corresponds to the fastest response time. The "Reference Power" value corresponds in practice to the baseline (in μCal/s). In the high gain mode, the instrument's reference power ranges from 0 to 12.25 μCal/s. The anticipated size and direction of the titration peaks will determine the baseline setting. For exothermic interactions, a large reference power setting is required (~5 μcal/s) and for endothermic interactions, a small reference power setting is required (0.5 μCal/s). It is essential to set the reference power so that the signal generated by each peak remains positive at any time. The stirring speed is set to 1000 × *g* by default (*see* **Note 5**).

Regarding the cell temperature, the instrument's theoretical operating range is 2–80 °C but most of the experiments are performed between 4 and 37 °C. The temperature of the experiment is often an overlooked parameter that might be very useful to exploit in order to improve the results through an optimization of the Wiseman c value by playing on two parameters: ΔH and K_d. Indeed, most biomolecular interactions are characterized by temperature dependence of the ΔH due to the heat capacity change of the event:

$$\Delta C_p = (\partial \Delta H / \partial T)$$

This temperature dependence is generally negative and associated with a local folding following binding [36]. It is a valuable tool for optimizing the study of any binding event: for an exothermic reaction, ΔH will be more negative at 35 °C than 25 °C and, consequently, the signal will be increased at 35 °C. Regarding the effect on the affinity, the K_d of an exothermic reaction should decrease at lower temperature (in agreement with van't Hoff's relationship), which might improve data in systems where the c value is too low.

3.3 Performing the ITC Experiment

1. Fill the reference cell with ultra pure water using a 0.5 ml glass Hamilton syringe.
2. In the instrument controls tab, set the temperature at 12 °C in "Thermostat Control," and click "Set Jacket Temp." This temperature is selected in order to improve the affinity of the aminoglycoside for the DIS RNA and to reduce any potential hairpin to duplex conversion of the RNA in presence of magnesium cations and rather high RNA concentration [31].
3. Rinse the sample cell, two or three times, with 1× ITC buffer.
4. Gently load the sample cell with 300 μl of DIS F BrU3 RNA at 0.120 mM. The small excess should be removed. The final

sample volume in the cell is ca. 280 μl. This step is delicate since no bubble should be present in the sample cell.

5. To load the titration syringe, place a microcentrifuge tube with at least 70 μl of lividomycin sample at 2.4 mM. Place the syringe titration above the microcentrifuge tube so that the tip of the syringe is immerged in ligand solution. Manual loading of the syringe is by far much safer than the semi-automatic procedure, especially if the sample is viscous (i.e. with a buffer containing some glycerol). Connect the BD syringe and tubing to the titration syringe and load the sample manually by aspiration into the titration syringe (before aspiration, make sure that the plunger is in the open position). When the sample gets to the plunger, click "Close Fill Port" in the instrument controls tab; this will move the plunger tip down so that it blocks the fill port. Perform the "Purge/ReFill" procedure to remove small air bubbles in the syringe. Unplug the BD syringe from the titration syringe.
6. Place the titration syringe into the sample cell.
7. Set the reference power to 5 μCal/s.
8. Set the initial delay to 100 s and the stirring speed to 1000×*g*.
9. Set the number of injections to 40, the sample cell concentration to 0.12 mM and the concentration in the syringe to 2.4 mM.
10. Set the titration parameters (the number and volume of the injections as well as the spacing between two injections) in the "Injection Parameters" window into the "Advanced Experimental Design" tab. Set the first injection volume to 0.3 μl and the next injections volume to 1 μl. Set the spacing between each injection to 180 s.
11. Press the "Start" button to begin the experiment. The experiment will start when both cells are thermostated at 12 °C.
12. At the end of the titration, recover the RNA/ligand sample, wash the sample cell, reload the sample cell with 1× ITC buffer, reload the syringe with lividomycin solution and perform a blank experiment using the same parameters. A signal (here endothermic) will be produced due to the dilution of the ligand into the buffer. This signal should be taken into account for subsequent data analysis.

3.4 ITC Data Analysis

ITC data analysis using the Microcal Origin7 software will be described here.

Start the Origin software. The program opens and displays the rawITC plot window.

1. Click on "Read Data" to open the titration experiment (file with .itc extension). Origin will integrate the injection peaks

and a new "Delta H" window will appear (Fig. 4, step 1). It is possible at this stage to modify the concentrations of your samples by using the "Concentration button."

2. The software will automatically generate a baseline important for peak integration. If required, it can be manually adjusted by using the "adjust integration" button in the "rawITC" window in the "window" menu.
3. Go to the "rawITC" window.
4. Click again "Read data."
5. Select the blank file generated by injection of lividomycin in the buffer.
6. Remove aberrant data points and/or adjust the baseline if necessary.
7. Go to the "Math" menu and select "linear regression." A red line will be drawn (Fig. 4, step 2).
8. Double click on the "1" button located on the top left corner of the window (red arrow in Fig. 4, step 1). A new window labeled "layer 1" appears, with two lists: "available data" on the left and "Layer content" on the right. The "Layer content" list contains two items: one "xxx_ndh" corresponding to integrated heats of the blank experiment and a "linearfit1_xxxndh" corresponding to the linear regression.
9. Select data corresponding to integrated heats ("yyy_ndh") generated by the injection of ligand onto the RNA and click on the right arrow placed between both lists. The data set will then appear in the "Layer content" list. Click "OK" on the "Layer 1" window and on the "Rescale" button on the button bar of the Origin window (red arrow in Fig. 4, step 2). Integrated heats of both the ligand/RNA and ligand/buffer experiments are displayed (Fig. 4, step 3).
10. Go to the "Math" menu and select "Simple math." A new "Math on/Between data Set" window appears.
11. In the "Available data" list, select the "yyy_NDH" item and click on the right arrow close to "Y1."
12. In "operator," write the minus "-" operation.
13. In the "Available data" list, select the "linearfit1_xxxndh" item and click on the right arrow close to "Y2" (Fig. 4, step 4). Press "OK" in the "Math on/Between data Set" window. This will subtract the blank experiment to the RNA/ligand titration.
14. Double click on the "1" layer button located on the top left corner of the window. In the "Layer content" list, select "xxx_ndh" and "linearfit1_xxxndh" items and click on the left window (Fig. 4, step 5). This will remove the linear regression and

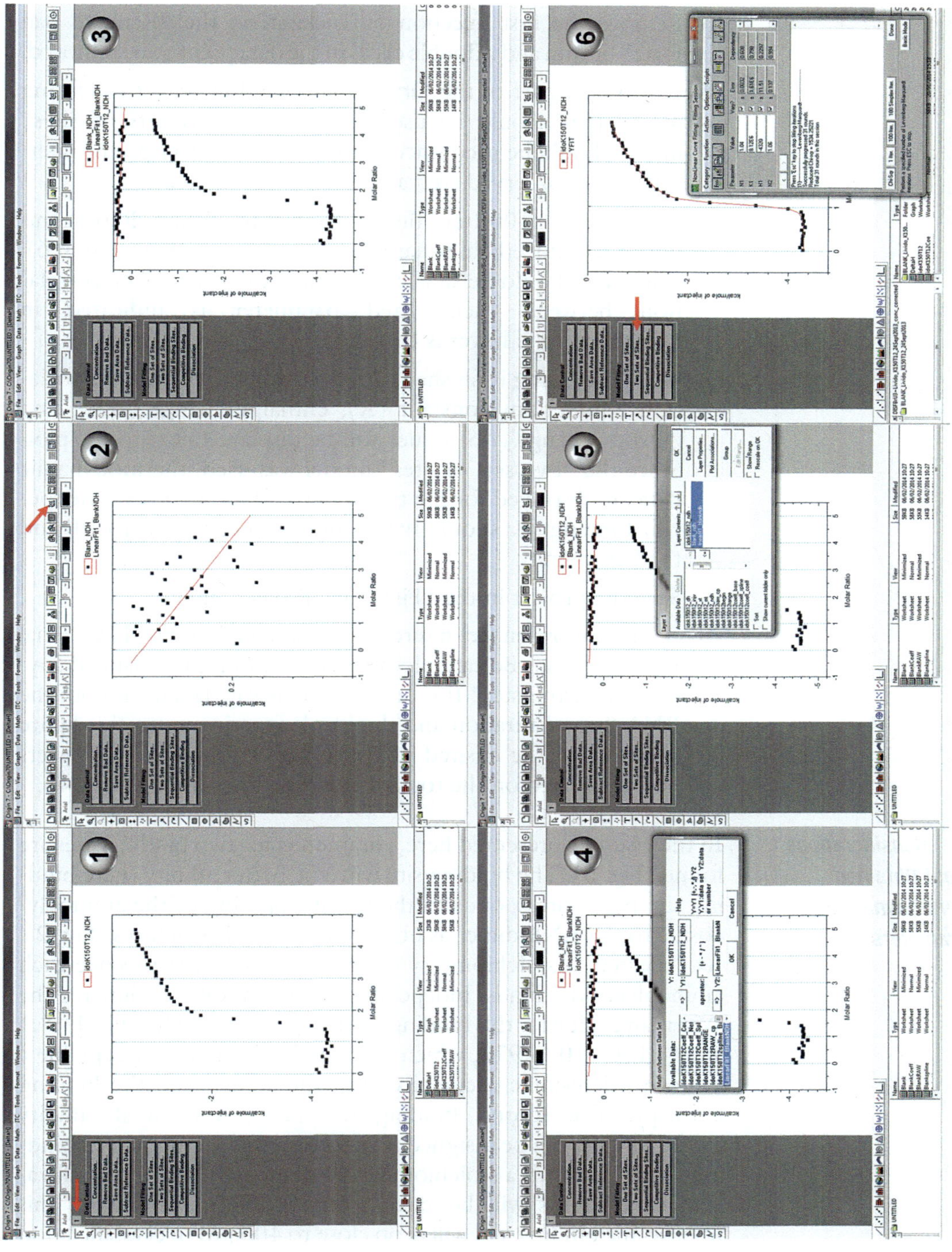

Fig. 4 Screenshots of six consecutive steps during ITC data processing with Microcal Origin

integrated heats of the blank experiment from graph in the main window. Click on the "Rescale" button. Now it is time to fit the data using an available model.

15. Remove the first injection by clicking on the "Remove Bad Data" button and a double click on the data point to be removed.
16. On the left part of the Origin main window, select the fitting model that should be used. Here the "Two Sets of Sites" model should be used since both specific and nonspecific interactions are observed (red arrow on Fig. 4, step 6).
17. Click on the "100 Iter" button to start the fitting. Repeat this until the reduced chi square remains constant (Fig. 4, step 6). The error should be smaller than the fitted value. The correlation between each fitted parameter is indicated in "Dependency": lower is better.
18. Once the fitting is finished, the stoichiometry (N), the association equilibrium constant (K), enthalpy change (ΔH), and entropy change (ΔS) values will be displayed next to the plot. Since the "two sets of sites" model was used, thermodynamic data are displayed for the two different binding sites. The specific binding site should have a stoichiometry close to 1.0 and the best affinity.
19. Save the project in the "File/Save Project" menu.
20. A publication-quality figure can be obtained in the "ITC/Final Figure" menu. Be careful: the baseline cannot be changed anymore after this step! It is possible to display data from both the RNA/ligand titration and the blank experiment on this figure by selecting the desired items on the "1" and/or "2" layers button located on the top left corner of the window (Fig. 5).

3.5 Considerations About Obtained Thermodynamic Parameters

In the example presented here, parameters for two binding sites are found (Fig. 5). The binding site with the better affinity (here binding site n°1) corresponds to the specific binding of the aminoglycoside to the RNA dimer as observed in crystal structures (Fig. 2). As expected, a stoichiometry close to 1.0 is observed since each aminoglycoside binds independently to the RNA dimer. The Wiseman c value for this binding site is $N_1 \cdot (1/K_{d1}) \cdot [\text{RNA}] = 1 \times 8.1 \times 10^6 \times 120 \times 10^{-6} = 972$, which is in the high range of acceptable c values to determine accurately both the affinity and the binding enthalpy. The second binding site has a very weak affinity (~140 μM). It can be assigned to a nonspecific component of binding (here also with a stoichiometry close to 1.0). Since crystallization experiments will be carried out at 37 °C, affinity of this nonspecific binding site will drop close to 400 μM in crystallization conditions (this estimation can be obtained using the known heat capacity change for the DIS RNA/lividomycin interaction [28] and the van't Hoff's relationship). It should therefore not interfere with co-crystallization of the specific RNA/ligand complex.

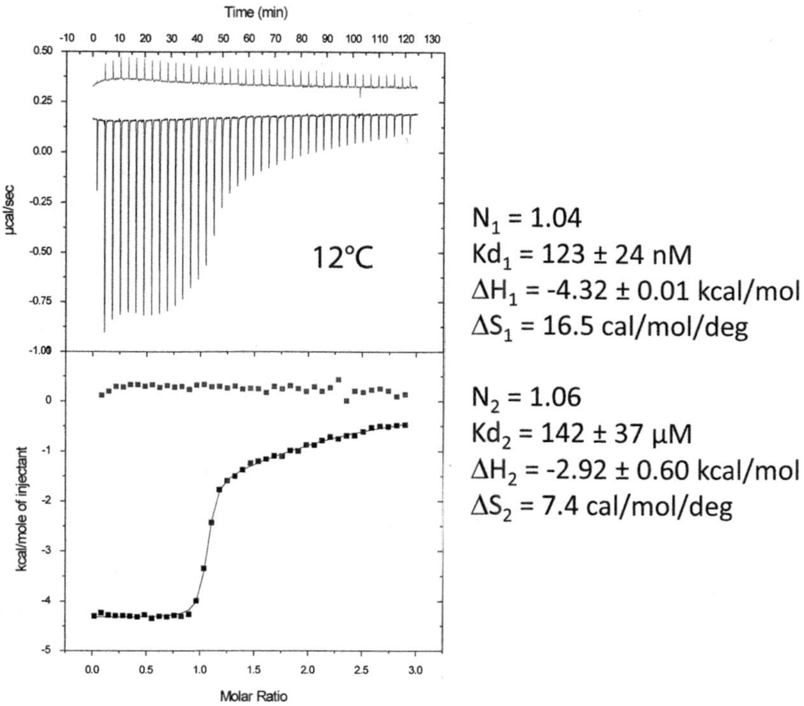

Fig. 5 Final figure obtained after the titration of DIS RNA with lividomycin at 12 °C. The blank experiment corresponding to injection of the aminoglycoside into ITC buffer is shown in *grey* and yield an endothermic dilution signal. Values obtained for the specific and the nonspecific binding sites after fitting the data using the "two set of sites" model are shown

3.6 RNA–Aminoglycosides Crystallization

The RNA–lividomycin complex obtained after the ITC experiment is concentrated to a final concentration of 0.33–0.40 mM with an Amicon Ultra 0.5 ml concentrator (10,000 MWCO) at 4 °C.

3.6.1 RNA/Lividomycin Co-crystallization

1. Prior the crystallization trials, the Natrix HT kit should be placed at room temperature for at least 1 h. Dispense 0.16 ml of Natrix HT crystallization kit solutions in each well of the 96-well CrystalEX™ microplate.
2. Using the Mosquito crystallization robot, mix 0.2 µl of the RNA–lividomycin sample with 0.2 µl of the crystallization buffer in the top left (elevated) drop wells of the CrystalEX™ microplate. Alternatively, this step can be performed manually on a larger scale using 1.0 + 1.0 µl drops.
3. Seal the crystallization plate with the Crystal clear sealing film.
4. Store the crystallization plates at 37 °C overnight.
5. Check for crystals ~16 h later. After 24 h at 37 °C, place the crystallization plates containing crystals at 20 °C. Large monocrystals suitable for diffraction should grow from Natrix HT conditions E4, E8, G3, G4, G6, G11, G12 and H9, which corresponds to a remarkable score of 16.6 % successful crystallization rate (Fig. 6). Many other conditions lead to microcrystals.

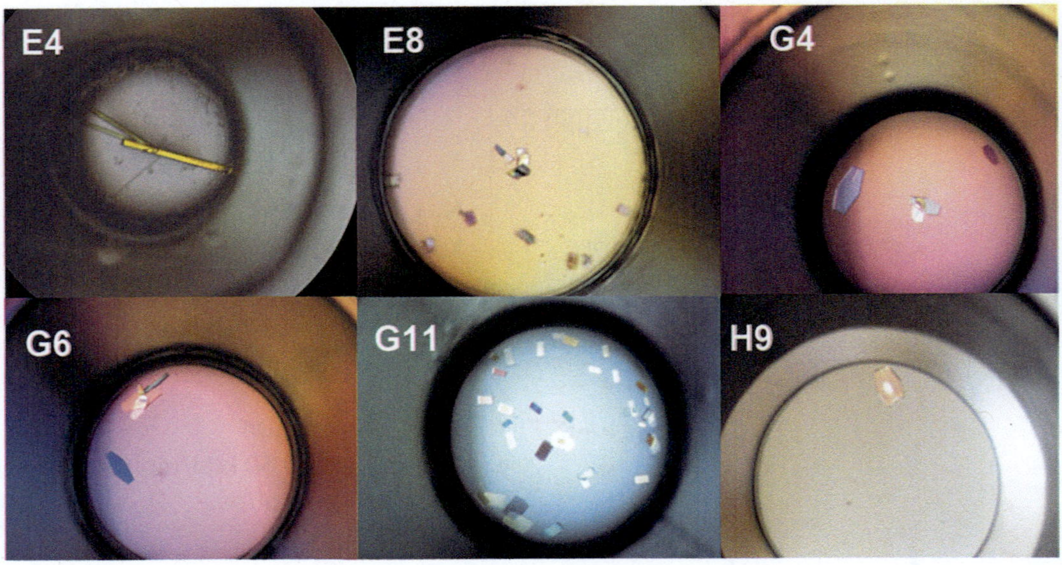

Fig. 6 Some of the crystals obtained using a Natrix HT kit with a DIS RNA loop–loop complex/lividomycin complex. The RNA/ligand complex was recovered from the ITC experiment shown in Fig. 5. The crystallization conditions is indicated in the top-left part of each picture

3.6.2 Crystals Harvest and Freezing

1. Harvest crystals that are deemed suitable for diffraction. Crystals grown in 2-methyl-2,4-pentanediol conditions do not require further cryoprotection. Otherwise glycerol can be used as a cryoprotectant for other precipitants.
2. Flash freeze crystals in liquid ethane or propane (not directly in liquid nitrogen!).
3. Store the frozen crystals in liquid nitrogen.

3.6.3 X-Ray Diffraction and Data Collection

Diffraction data for several crystals directly obtained from this screen were obtained on the X06DA beam line (bending magnet) at the SLS synchrotron. Complete and good data sets were collected at the bromine K-edge up to 1.7 Å resolution and the structure could be solved ab initio using the SAD approach.

4 Notes

1. Note that AFFINImeter offers many automatic procedures and is the only one allowing to obtain potential kinetic information by using *kinITC* [35].
2. It is extremely important of not being misled by situations where no heat signal can be detected. Indeed, two situations are possible: either there is no interaction at all (and a co-crystallization will be useless), or there is interaction, but the ΔH is null in the conditions of the experiment. The problem is thus to distinguish the two situations. Since the most important parameter influencing the ΔH of a reaction is the temperature

(which is characterized by the term $\Delta C_p = \partial \Delta H / \partial T$), one should immediately repeat the same experiment at, at least, another temperature or, preferably, at two temperatures, one above and the other one below the initial temperature T_0. In situations where $|\Delta C_p|$ is of order 0.5 kcal/mol/K, which is quite a common order of magnitude, an overall change by +10 K (from $T_0 - 5$ K to $T_0 + 5$ K) would induce an overall shift in ΔH by +(sign ΔC_p)×5 kcal/mol, which would be readily detectable. Furthermore, the sign of ΔH on either side of T_0 should be opposite. Very importantly, following the van't Hoff equation $[\partial \ln K_d / \partial T = -\Delta H / (RT^2)]$, a null ΔH at some temperature implies that $\ln K_d$, and thus K_d too, are extremum at that temperature. If ΔC_p is positive (i.e. if ΔH increases when the temperature increases), a null ΔH is the mark of a minimum affinity (K_d is maximum) at that temperature. Conversely, if ΔC_p is negative (i.e. if ΔH decreases when the temperature increases), a null ΔH is the mark of a maximum affinity (K_d is minimum) at that temperature.

3. The mismatch may be produced by pH difference between the ligand solution and the sample solution, as well as any difference in salt concentration, buffer concentration or additives. To minimize these differences, the interactants should be prepared in identical solutions. In addition, because of the high sensitivity of the microcalorimeter to any temperature change, it is highly recommended to place the instrument far from any air flow, especially air conditioning, and in a room exposed to minimal temperature changes (avoid heaters, large windows or other instruments generating lots of heat such as freezers). To obtain steady baselines, the instrument should be cleaned regularly and the syringe should be in a very good state (even very slightly bent syringe will generate a significant noise). Obviously, the sample cell should be kept RNase free for experiments involving RNA.
4. Usually, this first injection is not correct due to artifacts introduced by the mechanical backlash of the screw mechanism in the injector syringe and by a frequent looseness in the Teflon tip in the syringe.
5. Reducing the stirring speed generally slightly improve the baseline stability but might introduce artifacts due to insufficient mixing.

Acknowledgments

This work was supported by grants from the Agence Nationale pour la Recherche (grant ANR-12-BS07-0007-03 "ClickEnARN") and the Agence Nationale de Recherches sur le SIDA (ANRS). The authors would like to thank Vincent Olieric (Paul Scherrer Institute/Swiss Light Source, Villigen, Switzerland), Natalia Markova and Peter Gimeson (Microcal-Malvern, Uppsala, Sweden).

References

1. Feig AL (2007) Applications of isothermal titration calorimetry in RNA biochemistry and biophysics. Biopolymers 87:293–301
2. Feig AL (2009) Studying RNA-RNA and RNA-protein interactions by isothermal titration calorimetry. Methods Enzymol 468:409–422
3. Gilbert SD, Batey RT (2009) Monitoring RNA-ligand interactions using isothermal titration calorimetry. Methods Mol Biol 540: 97–114
4. Salim NN, Feig AL (2009) Isothermal titration calorimetry of RNA. Methods 47:198–205
5. Pilch DS, Kaul M, Barbieri CM, Kerrigan JE (2003) Thermodynamics of aminoglycoside-rRNA recognition. Biopolymers 70:58–79
6. Kulshina N, Edwards TE, Ferre-D'Amare AR (2010) Thermodynamic analysis of ligand binding and ligand binding-induced tertiary structure formation by the thiamine pyrophosphate riboswitch. RNA 16:186–196
7. Pikovskaya O, Polonskaia A, Patel DJ, Serganov A (2011) Structural principles of nucleoside selectivity in a 2′-deoxyguanosine riboswitch. Nat Chem Biol 7:748–755
8. Sokoloski JE, Dombrowski SE, Bevilacqua PC (2012) Thermodynamics of ligand binding to a heterogeneous RNA population in the malachite green aptamer. Biochemistry 51: 565–572
9. Ennifar E, Aslam MW, Strasser P, Hoffmann G, Dumas P, van Delft FL (2013) Structure-guided discovery of a novel aminoglycoside conjugate targeting HIV-1 RNA viral genome. ACS Chem Biol 8:2509–2517
10. Trausch JJ, Batey RT (2013) A disconnect between high-affinity binding and efficient regulation by antifolates and purines in the tetrahydrofolate riboswitch. Chem Biol 1(2): 205–216
11. Datta K, LiCata VJ (2003) Thermodynamics of the binding of Thermus aquaticus DNA polymerase to primed-template DNA. Nucleic Acids Res 31:5590–5597
12. Recht MI, Williamson JR (2004) RNA tertiary structure and cooperative assembly of a large ribonucleoprotein complex. J Mol Biol 344: 395–407
13. Bauer WJ, Heath J, Jenkins JL, Kielkopf CL (2012) Three RNA recognition motifs participate in RNA recognition and structural organization by the pro-apoptotic factor TIA-1. J Mol Biol 415:727–740
14. Freisz S, Mezher J, Hafirassou L, Wolff P, Nomine Y, Romier C, Dumas P, Ennifar E (2012) Sequence and structure requirements for specific recognition of HIV-1 TAR and DIS RNA by the HIV-1 Vif protein. RNA Biol 9:966–977
15. Bec G, Meyer B, Gerard MA, Steger J, Fauster K, Wolff P, Burnouf D, Micura R, Dumas P, Ennifar E (2013) Thermodynamics of HIV-1 reverse transcriptase in action elucidates the mechanism of action of non-nucleoside inhibitors. J Am Chem Soc 135:9743–9752
16. Lukavsky PJ, Daujotyte D, Tollervey JR, Ule J, Stuani C, Buratti E, Baralle FE, Damberger FF, Allain FH (2013) Molecular basis of UG-rich RNA recognition by the human splicing factor TDP-43. Nat Struct Mol Biol 20:1443–1449
17. Neuenfeldt A, Lorber B, Ennifar E, Gaudry A, Sauter C, Sissler M, Florentz C (2013) Thermodynamic properties distinguish human mitochondrial aspartyl-tRNA synthetase from bacterial homolog with same 3D architecture. Nucleic Acids Res 41:2698–2708
18. Vander Meulen KA, Davis JH, Foster TR, Record MT Jr, Butcher SE (2008) Thermodynamics and folding pathway of tetraloop receptor-mediated RNA helical packing. J Mol Biol 384:702–717
19. Reymond C, Bisaillon M, Perreault JP (2009) Monitoring of an RNA multistep folding pathway by isothermal titration calorimetry. Biophys J 96:132–140
20. Salim N, Lamichhane R, Zhao R, Banerjee T, Philip J, Rueda D, Feig AL (2012) Thermodynamic and kinetic analysis of an RNA kissing interaction and its resolution into an extended duplex. Biophys J 102:1097–1107
21. Zhang J, Ferre-D'Amare AR (2013) Co-crystal structure of a T-box riboswitch stem I domain in complex with its cognate tRNA. Nature 500:363–366
22. Ladbury JE, Chowdhry BZ (1996) Sensing the heat: the application of isothermal titration calorimetry to thermodynamic studies of biomolecular interactions. Chem Biol 3:791–801
23. Leavitt S, Freire E (2001) Direct measurement of protein binding energetics by isothermal titration calorimetry. Curr Opin Struct Biol 11:560–566

24. Velazquez Campoy A, Freire E (2005) ITC in the post-genomic era...? Priceless. Biophys Chem 115:115–124
25. Privalov PL, Dragan AI (2007) Microcalorimetry of biological macromolecules. Biophys Chem 126:16–24
26. Ennifar E, Paillart JC, Marquet R, Ehresmann B, Ehresmann C, Dumas P, Walter P (2003) HIV-1 RNA dimerization initiation site is structurally similar to the ribosomal A site and binds aminoglycoside antibiotics. J Biol Chem 278:2723–2730
27. Ennifar E, Paillart JC, Bodlenner A, Walter P, Weibel JM, Aubertin AM, Pale P, Dumas P, Marquet R (2006) Targeting the dimerization initiation site of HIV-1 RNA with aminoglycosides: from crystal to cell. Nucleic Acids Res 34:2328–2339
28. Bernacchi S, Freisz S, Maechling C, Spiess B, Marquet R, Dumas P, Ennifar E (2007) Aminoglycoside binding to the HIV-1 RNA dimerization initiation site: thermodynamics and effect on the kissing-loop to duplex conversion. Nucleic Acids Res 35:7128–7139
29. Ennifar E, Paillart JC, Bernacchi S, Walter P, Pale P, Decout JL, Marquet R, Dumas P (2007) A structure-based approach for targeting the HIV-1 genomic RNA dimerization initiation site. Biochimie 89:1195–1203
30. Keller S, Vargas C, Zhao H, Piszczek G, Brautigam CA, Schuck P (2012) High-precision isothermal titration calorimetry with automated peak-shape analysis. Anal Chem 84: 5066–5073
31. Bernacchi S, Ennifar E, Toth K, Walter P, Langowski J, Dumas P (2005) Mechanism of hairpin-duplex conversion for the HIV-1 dimerization initiation site. J Biol Chem 280:40112–40121
32. Wiseman T, Williston S, Brandts JF, Lin LN (1989) Rapid measurement of binding constants and heats of binding using a new titration calorimeter. Anal Biochem 179:131–137
33. Tellinghuisen J (2005) Optimizing experimental parameters in isothermal titration calorimetry. J Phys Chem B 109:20027–20035
34. Tellinghuisen J (2008) Isothermal titration calorimetry at very low c. Anal Biochem 373: 395–397
35. Burnouf D, Ennifar E, Guedich S, Puffer B, Hoffmann G, Bec G, Disdier F, Baltzinger M, Dumas P (2012) kinITC: a new method for obtaining joint thermodynamic and kinetic data by isothermal titration calorimetry. J Am Chem Soc 134:559–565
36. Spolar RS, Record MT Jr (1994) Coupling of local folding to site-specific binding of proteins to DNA. Science 263:777–784

Part III

Data Collection, Phasing and Refinement

Chapter 10

Crystallographic Data and Model Quality

Kay Diederichs

Abstract

This article gives a consistent classification of sources of random and systematic errors in crystallographic data, and their influence on the averaged dataset obtained from a diffraction experiment. It discusses the relation between precision and accuracy and the crystallographic indicators used to estimate them, as well as topics like completeness and high-resolution cutoff. These concepts are applied in the context of presenting good practices for data processing with a widely used package, XDS. Recommendations are given for how to minimize the impact of several typical problems, like ice rings and shaded areas. Then, procedures for optimizing the processing parameters are explained. Finally, a simple graphical expression of some basic relations between data error and model error is suggested.

Key words X-ray crystallography, Accuracy, Precision, Random errors, Systematic errors, Merged data, Unmerged data, Indicators

1 Introduction

In the last decades, crystallography has been highly successful in delivering structural information about proteins, DNA, and RNA, the substrates of life on earth. The resolution of the method is good enough to discern the three dimensional structure of these macromolecules at the atomic level, which is essential to understand their diverse properties, functions and interactions. However, although it is easy to calculate the diffraction pattern for a given structure, the reverse task of deriving a molecular structure from just a single set of unique diffracted intensities is difficult, as the mathematical operation governing the former direction cannot be inverted in a unique way. To be solved experimentally, this "inverse problem," or more specifically "phase problem," requires more than just a single set of unique diffraction data. High quality of the data is a requirement for the experimental solution of the problem, but also for the refinement of the macromolecular structure, as discussed in Subheading 4.

Eric Ennifar (ed.), *Nucleic Acid Crystallography: Methods and Protocols*, Methods in Molecular Biology, vol. 1320,
DOI 10.1007/978-1-4939-2763-0_10, © Springer Science+Business Media New York 2016

The correct biological interpretation requires the best possible model of the macromolecule. To obtain the best model, every step of the structure determination procedure has to be performed in a close-to-optimal way. This means that the purification of the macromolecule, its crystallization, crystal handling, measurement of diffraction data, processing of the resulting datasets, and downstream steps such as structure solution, refinement, and validation each constitute scientific tasks that deserve specific attention, and have been undergoing continuous enhancements throughout the history of macromolecular crystallography.

Two kinds of numerical data are the result of a crystallographic experiment and usually deposited as such in the Protein Data Bank: the diffraction intensities as a reduced representation of the diffraction experiment, and the atomic coordinates resulting from the visual inspection and interpretation of electron density maps, and subsequent refinement. A third kind of numerical data, the raw data (frames) obtained in the diffraction experiment, have so far not been usually deposited in long-term archives, mainly due to (disk) space concerns. This is unfortunate since archiving of raw data would enable reprocessing of incorrectly processed data as well as enabling and taking advantage of future improvements in methodology, like extracting the diffuse scattering information.

The discussion focuses on data that correspond to a *single* atomic model. This rules out all the complications that arise from merging of non-isomorphous datasets, where each individual dataset corresponds to a different model—in this situation, a merged dataset would represent something like an average model, which violates the physicochemical requirements, and may not be biologically meaningful.

This chapter first presents the principles and concepts that need to be understood in the context of the rather broadly used term "data quality"; similar presentations may be found for example in refs. [1–3]. Second, the application of these principles to data processing with the XDS program package [4, 5], which the author is most familiar with, is explained. Third, data and atomic model are related in a graphical way, which allows some important and nontrivial conclusions to be drawn about how the former influence the latter.

2 Errors and Crystallographic Indicators

The goal of a crystallographic experiment is to obtain accurate intensities $I(hkl)$ for as many Bragg reflections hkl as possible. I discuss two kinds of errors, random and systematic error, which exist in any experiment. A major difference between them is that the relative error arising from the random component *decreases* with increasing intensity, whereas the relative error in intensity from the systematic component is (at least on average) *constant*, often in the range between 1 and 10 %. A more specific description for systematic error would thus be "fractional error," but this name is not in common use. Nonlinear errors also exist, but play a minor role.

A well-designed crystallographic experiment has to strike an appropriate compromise between the two kinds of error. For example, a reduction of random error (see below) can be obtained by longer or stronger exposure of the crystal, but this will inevitably increase the systematic error from radiation damage to the crystal. Ideally, the sum of both errors should be minimal, and programs, e.g., "BEST" [6], exist that suggest a compromise, in the form of a proposed "strategy" for the experiment. Fortunately, the gradient of the sum of both errors is close to zero at and near the optimal strategy, which means that small deviations from the optimal strategy do not substantially decrease data quality.

The discussion of errors has to take the distinction between precision and accuracy into account. The term "precision" refers to the reproducibility of an experiment, and to the internal consistency or relative deviation of the values obtained. For example, if the number $e = 2.718...$ should be determined in an experiment, and two measurements would yield the values 3.217 and 3.219, then these measurements are considered precise, because they agree well with each other—their relative deviation is small. However, they are not close to the true value—the error (or inaccuracy) in their measurement amounts to about 0.5.

The term "accuracy," on the other hand, refers to the deviation of measured values from the true values. In this example, if two measurements would yield the values 2.6 and 2.8, then the results from this experiment are more accurate than that from the previously mentioned experiment, although they are not as precise.

Optimizing an experiment for precision alone therefore does not ensure accuracy; rather, equating accuracy with precision also requires the absence of any kind of error that has not been taken into account in the precision estimate. To estimate accuracy, we thus need to quantify both the precision of the data, and the undetected error (which usually requires some knowledge about the true value obtained by other means). If both can be quantified, we can estimate the accuracy as the absolute or relative error of a measurement.

2.1 Random Error

The crystallographic experiment measures the number of photons contributing to each detector pixel. These photons arise from Bragg reflections, but also from background scatter. The number of photons in each pixel is subject to random fluctuations. These are due to the quantum nature of photons; there exists a certain probability of emission of a given photon by the crystal into a given pixel in a unit of time, and each photon's emission into that pixel is independent from that of other photons. As a result, photon counts are governed by Poisson (counting) statistics, which mathematically means that the variance of the photon number is equal to the photon number itself. Furthermore, a CCD detector may contribute a random component ("read-out noise") to the total photon count (pixel detectors are almost noise-free), which is also due to quantum fluctuations in the detector hardware and may be considered as additional background.

Data processing software essentially adds the counts of the pixels belonging to each reflection, and subtracts an estimate of the background in each pixel, to give $I(hkl)$, the intensity of the Bragg reflection. The variance of $I(hkl)$ may be calculated, using the rules of error propagation, from the known variances of each contribution to $I(hkl)$; its square root will be called $\sigma_0(hkl)$ in the following. For strong reflections, where the background is negligible, this procedure gives a precision, expressed as relative random error in $I(hkl)$, of $\sigma_0(hkl)/I(hkl) \sim 1/\sqrt{I(hkl)}$.

The relative amount of the random error may be reduced by repeating the experiment, and averaging the results of the individual experiments. As the laws of error propagation show, the precision of the estimate of the averaged intensity is improved by a factor of $\sqrt{n}$ over that of an individual measurement, if n is the number of repeated experiments with independent errors. Thus, the precision of the averaged (also called "merged") data may be high even if the precision of each individual observation is low.

The square root function appears both in the relative error of a photon count arising from Poisson statistics, and in the improvement of precision from averaging of multiple measurements. It is important to realize that the mathematical reasons for the occurrence of the square root differ. Nevertheless, the fact that the square root occurs in both situations means that the relative error is in principle the same whether a reflection is measured ten times and averaged, or measured just once, but with ten times stronger exposure or ten times as long.

A high number (multiplicity, sometimes called redundancy) of observations of each unique reflection, together with low exposure of each observation, is therefore equivalent in terms of the precision of the merged data, to an experiment in which each unique reflection is just measured once, but exposed proportionally stronger. Thus, if only random error is considered, there would be no reason to perform experiments with high multiplicity.

2.2 Systematic Error

The term "systematic error" summarizes all types of error that are not purely random in nature, and these are due to macroscopic physical or technical properties of the experimental setup, the crystal, and the processing of its data. For instance, systematic errors may arise from imperfect spot shapes (split crystal), radiation damage, absorption differences due to crystal shape and mounting, shutter synchronization problems, imperfect detector calibration and inhomogeneity of detector sensitivity, shadowed parts of the detector, nonlinear or overloaded detector, vibrations for example due to the cryo stream or fluctuations of the primary X-ray beam, imperfect or inaccurate assumptions about geometric parameters and computational models applied in the data processing step, and to other problems that may be significant for a given experiment.

Systematic error may appear to be random if its cause is unknown or cannot be fully described or modelled, but contrary to random counting error, the change of a reflection's intensity is usually (at least on average) proportional to the intensity itself—thus the term "fractional error." For example, a fluctuation in beam intensity changes all intensity values by the same percentage; absorption in the crystal or loop changes intensities in proportion to their original value; and using, during data processing, a mosaicity value that is too low, or a summation (integration) area that is too small, will chop off a certain fraction of the intensity.

Contrary to random error, the relative error of a single observation is *not* decreased by higher flux or longer exposure. However, many kinds of systematic errors in a crystallographic experiment at least partially cancel out if multiple measurements are averaged. This is the case if the experiment samples the possible values of the error term multiple times, in an even (or at least random) and unbiased way. Examples are beam instability, shutter problems, and most aspects of detector non-ideality, except those that result in nonlinear response (e.g., overload). Their influence on the final averaged data is decreased by averaging of n independent observations, and indeed the reduction of error then follows the same $\sqrt{n}$ rule as applies to random error.

These kinds of systematic error may thus be considered as benign: their influence on the merged data may be mitigated by distributing the total experimental time and dose over many observations, or by collecting multiple datasets [7]. It is therefore systematic error, not random error, that mandates the collection of data over more than the absolute minimum of rotation range required for obtaining complete data. However, it is important to realize that after a full turn of the spindle, all those systematic errors that depend on the geometry of the experiment will be exactly repeated. It is thus highly advisable to change the crystal setting on the goniostat after at most 360°.

If all or most observations of a unique reflection are systematically affected in the same or a similar way, their systematic errors are not independent, and averaging may not necessarily decrease the systematic difference between true and estimated intensity (the accuracy). Known or well understood effects may often be modelled by analytical or empirical formulas. If a model for the specific error type is available and appropriate, the systematic difference is accounted for, and any remaining difference between intensities may become a useful signal. In this way, a systematic effect may become a part of an extended description of the experiment, and does no longer contribute to the experimental error.

An example for this is absorption by the crystal and its environment (loop, mother liquor)—if it can be properly modelled, its influence is compensated. However, in low-symmetry space groups, all symmetry-related reflections may systematically be weakened or strengthened in the same way. Since only those systematic errors that lead to systematic differences can be corrected, no information about the proper absorption correction is available in this case. Therefore, at least one additional dataset should be measured in a different orientation of the crystal. The systematic absorption difference between the two resulting datasets may then be detected and corrected in software. It should be noted that even if absorption is not corrected in the data processing stage, it can be approximately compensated by an overall anisotropic overall displacement parameter in the refinement stage. This parameter then should not be interpreted as its name suggests, but rather as a compensation factor for an experimental property.

Importantly, for strong reflections (low resolution), systematic error is usually higher than random error; the converse is true for weak reflections (high resolution), where the signal-to-noise ratio is usually dominated by the random error term. However, radiation damage, the most devastating kind of systematic error, is an exception to this rule. Radiation damage, which changes (and ultimately destroys) the structure of the macromolecule during the measurement, induces a systematic error that is not mitigated by averaging of multiple observations, because it results in intensity measurements that do not scatter around a true value, but rather, with increasing dose, deviate further and further from the true value—the intensity at the beginning of the experiment.

The detrimental influence of radiation damage has to be avoided to a degree that depends on the kind of experiment, and its desired goal. In recent years, there has been some progress in describing the relation between dose and its footprint on the macromolecule [8, 9]. Furthermore, the influence of radiation damage may be partially compensated by zero-dose extrapolation, a computational technique [10]. However, it should be noted that the relative change of intensities by radiation damage is biggest at high resolution, where the signal may be so weak (i.e., the individual measurements so imprecise) that zero-dose extrapolation becomes inaccurate.

2.3 An Indicator for the Systematic Error

The error estimates $\sigma_0(hkl)$ obtained during integration of observations, being only based on counting statistics, are lower than the actual differences between intensities of symmetry-related observations, because the latter include the differences due to systematic errors. To account for the full difference, and thus to arrive at more useful error estimates $\sigma(hkl)$, the scaling and merging procedures operating on multiple reflections inflate the estimated error using an empirical formula [3, 11] that employs separate scale factors for the random and the systematic error. This modification of error estimates works quite well in practice, and the coefficients of the formula (the "error model") can be used to obtain an estimate of the systematic error of the dataset. The resulting estimate can be expressed as an "asymptotic signal-to-noise ratio," abbreviated ISa [11]. ISa gives the numerical value of I/σ for a hypothetical, infinitely strong reflection in the dataset, after adjustment of its σ by the new error model. If no systematic errors would exist, ISa would be infinite, since σ would be, from Poisson counting statistics, just the square root of I. However, real experiments are never ideal, which is why ISa is finite. ISa does not depend on the random error, making it insensitive to for example crystal size, mosaicity, exposure time, and flux, and is thus on an absolute scale.

The author has processed many datasets from different synchrotron beamlines and detectors. Empirically, it is found that datasets from CCD detectors rarely have ISa bigger than 30; datasets from pixel array detectors may have ISa values about twice as high compared to datasets from the same crystals collected on CCD detectors. This means that the use of a pixel array detector at a very stable beamline in an experiment with a good crystal may result in down to half the systematic error, compared to a CCD detector, and demonstrates the importance of detector technology. Conversely, if a split crystal, or a dataset with strong radiation damage, is measured on a perfect beamline, ISa may be as low as 10 (i.e., meaning that even the strongest reflections in the dataset will have signal-to-noise less than 10). Likewise, a good crystal may give a low ISa when the experimental setup or beam suffers from instability, or the cryo stream makes the crystal vibrate.

Data processing programs implement simplified or idealized assumptions about the experiment, and thus may themselves contribute some systematic error. To investigate the magnitude and properties of the systematic error from data processing, the author wrote the program SIM_MX [12] that allows to simulate complete datasets with specified amounts of random and systematic errors. A simulated dataset with only random error should ideally produce an ISa of infinity. However, the data processing program that the author is most familiar with (XDS), in this situation gives ISa well above 100. This means that if a good crystal is used to obtain a dataset, the overall systematic error, as measured by ISa, is mostly due to deficiencies of the experimental setup, and not due to shortcomings of the data processing program. Unfortunately, a general weakness of the analysis of systematic error is that the simple error model usually employed does not allow to identify and thus correctly model the exact source of systematic error.

Conceptually, it is almost always impossible to accurately measure the systematic error, since the true intensities are unknown and not measurable. Obviously, ISa *underestimates* the overall systematic error, as only those systematic errors that lead to systematic differences enter its calculation. In my research group, we usually find that upon scaling two or more datasets together, the new error model for each dataset, which is calculated by the scaling program (XSCALE), almost always leads to a reduction of its ISa. This is due to the fact that only by comparing datasets can some systematic errors be detected.

For isomorphous datasets (e.g., same crystal but different orientation) which are >90 % complete, have average multiplicity of 3 and more, and have higher symmetry than triclinic, we find that the reduction of ISa is usually less than 10 %. Therefore, the underestimation of the systematic error by ISa is usually minor.

If non-isomorphous crystals are scaled and merged and the error model is recalculated, the newly determined ISa values have to account for the systematic differences arising from differences in unit-cell parameters and crystal contents, and may then be much reduced.

2.4 Indicators for the Precision of Unmerged Data

In crystallography, indicators to describe aggregated properties of the data are necessary because the number of reflections is so large that it is prohibitive to inspect individual reflections. The availability of multiple observations (called "multiplicity" or "redundancy") allows their precision to be measured.

Historically, R_{sym} was proposed by Arndt [13] for analysis of data from the first electronic area detectors. At the time, his interest was to create an indicator for how reproducibly the data are measured with these devices. This lead him to use a formula similar to the "R-factor," which has been in use since before the middle of last century (for example in [14]), and compares the experimental amplitudes with those derived from a model. His formula

$$R_{sym} = \frac{\sum_{i}\sum_{j=1}^{n_i}\left|I_j(hkl) - \bar{I}(hkl)\right|}{\sum_{i}\sum_{j=1}^{n_i} I_j(hkl)}$$

calculates the relative absolute deviation of intensity measurements from their mean value. R_{sym} has been in use since then, but renamed to R_{merge} probably because the formula can also be applied when merging symmetry-equivalent observations obtained from one or more crystals. R_{merge} measures the *precision* of the individual measurements (observations) of the intensities, and takes both the random and systematic error into account, in as far as the latter leads to differences in symmetry-related reflections.

It turned out that R_{merge}, as originally defined, has the flaw that for low multiplicity datasets it makes the precision appear to be better than it really is (by up to a factor of $\sqrt{2}$; [15]). This can be fixed by including a factor $\sqrt{\frac{n_i}{n_i - 1}}$ in the numerator [15], and the resulting precision indicator is called R_{meas} (or $R_{r.i.m.}$; [16]). Even though it more accurately reflects the precision of the data, it has unfortunately not been fully adopted by the crystallographic community; mainly, I assume, for the psychological reason that R_{meas} has a higher numerical value than R_{merge}.

Another measure of precision of intensities is the average I/σ ratio of the observations, $\langle I/\sigma \rangle_{obs}$. R_{meas} and $\langle I/\sigma \rangle_{obs}$ obey $R_{meas} \sim 0.8/\langle I/\sigma \rangle_{obs}$. This approximate relation is valid provided that the error model has been adjusted such that χ^2, an indicator of the agreement between estimated and observed differences between symmetry-related observations, is near 1, and provided that $\langle I/\sigma \rangle$ is a good approximation of $\langle I \rangle / \langle \sigma \rangle$. For a given dataset, these approximations are usually well fulfilled at high resolution. Nevertheless, the error models of different data processing programs usually yield quite different estimates of the $\sigma(hkl)$ and $\langle I/\sigma \rangle_{obs}$ values [17].

A precision indicator like R_{meas} or $\langle I/\sigma \rangle_{obs}$ is only useful in comparisons of datasets if the multiplicity of observations in the compared datasets is approximately the same. However, neither measure is useful for defining, for instance, a high-resolution cutoff since it does not take into account the obvious fact that multiple observations increase the precision.

The low-resolution R_{meas} value of a strongly exposed crystal mainly measures detectable systematic error and may therefore be considered another indicator of systematic error. This may explain the historical popularity of the related (but less suitable) R_{sym} value for (broadly) characterizing the "quality" of a dataset.

2.5 Indicators for the Precision of Merged Data

With one exception [18], merged data are used in all crystallographic calculations after the data processing step. Statistics referring to merged data are thus much more important than those referring to unmerged data, and ignorance or misunderstanding of this fact has led to common misconceptions about for example the choice data collection strategy, the choice of dataset to refine against, the possibility of merging of datasets, and about a suitable high-resolution cutoff [19].

In 1997, Diederichs and Karplus [15] therefore introduced a specially defined R-value which takes the multiplicity of observations into account, and calculates the precision of the merged data. This quantity, $R_{mrgd\text{-}I}$, measures the differences between merged intensities from two randomly selected subsets of the data. It is not in common use, but R_{split},

$$R_{split} = \frac{1}{\sqrt{2}} \frac{\sum_i |I_{even} - I_{odd}|}{\frac{1}{2}\sum_i I_{even} + I_{odd}}$$

which is the same as $R_{mrgd\text{-}I}$ except for a factor of $1/\sqrt{2}$ and a simplified assignment of measurements to subsets, is in use by the Free Electron Laser community [20]. An equivalent quantity, $R_{p.i.m.}$ ("precision indicating merging R-factor"; [16])

$$R_{pim} = \frac{\sum_i \frac{1}{\sqrt{n_i - 1}} \sum_{j=1}^{n_i} \left| I_j(hkl) - \bar{I}(hkl) \right|}{\sum_i \sum_{j=1}^{n_i} I_j(hkl)}$$

takes the $\sqrt{n}$ improvement of precision directly into account, and has been shown to be useful when compared to R_{anom} which measures the precision of the anomalous signal.

High-resolution R-values go to infinity when the signal vanishes [19]. This is obvious from the fact that the mean intensity, in the denominator of the formula, approaches zero in this situation, whereas the numerator approaches a constant which is determined by the variance of the background. This prevents the aforementioned data R-values from being useful for comparisons with model R-values at vanishing signal, where the latter approach a constant value [21]. As a consequence, data R-values are not suitable for defining a high-resolution cutoff, a little-known fact that has led to wrong conclusions for numerous datasets.

One of the oldest and more useful estimators of precision is $\langle I/\sigma \rangle_{mrgd}$ (the subscript "mrgd" is added here only to distinguish it from $\langle I/\sigma \rangle_{obs}$; the subscript is not in common use) of the averaged data, which most data processing programs print out. There exists a reciprocal relationship between $\langle I/\sigma \rangle_{mrgd}$ and $R_{mrgd\text{-}I}/R_{split}/R_{p.i.m.}$ similar to the relation between $\langle I/\sigma \rangle_{obs}$ and R_{meas}. Unfortunately, the value of $\langle I/\sigma \rangle_{mrgd}$ depends on the error model, which usually varies significantly between different data processing programs [17]. Furthermore, for a given error model, $\langle I/\sigma \rangle_{mrgd}$ rises monotonously with higher multiplicity, even if the additional data are bad, e.g., in case of radiation damage. Nevertheless, historically a value of $\langle I/\sigma \rangle_{mrgd} > 2$ has been and continues to be used by many crystallographers as indicating the highest resolution shell that should be used for refinement [3, 22].

The latest, statistically justifiable and so far most useful addition to the crystallographic data precision indicators is $CC_{1/2}$, which is derived from mainstream statistics and measures the correlation coefficient between merged intensities obtained from two random subsets of the data. Its properties have been investigated recently [19, 23, 24]. It does not depend on estimated standard deviations of intensities, and its value is not misleadingly increased by important types of systematic errors [23]. Being a correlation coefficient, its value can be assessed for significance by a *t*-test, and most importantly, it offers the possibility to define a high-resolution cutoff based on the question "where do the data still have significant signal."

Furthermore, from $CC_{1/2}$ we can calculate CC*, a quantity on the same scale as correlation coefficients between measured intensities and F_{calc}^2 that are obtained from a model [19]. The latter, termed CC_{work}/CC_{free}, are defined for the "working" and the "free" set of reflections used in refinement, and should converge towards CC* in the course of model completion and correction.

2.6 Accuracy of the Merged Data

Since the true intensity values are usually unknown and not measurable, in a strict sense it is impossible to estimate the accuracy of the merged intensity values, because undetected systematic error may be present which has to be added to the error estimate corresponding to the precision of the merged data.

As discussed, a few sources of systematic error remain potentially undetected; most notable are absorption, diffuse scattering and detector nonlinearity. Experience suggests that the undetected systematic error in the merged data may be on the order of a few percent for a good experiment; this is the relative difference between observed data and calculated intensities seen in small-molecule experiments where a complete and accurate model of the structure is available. A more quantitative, but still conservative upper limit is the reciprocal of ISa: this estimate asserts that the undetected systematic error is unlikely to be higher than the detected systematic error.

It is important to realize that when adding independent errors or error estimates, error propagation tells us that we have to add their squares, and finally take the square root. To give an example: suppose we expect an undetected relative systematic error of 3 % (conservative upper limit at ISa = 33), and a detected relative systematic error of 1.5 % (corresponding to ISa = 33 and fourfold multiplicity). In a low-resolution shell of a crystal, the random error in the merged data may amount to 2 %. We then have a relative accuracy estimate of about 4 % ($\sqrt{(0.03^2 + 0.015^2 + 0.02^2)} = 0.039$). In a resolution shell with 20 % relative random error in the merged data, we have a relative accuracy estimate of slightly more than 20 % ($\sqrt{(0.03^2 + 0.015^2 + 0.20^2)} = 0.203$). Thus, in practice, the estimate of the undetected error dominates the accuracy estimate of the strong low-resolution data, whereas for weak high-resolution data, the accuracy estimate is determined mostly by the precision of the merged data.

2.7 Completeness

All reflections in a dataset contribute to any place in the Fourier synthesis of the electron density. In principle, this means that the quality of the electron density map is compromised if not all reflections are measured. Quantitatively, since the reflections contribute to the map in proportion to their amplitude, it is clear that the strongest reflections are most important. Strong reflections are found mainly at low resolution, where completeness fortunately is favoured by the geometry of the diffraction experiment, i.e., the low-resolution shells are usually more complete than the high-resolution shells. Then again, the strongest reflections are those that are most easily lost due to detector overload, which means that another dataset ("low resolution pass") may be needed to fill in the missing (that is, saturated and thus inaccurate) reflections.

There exist no hard rules or studies about how incomplete data may be to be still useful. If non-crystallographic symmetry is present, averaging of the electron density maps of the copies of the molecule may be performed, which partly substitutes for missing completeness by virtue of redundancy in the unique dataset. If only a single copy of the molecule resides in the asymmetric unit, a low-resolution completeness of less than 75 % can be expected to lead to quite noticeable degradation (artifacts) of maps (for an example, see http://ucxray.berkeley.edu/~jamesh/movies/completeness.mpeg); also, data missing systematically in a region of reciprocal space leads to more noticeable defects in the electron density than randomly missing data (http://www.ysbl.york.ac.uk/~cowtan/fourier/duck4.html). On the other hand, if the low resolution is almost complete, there is no reason to discard high-resolution shells just for lack of completeness. To the contrary: all measured reflections are valuable as they mitigate Fourier ripples, contribute to the fine details in the electron density map, and constitute useful restraints in refinement. The common practice of discarding high-resolution data if their completeness is not "high enough" is questionable, and has never been carefully tested.

3 How to Obtain the Best Data from XDS

The procedures for processing data with XDS have been described [4, 5] and are not repeated here. Instead, based on first hand experiences when processing datasets from my own group and helping others with their challenging datasets, I focus on those steps that are critical for data quality. For simplicity, we assume that a given dataset can be indexed in the correct space group.

The overarching rules for data processing, in the order of their importance, are that

(a) Sources of systematic error should be excluded if possible.

(b) The impact of any remaining sources of systematic error on the data should be minimized.

(c) The random error should be minimized.

(d) The completeness of the data should be maximized.

Experience shows that goals (a), (b), and (c) are not conflicting, but can be met with the same set of processing parameters. Topic (d), however, requires a compromise. For instance, rejecting the final frames of a dataset, in order to minimize the impact of radiation damage, will reduce the completeness, or at least the multiplicity of the data. Likewise, too generous masking of shadowed detector regions might lead to rejection of well-measured reflections.

Analysis of the information provided by XDS (see below) may lead to deeper insight about the data collection experiment itself. Designing and evaluating an experiment is a genuinely scientific approach and can and should not be left to automatic procedures.

The goal of data processing is to best parameterize the data collection experiment. If the data processing is repeated with changed parameters, the magnitude of the systematic error should be monitored, using ISa. By *optimizing* (maximizing) ISa, indicators of data precision are usually enhanced along the way, mainly because the location and shape of the reflections on the frames can be predicted more accurately. Generally, if the systematic error in the data is reduced, the noise associated with it is converted to signal. In case of doubt about any specific aspect of data processing, the parameter value that maximizes ISa is usually the correct one.

3.1 General Approach

To discover problems associated with data processing, it is essential that in particular the files FRAME.cbf, INTEGRATE.LP, XDS_ASCII.HKL and CORRECT.LP are analyzed.

FRAME.cbf should be inspected (Fig. 1) to find out whether spot shapes are regular, or whether there is indication of splitting and multiple lattices. Irregular and split spots indicate problems in crystal growth or handling, and always compromise data quality due to higher random noise (because spots extend over more pixels) and higher systematic error (because the reflection profiles differ from the average). Furthermore, FRAME.cbf allows finding out if predicted and observed diffraction patterns match. If they do not, the space group or geometric parameters may be wrong which may either prevent data processing from giving useful data, or may lead to downstream problems in phasing and refinement. However, this is beyond the scope of this article. Finally, FRAME.cbf, which visualizes the last frame processed by INTEGRATE, should be checked for the presence of ice rings (see below).

The tables in INTEGRATE.LP should be inspected for jumps or large changes in frame-wise parameters like scale factors, mosaicity, beam divergence, or refined parameters like unit cell parameters, direct beam position, and distance (Fig. 2). Such changes

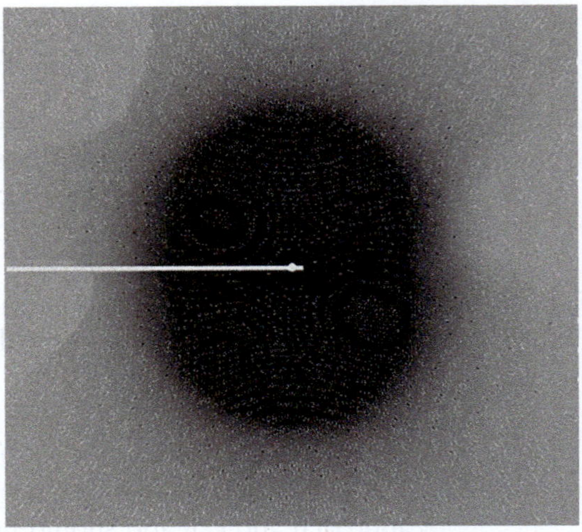

Fig. 1 Visualization of FRAME.cbf with xds-viewer. Predicted reflections are encircled. Two areas that are *shaded* by the cryo nozzle are visible (*left*, and *upper left*) that the user has not masked, which compromises data quality. The raw data used to prepare the figures are from a sulfur-SAD experiment with cubic insulin [25]; they may be obtained from http://www.helmholtz-berlin.de/forschung/funkma/soft-matter/forschung/bessy-mx/tutorial/experiment-1_en.html

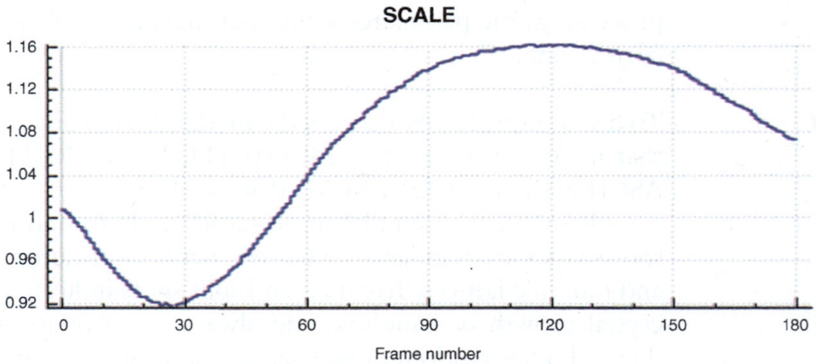

Fig. 2 Plots of some tabular quantities given in INTEGRATE.LP. Plot (**a**) shows scale factors, based on background pixels, of each frame. The plot is smooth which attests to the stability of the beam. Plot (**b**) shows different mosaicity estimates: for each frame (*blue*); for every 5° batch of data (*green*); for whole dataset (*red*). Due to the high symmetry, the curves are smooth. Plot (**c**) shows refined cell parameters, and (**d**) shows refined crystal setting angles. The variations in cell parameters do not follow a trend, which suggests that CELL should not be refined in INTEGRATE. The variations in setting angles are small; whether refinement of ORIENTATION in INTEGRATE improves the data should be tested, and the decision should be made based on Isa

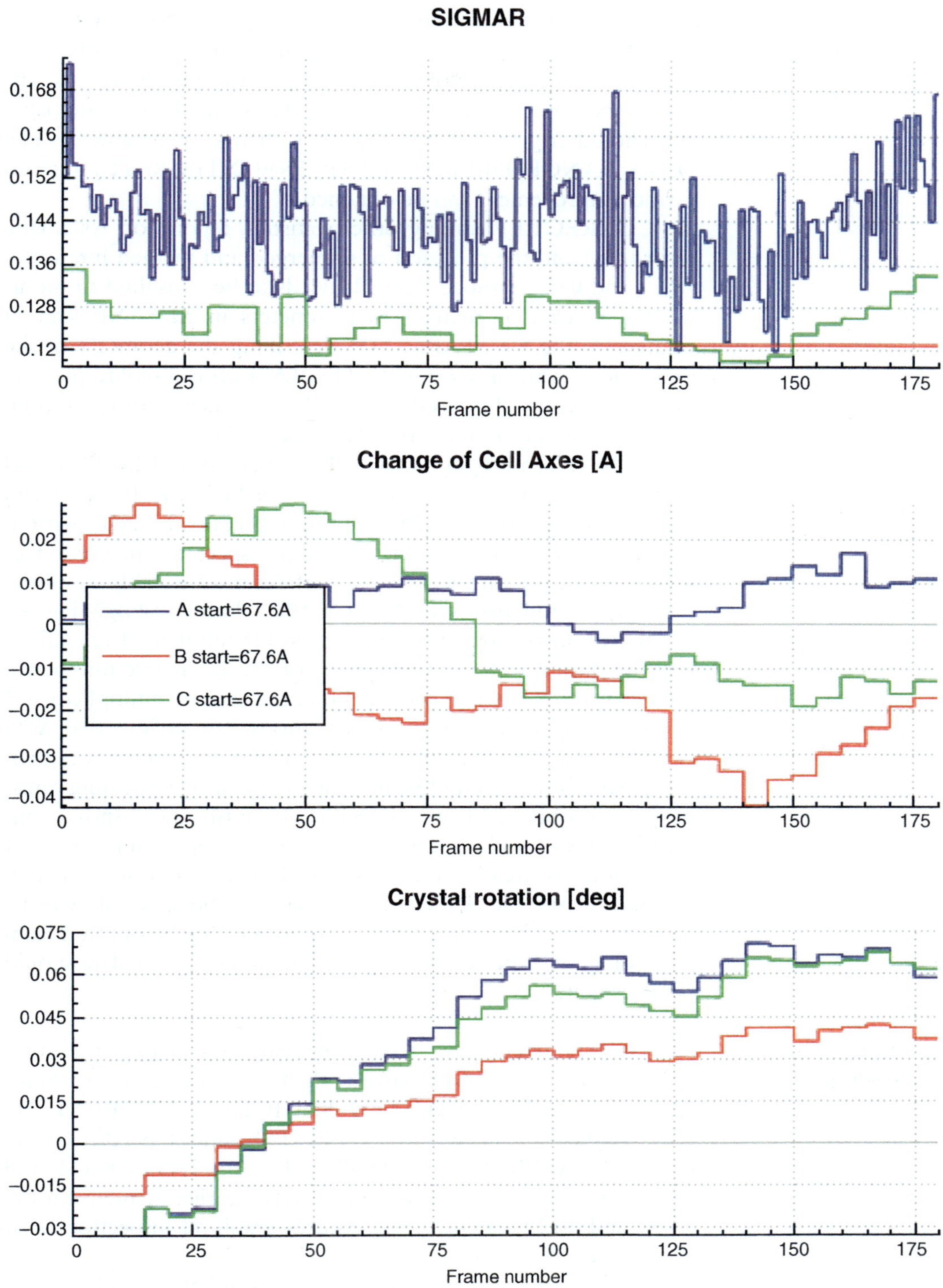

Fig. 2 (continued)

should be understood as indicating a potential source of systematic error. Scale factor jumps should be brought to the attention of the beamline manager; the other changes point to problems concerning the experiment parameterization, like crystal decay or slippage, and should trigger reprocessing after change of parameters like DATA_RANGE, DELPHI, and REFINE(INTEGRATE) until no further improvement can be obtained.

CORRECT.LP, among other statistics, reports on systematic error (ISa) and the precision of unmerged and merged intensities (R_{meas} and $CC_{1/2}$, respectively). It needs to be consulted to monitor the success of changes to parameters in XDS.INP, and of changes to the file XPARM.XDS describing the geometry of the experiment, which is used by INTEGRATE. It is useful to plot the quantities reported in CORRECT.LP as a function of resolution, and as a function of the upper frame range (Fig. 3).

XDSSTAT, a program that analyzes XDS_ASCII.HKL, should be run and its output diverted to XDSSTAT.LP, to be visualized with a plotting program. In addition, the control images written by XDSSTAT offer a graphical way to inspect the projection of several quantities on the detector surface, most notably R-values, scale factors, and misfits (outliers identified during scaling) (Fig. 4).

Better processing may lead to a lower number of reflections rejected during scaling. A guideline for the acceptable number of outliers is the following: provided that the average multiplicity is 2 or higher, up to 1 % of the observations (the default that XDS employs) may be rejected as outliers. If the percentage is higher, the reason for this should be investigated, first by inspecting "misfits.pck" as obtained from XDSSTAT. If "misfits.pck" shows concentric rings of outliers, the high percentage appears justified, but the options for treating ice rings (see above) should be evaluated. Second, if specific frames have many outliers, as shown by XDSSTAT.LP, then these frames should possibly be omitted from processing, and the reason why they delivered outlier data should be investigated.

3.2 Shaded Areas of the Detector

Several parameters have to be manually set before the integration step of XDS to mask shaded detector areas. Since the keywords TRUSTED_REGION, UNTRUSTED_RECTANGLE, UNTRUSTED_ELLIPSE, and UNTRUSTED_QUADRILATERAL are not evaluated by the INTEGRATE and CORRECT steps, they have to be specified earlier, namely, for the INIT or DEFPIX steps. This requires graphical inspection of at least a single data frame.

The low resolution limit of the data should be set such that the shadow of the beam stop is completely excluded, using INCLUDE_RESOLUTION_RANGE. Contrary to the keywords mentioned before, this keyword can be specified at a later step (CORRECT). If the lower resolution limit is too optimistic (i.e., too low), many

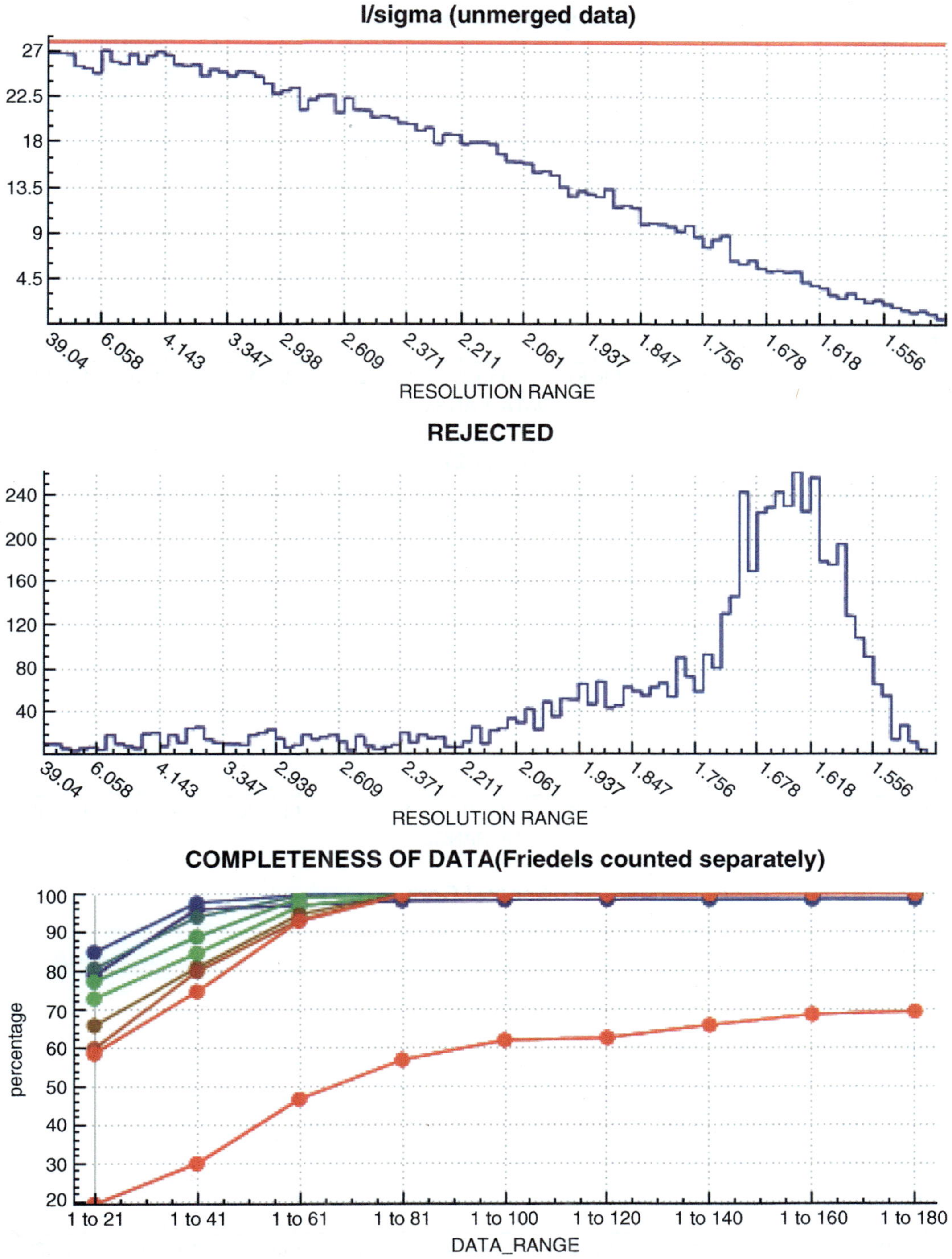

Fig. 3 Plots of some tabular quantities given in CORRECT.LP. Plot (**a**) shows $\langle I/\sigma \rangle_{obs}$ (*blue*) and ISa (*red*), (**b**) the number of rejected observations. Both quantities are given as a function of resolution. Rejections peak at high resolution, due to the user's neglect of masking the *shaded regions* of the detector. The remaining plots show different quantities as a function of the number of frames, and thus of the multiplicity; the *coloured curves* (*blue* to *red*) correspond to different resolution ranges (low to high resolution): (**c**) completeness, (**d**) $\langle I/\sigma \rangle_{mrgd}$, (**e**) $CC_{1/2}$, and (**f**) CC_{anom}, the correlation coefficient between the anomalous signals obtained from half-datasets [26]

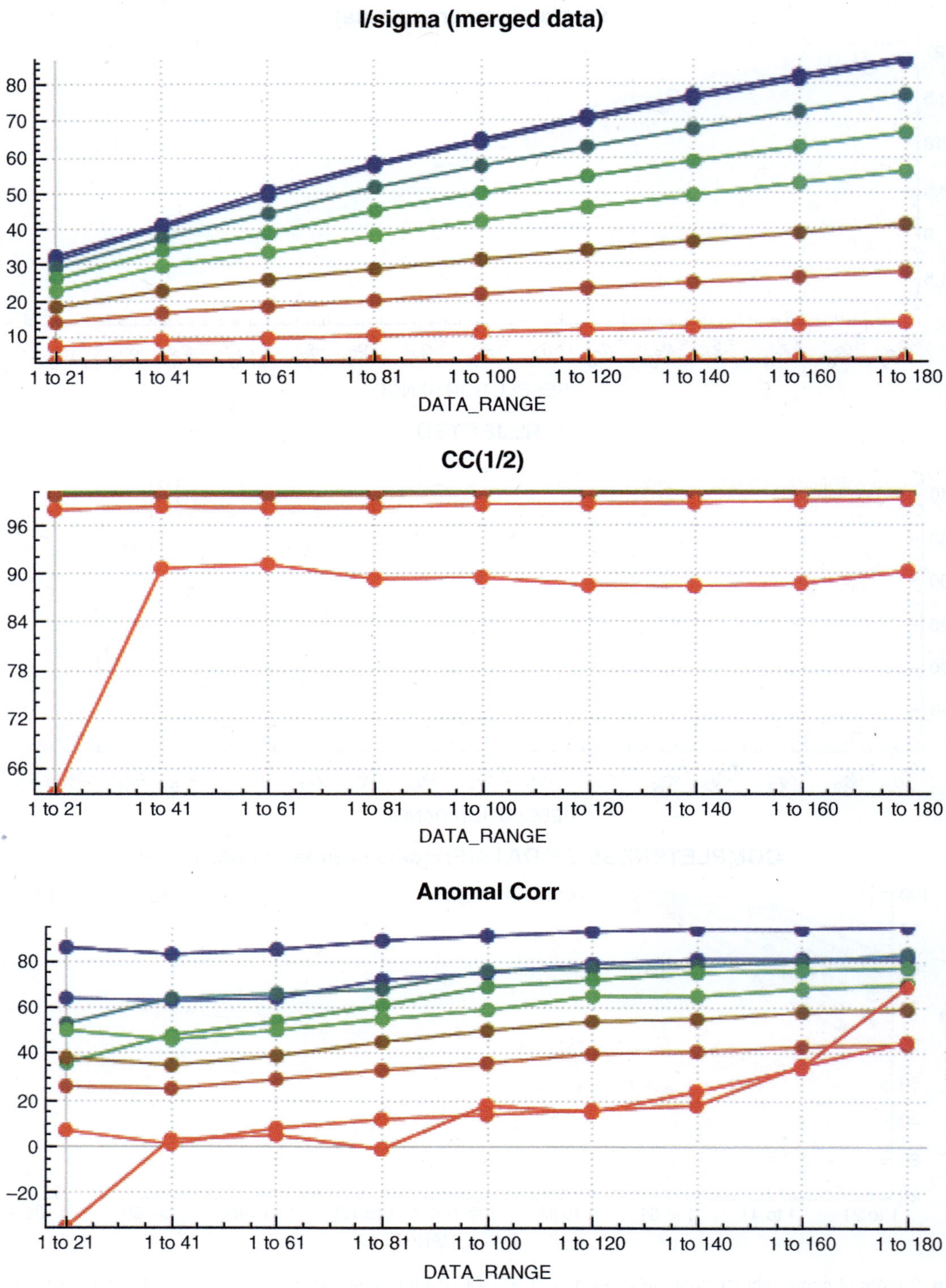

Fig. 3 (continued)

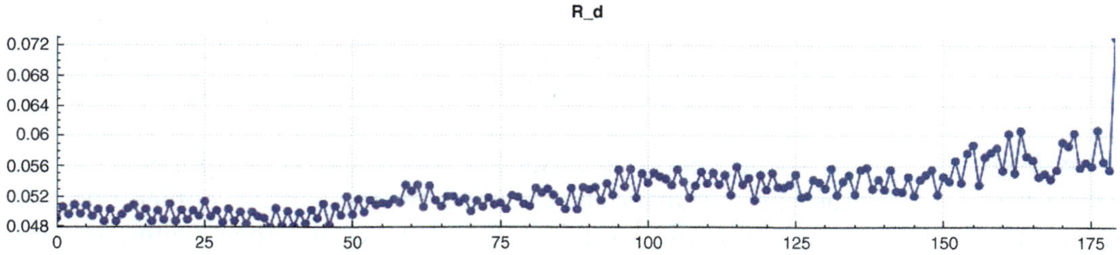

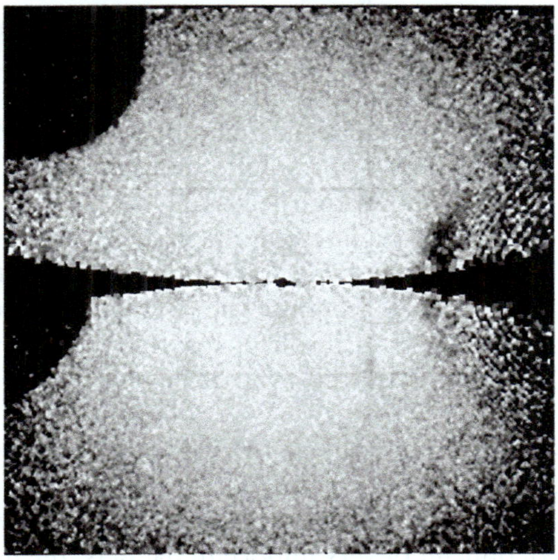

Fig. 4 Plots of some quantities obtained from XDSSTAT. Plot (**a**) shows a measure of radiation damage, R_d [27], revealing an increase in R_{meas} from 5.0 to 5.8 %. This corresponds to a 3 % contribution by radiation damage for the final frames of the dataset ($0.03^2 + 0.05^2 \sim 0.058^2$). Plot (**b**) shows the outliers projected on the detector; consistent with the high number of outliers revealed in Fig. 3b. Plot (**c**) displays R_{meas} projected on the detector, which reveals high values (*dark areas*) in the *shaded areas*

rejections and high χ^2 values result in the low-resolution shell of the first statistics table available from CORRECT. If this is indeed observed, the lower resolution limit should be raised.

3.3 Ice Rings, Ice Reflections, and "Aliens"

Single ice reflections, which fall onto a predicted spot position, are usually automatically excluded by the default outlier rejection mechanisms in CORRECT, either because their symmetry does not obey that of the macromolecular crystal, or because they are much stronger ("aliens" in CORRECT.LP) than the other reflections in their resolution range. The positions of rejected reflections can be visualized by inspecting the file "misfits.pck" using XDS-Viewer or adxv.

Strong ice rings should be manually excluded using EXCLUDE_RESOLUTION_RANGE; weak ice rings should be left to the automatic mechanisms for outlier rejection, because that results in higher completeness. To decide whether an ice ring should be considered strong or weak, the user should inspect the first statistics table in CORRECT.LP ("STANDARD ERROR OF REFLECTION INTENSITIES AS FUNCTION OF RESOLUTION"); ice rings are easily identified by a large number of rejections at resolution values near those of ice reflections (3.897, 3.669, 3.441, 2.671, 2.249, 2.072, 1.948, 1.918, 1.883, 1.721 Å for hexagonal ice, the form most often encountered). If the χ^2 and R-values in these resolution ranges are much higher than in the other ranges, the user should consider to reject the ice rings, using EXCLUDE_RESOLUTION_RANGE. This should also be done if the control image "scales.pck" (written by XDSSTAT) shows a significant deviation of scale factors from the value of 100 % at resolution values close to those of ice rings, or if "rf.pck" shows high R-values.

At very high resolution, in shells with mean intensity approaching zero, the "alien" identification algorithm sometimes rejects very many reflections when using its default value of REJECT_ALIEN=20. If this happens, the default should be raised to, say, 100 to prevent this from happening.

3.4 Specific Procedures for Optimizing Data Quality

Since the defaults in XDS.INP are carefully chosen and XDS has robust routines, very good data are usually obtained from a single processing run, in particular from good crystals. However, in case of difficult or very important datasets, the user may want to try and optimize the data processing parameters. This can be understood as minimizing or eliminating the impact of systematic errors introduced by the data processing step.

Three simple options should be tried:

(a) The globally optimized geometric parameter file GXPARM.XDS (obtained from CORRECT) may be used for another run of INTEGRATE and CORRECT. This operation may reduce the systematic error which arises due to inaccurate geometric parameters. It requires that the values of "STANDARD DEVIATION OF SPOT POSITION" and "STANDARD DEVIATION OF SPINDLE POSITION" in CORRECT.LP are about as high as the corresponding values printed out multiple times in INTEGRATE.LP, for each batch of frames. This option is particularly successful if the SPOT_RANGE for COLSPOT was chosen significantly smaller than the DATA_RANGE, because in that case the accuracy of geometric parameters from IDXREF may not be optimal.

(b) In XDS.INP, the averages of the refined profile-fitting parameters as printed out in INTEGRATE.LP, may be specified for another run of INTEGRATE and CORRECT. Essentially, this option attempts to minimize the error associated with poorly determined spot profiles. This is most effective if there are few strong reflections and/or large frame-to-frame variations between estimates of SIGMAR (mosaicity) and SIGMAB (beam divergence) as listed in INTEGRATE.LP.

(c) In XDS.INP, one may specify the keyword REFINE (INTEGRATE) with fewer (e.g., only ORIENTATION) or no geometric parameters, instead of the default parameters DISTANCE BEAM ORIENTATION CELL. This approach, which also requires at least one more run of INTEGRATE and CORRECT, is most efficient if the refined parameters, as observed in previous INTEGRATE runs, vary randomly around a mean value. Of course, preventing refinement of a parameter is not the correct approach if its change is required to achieve a better fit between observed and predicted reflection pattern. If removal of certain geometric parameters from geometry refinement in INTEGRATE indeed improves ISa, this indicates that the geometry refinement is not well enough determined to improve them beyond those obtained by the global refinement in IDXREF or CORRECT. This option thus reduces the systematic error due to poorly determined geometry. An alternative to switching refinement off is to specify a larger DELPHI than the default (5°).

Ideally, each of the three options (a–c) should be tried separately. Those options that improve ISa can then be tried in combination, and the optimization procedure may be iterated as long as there is significant improvement (of, say, a few percent) in ISa.

In my experience, optimization may lead to significantly better data, as shown by improved high-resolution $CC_{1/2}$ and improved merging with other datasets, particularly for poor datasets with high mosaicity and/or strong anisotropy.

3.5 Don'ts

Two possible ways of misusing XDS parameters should be mentioned.

First, it may be tempting to increase the number of outliers and thereby to "improve" (or rather "beautify") the numerical values of quality indicators. This could in principle be achieved by lowering the WFAC1 parameter below its default of 1. However, the goal of data processing is to produce an accurate set of intensities for downstream calculations, not a set of statistical indicators that have been artificially "massaged." Experience shows that reducing WFAC1 below its default almost always results in data with worse accuracy; conversely, raising WFAC1 may sometimes be a way to prevent too many observations to be rejected as outliers. Only if there is additional evidence for the validity of reducing WFAC1 should this quantity be lowered.

The second way to misuse XDS is to consider all the reflections listed as "aliens" in CORRECT.LP as outliers, and to place them into the file REMOVE.HKL to reject them in another CORRECT run. This is not appropriate; it should only be done if there is additional evidence that these reflections are indeed outliers. Such evidence could be the fact that the "aliens" occur at resolution values corresponding to ice reflections (see above).

3.6 High-Resolution Cutoff

The correct choice of high-resolution cutoff need not be made just once, but can be made at various times during a crystallographic study. The first is during data processing, and the additional times are when the data are used for calculations such as molecular replacement or model refinement, or calculating anomalous difference maps.

At the data processing stage, $CC_{1/2}$ should be used as the sole indicator to determine a generous cutoff—one that avoids rejecting potentially useful data. It appears prudent not to discard resolution shells with $CC_{1/2}$ larger than, say, 10 %, and it would appear useful to deposit all of these data into the PDB, to enable later re-refinement with refinement programs that can extract more information from weak data.

For further crystallographic calculations, one must decide upon the best cutoff to use for each application. Sometimes, as for molecular replacement, all one desires is a successful solution and a variety of choices may all work well. During the final model refinement stage, when the goal is to get the most accurate model possible, a recent suggestion is that one need not make this decision blindly, but that several high-resolution cutoffs can be compared using the "pairwise refinement technique" [19, 23], to find the high-resolution cutoff that delivers the best model under the given circumstances: data, starting model, refinement strategy, and refinement program.

4 The Relation of Data and Model Errors

An atomic model of a macromolecule has to fulfil certain geometric restraints (bond lengths, angles, dihedrals, planes, van der Waals distances) because all macromolecules obey the same physico-chemical principles and consist of the same building blocks whose stereochemistry and physical properties are well known from high-resolution structures. Given a suitable starting model, these restraints leave several degrees of freedom that can be used, by a refinement program, to fit the experimental data.

Only recently has it been possible to connect data quality to model quality [19], which requires definition of suitable indicators, most notably CC*, CC_{work}, and CC_{free} (*see* Subheading 2.5). The advantage of using correlation coefficients on intensities to measure both, the agreement of the observed intensities with the (unmeasurable!) true (ideal) intensities using CC*, and the agreement of the observed intensities with the model intensities using CC_{work} and CC_{free}, lies in the fact that these correlation coefficients are comparable since they are defined in a consistent way—other than is the situation with R_{work}/R_{free} and (e.g.,) R_{meas} or $R_{p.i.m.}$.

Importantly, refinement should not result in CC_{work} being numerically higher than CC* since that would mean that the model intensities agree better with the measured intensities than the ideal (true) intensities do. Since the measured intensities differ from the true intensities by noise, that would mean that the model fits the noise in the data, a situation that is called “overfitting.”

Since refinement makes CC_{work} approach CC*, CC* is a meaningful upper limit for CC_{work}. Any improvement in the data (from better processing or a new experiment) that results in higher CC* allows obtaining a better model, with a higher CC_{work} (and of course better R_{work}/R_{free}).

In the following, we introduce a simple graphical representation for the relation between experimental data, true data, and the data corresponding to several models. If n is the number of reflections in a dataset, an *n*-dimensional space can represent all possible combinations of intensities. The set of true intensities T is represented by a point in this space, and so is the set of starting model intensities M and the set of measured unique intensities D. The three points T, M, and D can be conveniently represented as points in a two-dimensional subspace (plane) of this *n*-dimensional space, and other datasets may be represented as projections on this plane. It is this two-dimensional plane which is shown in Fig. 5.

A possible distance measure in this space may be established by considering 1-CC (“Pearson distance”), where CC is the correlation coefficient between intensities of a pair of datasets. Any relevant value of CC yields a Pearson distance less than 1; values of CC lower than 0, giving a Pearson distance greater than 1, are not meaningful because they correspond to unrelated data and models.

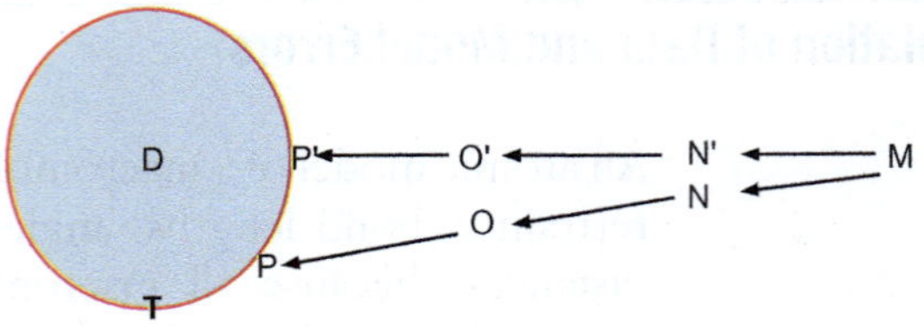

Fig. 5 Sketch of the relation between the intensities of the experimental data (D), the true (unmeasurable) data T, and those corresponding to various models. *Arrows* indicate the progression from a first starting model (M) to a final model P′ (without restraints) or P (with restraints). Local minima of the refinement target give rise to the intermediate models N′, O′ and N, O, respectively

In principle, there are infinitely many atomic models which fulfil the geometric restraints and could be used to calculate intensities. This means that the density of points in the plane that correspond to potential model structure factors is high everywhere. However, there is no smooth transition path, like that produced by refinement, between all these points. Nevertheless, they create local minima of the target function in refinement because these models fulfil the physicochemical restraints. In contrast, the subset of these local minima that are actually *biologically meaningful* is low overall, but high near T.

Since CC* estimates the correlation of the data with the true intensities, we realize that all points on the circle with radius 1-CC* around D denote potential positions of the true intensities T. For the purposes of this discussion, one particular position of T at the lower edge of the circle has been marked. For the starting model M, a reasonable assumption is that the differences between M and D are not correlated with the error in D—after all, a model obtained by Molecular Replacement is oriented and translated based on the signal, not the error in D. Likewise, a map calculated from experimental phases is based on the signal in D; the error in D just produces noise in the map. As represented in the Figure, this means that the vector from D to M is approximately at right angle to the vector between D and T.

If no or weak restraints were applied, refinement of the starting model M would produce the sequence of models N′, O′, and P′—in other words, the intensities of the model would almost linearly approach those that were measured. However, applying the proper restraints adds information to the refinement which biases the model towards the truth; thus instead of N′, O′, and P′ the model intensities are represented successively by (say) the local minima N, O, and P. The model, depending on the radius of convergence of the refinement protocol, needs to be manually adjusted to escape from these minima, and to progress towards T. Importantly, this only works if the starting model M is "close enough" to T; if it is not, manual adjustment becomes impossible as the electron density maps are too poor, and at the same time there is too little biological meaning in the model to guide its manual improvement.

As soon as the circle around D is reached (near P) after manual corrections and restrained refinement, the desired change of model intensities further towards T is almost orthogonal to the direction towards D; thus, the model may easily become stuck in one of the many local minima on the arc, which fulfil the geometric restraints, are biologically meaningful, and represent similarly good CC_{work} values. This means that it becomes increasingly difficult to improve the model any further. After a few iterations without clear progress, crystallographers—subject to individual levels of experience and ambition—tend to abandon manual model correction and refinement. This explains why different crystallographers obtain different models from the same data. In any case, T is never reached, i.e., a residual error remains, but its amount depends on details of the refinement protocol and program, as well as on the amount of time and dedication that is invested into improvement of the model.

The following points are also noteworthy. First, if overfitting is avoided, the refined model P is outside or on the circle around D, because $1\text{-}CC_{work} \geq 1\text{-}CC^*$ due to $CC_{work} \leq CC^*$. If $CC_{work} = CC^*$, P lies on the arc between P′ and T. One could argue that some overfitting could be tolerated as long as it reduces the distance between the refined model and T. Unfortunately, the latter distance cannot be measured, which is why it appears prudent to accept only little overfitting.

Second, the length of the arc between P′ and T is proportional to $1\text{-}CC^*$ which means that there are more local minima available for the refined model if the error in the data is higher. In reality, the space depicted as a one-dimensional arc in Figure N is a multidimensional one, and the number of local minima grows not only proportionally with $1\text{-}CC^*$, but rather with a large exponent. Thus, a large family of similar models with indistinguishable quality may be obtained, simply by varying some refinement parameters, or displacing the coordinates a few tenths of an Angström during manual adjustment.

Third, larger random and systematic errors will lead to a larger radius of the circle. The average distance between T and those points on the circle that correspond to refined models depends linearly on the radius, which emphasizes that better data produce better models. Undetected systematic errors may lead to T being outside the circle, which means that refinement will not be able to push the model as close to T as when T is on the circle, demonstrating that it is important to detect and minimize systematic errors.

Fourth, the refined model P can be closer to T than to D which means that—somewhat counter intuitively at first!—the final model is actually *better* than the data. This is trivially true if the starting model M happens to be close to T, but actually it is even the expected result, because judicious refinement and manual adjustment of a model takes sources of information beyond the mere experimental data restraints into account.

These considerations may illuminate the relation between data and model, and demonstrate that understanding and eliminating the sources of errors in the data helps in improving the atomic models on which our biological insight relies.

Acknowledgement

The author wishes to thank P. Andrew Karplus and Bernhard Rupp for critically reading and commenting on the manuscript.

References

1. Borek D, Minor W, Otwinowski Z (2003) Measurement errors and their consequences in protein crystallography. Acta Crystallogr D 59:2031–2038
2. Evans PR (2006) Scaling and assessment of data quality. Acta Crystallogr D 62:72–82
3. Evans PR (2011) An introduction to data reduction: space-group determination, scaling and intensity statistics. Acta Crystallogr D 67:282–292
4. Kabsch W (2010) Integration, scaling, space-group assignment and post-refinement. Acta Crystallogr D 66:133–144
5. Kabsch W (2010) XDS. Acta Crystallogr D 66:125–132
6. Bourenkov GP, Popov AN (2010) Optimization of data collection taking radiation damage into account. Acta Crystallogr D 66:409–419
7. Liu Z-J, Chen L, Wu D, Ding W, Zhang H, Zhou W, Fu Z-Q, Wang B-C (2011) A multi-dataset data-collection strategy produces better diffraction data. Acta Crystallogr A 67: 544–549
8. Ravelli RBG, McSweeney SM (2000) The 'fingerprint' that X-rays can leave on structures. Structure 8:315–328
9. Burmeister WP (2000) Structural changes in a cryo-cooled protein crystal owing to radiation damage. Acta Crystallogr D 56:328–341
10. Diederichs K, McSweeney S, Ravelli RBG (2003) Zero-dose extrapolation as part of macromolecular synchrotron data reduction. Acta Crystallogr D 59:903–909
11. Diederichs K (2010) Quantifying instrument errors in macromolecular X-ray data sets. Acta Crystallogr D 66:733–740
12. Diederichs K (2009) Simulation of X-ray frames from macromolecular crystals using a ray-tracing approach. Acta Crystallogr D 65: 535–542
13. Arndt UW, Crowther RA, Mallett JFW (1968) A computer-linked cathode-ray tube microdensitometer for x-ray crystallography. J Phys E Sci Instrum 1:510–516
14. Wilson AJC (1950) Largest likely values for the reliability index. Acta Crystallogr 3:397–398
15. Diederichs K, Karplus PA (1997) Improved *R*-factors for diffraction data analysis in macromolecular crystallography. Nat Struct Biol 4:269–274
16. Weiss MS (2001) Global indicators of X-ray data quality. J Appl Crystallogr 34:130–135
17. Krojer T, von Delft F (2011) Assessment of radiation damage behaviour in a large collection of empirically optimized datasets highlights the importance of unmeasured complicating effects. J Synch Rad 18:387–397
18. Schiltz M, Dumas P, Ennifar E, Flensburg C, Paciorek W, Vonrhein C, Bricogne G (2004) Phasing in the presence of severe site-specific radiation damage through dose-dependent modelling of heavy atoms. Acta Crystallogr D 60:1024–1031
19. Karplus PA, Diederichs K (2012) Linking Crystallographic Model and Data Quality. Science 336:1030–1033
20. White TW, Barty A, Stellato F, Holton JM, Kirian RA, Zatsepin NA, Chapman HN (2013) Crystallographic data processing for free-electron laser sources. Acta Crystallogr D69:1231–1240
21. Murshudov GN (2011) Some properties of crystallographic reliability index – Rfactor:

effects of twinning. Appl Comput Math 10:250–261

22. Wlodawer A, Minor W, Dauter Z, Jaskolski M (2008) Protein crystallography for non-crystallographers, or how to get the best (but not more) from published macromolecular structures. FEBS J 275:1–21
23. Diederichs K, Karplus PA (2013) Better models by discarding data? Acta Crystallogr D 69:1215–1222
24. Evans PR, Murshudov GN (2013) How good are my data and what is the resolution? Acta Crystallogr D 69:1204–1214
25. Faust A, Puehringer S, Darowski N, Panjikar S, Diederichs K, Mueller U, Weiss MS (2010) Update on the tutorial for learning and teaching macromolecular crystallography. J Appl Crystallogr 43:1230–1237
26. Schneider TR, Sheldrick GM (2002) Substructure solution with SHELXD. Acta Crystallogr D 58:1772–1779
27. Diederichs K (2006) Some aspects of quantitative analysis and correction of radiation damage. Acta Crystallogr D 62:96–101

Chapter 11

Advanced Crystallographic Data Collection Protocols for Experimental Phasing

Aaron D. Finke, Ezequiel Panepucci, Clemens Vonrhein, Meitian Wang, Gérard Bricogne, and Vincent Oliéric

Abstract

Experimental phasing by single- or multi-wavelength anomalous dispersion (SAD or MAD) has become the most popular method of de novo macromolecular structure determination. Continuous advances at third-generation synchrotron sources have enabled the deployment of rapid data collection protocols that are capable of recording SAD or MAD data sets. However, procedural simplifications driven by the pursuit of high throughput have led to a loss of sophistication in data collection strategies, adversely affecting measurement accuracy from the viewpoint of anomalous phasing. In this chapter, we detail optimized strategies for collecting high-quality data for experimental phasing, with particular emphasis on minimizing errors from radiation damage as well as from the instrument. This chapter also emphasizes data processing for "on-the-fly" decision-making during data collection, a critical process when data quality depends directly on information gathered while at the synchrotron.

Key words Macromolecular crystallography, Multi-axis goniometry, Diffraction data collection strategies, Experimental phasing, Anomalous scattering

1 Introduction

1.1 Advanced Experimental Phasing Data Collection Protocols

De novo macromolecular structure determination is now primarily achieved by single- or multi-wavelength anomalous diffraction (SAD/MAD) experiments [1, 2] that have largely replaced traditional isomorphous replacement methods. These two methods, which offer the distinct advantage of requiring only one crystalline entity, were boosted by progress in synchrotron instrumentation, detectors and software, as well as by the extraordinary success of selenomethionine labeling of recombinant proteins [3, 4]. The requirements of data collection for experimental phasing by anomalous dispersion are different from those for native data [5] and great care must be taken to achieve the best possible data quality. As is the case with proteins, nucleic acid crystals can give rise to an anomalous signal coming either from endogenous light elements

Eric Ennifar (ed.), *Nucleic Acid Crystallography: Methods and Protocols*, Methods in Molecular Biology, vol. 1320,
DOI 10.1007/978-1-4939-2763-0_11, © Springer Science+Business Media New York 2016

(P and metal ions) or, more commonly, from derivatization with heavy atoms (HA). The latter is achieved either by covalent incorporation of anomalous scatterers (typically bromine, iodine, or selenium) at various positions on certain nucleotides or by classical HA soaking (hexammine ions of cobalt, iridium, osmium, and rhodium being among the most popular when working with nucleic acids). Specific base pairs can ease the binding of certain metal ions [6] and their insertions in the DNA/RNA sequence can increase the success rate of soaking [7].

The anomalous differences measured during a MAD/SAD experiment are typically weak (1–5 % of the total scattering) and are therefore subject to many systematic errors. The most detrimental source of error, X-ray radiation damage, will for instance affect specifically and irreversibly the very sensitive halogen derivatives at doses that are about 50 times lower than the Henderson limit of 20 MGy [8], a typical dose used for "native" crystals phased by molecular replacement. The early uses of SAD/MAD experimental phasing were therefore characterized by careful data collection strategies to optimize the delivery of the X-ray dose to the crystal in order to minimize the measurement error of the anomalous signal [9–13]. They involved collecting Bijvoet-related reflections [14], or the same reflections at different wavelengths, with almost identical X-ray dose and absorption effects, i.e., with strategies involving crystal alignment, inverse beam and energy interleaving as described below.

1.1.1 Crystal Alignment

One can take advantage of the symmetry present in a crystal to align an even-fold symmetry axis along the spindle, so that Bijvoet pairs are measured on the same image, thereby eliminating errors caused by a difference in radiation dose between the two members of such pairs. A multi-axis goniometer, combined with an alignment program such as XOalign [15], is required to perform this alignment accurately.

1.1.2 Inverse Beam

When crystal alignment as just described is not possible, the nature and orientation of the crystal's symmetry elements can lead to highly variable and uncontrollable dose differences between the members of the various Bijvoet pairs. In the extreme case of a triclinic crystal, where only Friedel pairs (consisting of a reflection (h, k, l) and its Friedel mate $(-h, -k, -l)$) can be used to measure anomalous differences, a continuous ω-scan, as is typically done, would lead to the members of such pairs being collected 180° apart from each other, i.e., with potentially very unequal X-ray doses. This can have a hugely detrimental affect on the anomalous difference estimation from the difference between their intensities. A simple and effective solution to this problem is the "inverse beam" method, which consists in collecting thin wedge of consecutive images 180° apart (a wedge being a scan of typically

10–30° around ω when working with a detector operating in shutterless mode). This procedure guarantees that both members of every Friedel pair are collected at most one wedge width apart, so that the difference in dose between them is smaller and more uniformly distributed than would be the case with a conventional continuous sweep on an unaligned crystal.

1.1.3 Energy Interleaving

In a MAD experiment, data are collected at multiple wavelengths. The simplest procedure is to collect separate complete data sets sequentially, one for each wavelength. However, this results in a large difference in dose between the measurements of the same reflection at the various wavelengths, so that the dispersive differences between them will be strongly affected by the intervening radiation damage. Collecting data by interleaving energies in thin wedges, in the same way as the inverse beam method interleaves ω angles by a 180° "Friedel flip," can minimize the dose difference between wavelengths and lead to better anomalous data being collected. Energy interleaving can be done with two or more wavelengths.

1.2 Advanced Experimental Phasing Data Collection Protocols at Modern Synchrotron Beamlines

Prior to the advent of fast pixel-array detectors [16], the above-mentioned advanced SAD/MAD protocols used to require hours of X-ray beam time. In addition, they were not compatible with the strong emphasis on speed emanating from high-throughput third-generation synchrotron macromolecular crystallography (MX) beamlines and thus rapidly fell into disuse after it was shown that simpler strategies involving the collection of full data sets, for one wavelength at a time, around a single axis, were capable of yielding experimentally phased electron-density maps in remarkably short times in favorable cases [17, 18]. As a consequence, the use of multi-axis goniometers at synchrotrons was severely impacted and rotation around a single axis was, until recently, the only option for data collection at most synchrotron MX beamlines.

With the challenges in structural biology steadily increasing, these older but more sophisticated data collection strategies have witnessed a resurgence of interest and have been successfully applied to large macromolecules such as ribosomes [19] and viruses [20]. Recently, new protocols involving inverse beam, multi-axis goniometer, as well as multi-crystal merging, have revived native SAD phasing, which exploits the weak intrinsic anomalous scattering from native macromolecules (i.e., from light atoms such as P, S, Cl^-, K^+ or Ca^{2+}), to solve challenging targets such as membrane proteins, multiprotein–ligand or protein–DNA complexes [21, 22]. Their implementations at beamlines, now greatly simplified by hardware developments such as noise-free single-photon counting detectors and accurate multi-axis goniometers, as well as software developments such as dedicated, integrated workflows and user friendly graphical interfaces, rarely exceed 30 min in the longest phasing scenario, thus matching the limitations of synchrotron beamline availability and making high-throughput phasing a reality [23].

In this chapter, we describe optimized synchrotron data collection protocols for SAD/MAD phasing that are available at beamline X06DA-PXIII at the Swiss Light Source (SLS) at the Paul Scherrer Institute.

2 Materials

2.1 X-Ray Source

The main requirement for collecting high-quality MAD/SAD phasing data is wavelength tunability. Most third-generation synchrotron light sources around the world have MX beamlines with a typical spectral range between 5 and 20 keV, suitable for eliciting anomalous scattering from many elements.

The super-bend magnet source for beamline X06DA-PXIII at SLS possesses a duo channel-cut monochromator capable of performing fast and stable energy change while keeping the same X-ray beam position across the whole energy range of 5.5–17.5 keV. The beam has a size of 90×45 μm^2 with a flux of 5×10^{11} photons/s at 12.398 keV (1 Å wavelength). This setup is very well suited for energy-interleaved data collection protocols, which is critical for high-quality MAD data. The typical time for an energy change at X06DA-PXIII is ~30 s.

2.2 Multi-axis Goniometers

The ability to freely orient crystals with respect to the directions of the X-ray beam and of the rotation axis can provide huge advantages for maximizing the anomalous signal while minimizing error. The use of multi-axis goniometers enables orientation of a crystallographic even-fold symmetry axis of the crystal (if available) along the rotation axis, allowing the collection of Bijvoet pairs on the same image as described in Subheading 1.1.1. Additionally, data collection in multiple crystal orientations gives access to true high multiplicity of measurement rather than just high repetition, and to the possibility of better absorption correction.

The most widespread multi-axis goniometer at synchrotron MX beamlines is the ESRF/EMBL mini-κ, which has partially solved collision and self-shadowing issues [24]. The PSI-developed multi-axis goniometer PRIGo [25] (Parallel Robotics Inspired Goniometer) that is installed on beamline X06DA-PXIII features a very compact design and micron precision for rotations around the main (ω) axis, a large collision-free angular range, as well as reduced self-shadowing. PRIGo emulates an Eulerian goniometer and is therefore described with ω, χ, and φ angles (Fig. 1).

2.3 Cryostats

Cryo-cooling is one of the most powerful devices to combat radiation damage. Thus, all modern stages for MX have a cryostream attachment to keep the sample at a consistently low temperature,

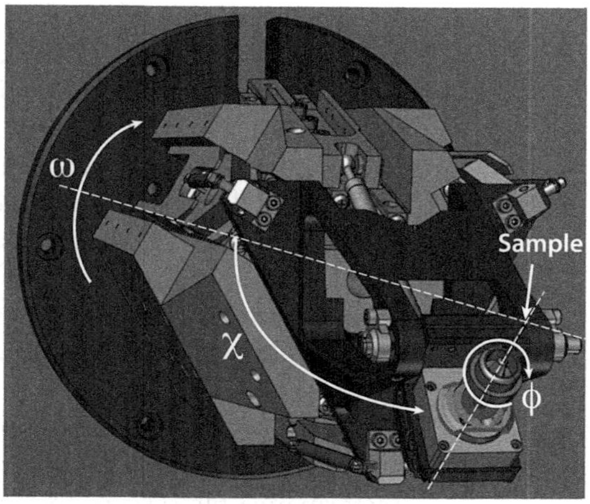

Fig. 1 Drawing of the multi-axis goniometer PRIGo with the ω, χ, and φ angles overlaid

typically around 100 K. A properly aligned cold stream with controlled flow is critical to the success of data collection, as sample vibration can severely compromise data collection as indicated by poorer statistics [26].

Beamline X06DA-PXIII has a Cryojet5 from Oxford instruments.

2.4 Detectors

Most modern MX beamlines have one of two kinds of X-ray detectors installed: a CCD detector or a single-photon-counting pixel detector. Both allow reliable reflection intensity measurements and are thus suitable for data collection for MAD/SAD experiments. However, the second detector type, as exemplified by the success of the PILATUS detector [16], has radically changed data collection practices. Such detectors feature the absence of both readout noise and dark current noise, as well as a sharp point-spread function, maximizing the signal-to-noise ratio and improving data quality. In addition, the extremely fast readout time (<3 ms), enabling shutterless and fine-slicing protocols [27], has led to the collection of complete data sets being performed in a matter of minutes. An added benefit of shutterless data collection is the exclusion of shutter jitter as a source of error. A further feature of PILATUS detectors is their energy discrimination capability, which can be used to decrease the background level by filtering out the contribution from fluorescence at K-edges.

Beamline X06DA-PXIII has a PILATUS 2M-F detector with 60 Hz frame rate.

2.5 X-Ray Absorption Spectroscopy

Critical to the success of HA-SAD/MAD is the correct choice of the wavelength(s) at which diffraction images will be collected. The absorption edges for the anomalous scatterers must be scanned, and therefore, it is crucial that the beamline has an X-ray fluorescence detector installed. While the energies for absorption edges found in the literature can be used as a starting point, the actual location of each absorption edge can vary depending on the chemical environment and oxidation state of the element of interest. Energy scan data can then be fed into programs such as CHOOCH [28] to determine the peak, inflection, and remote wavelengths to be used, as based on each wavelengths' computed f' and f'' values. The X-ray dose required to obtain such a fluorescence spectrum is not negligible but nonetheless small.

Beamline X06DA-PXIII has a Ketek AXAS system.

2.6 Software

A wide variety of software programs facilitate all aspects of data collection and strategy design. In addition, dedicated high-performance file servers and computing clusters now enable close to "real-time" data analysis. Here is a list of recommended programs for data collection and processing, with which a user should have a good level of familiarity before heading for the beamline.

2.6.1 Beamline Data Acquisition Graphical User-Interface

DA+ (Panepucci et al., unpublished), which features the advanced data collection protocols described in this chapter, is the data collection software in use at beamline X06DA-PXIII at SLS.

2.6.2 Anomalous Scattering Curves from a Fluorescence Energy Scan

CHOOCH [28] calculates values of the anomalous scattering factors f' and f'' from experimentally measured X-ray fluorescence data. This information is critical for HA-SAD and MAD experiments.

2.6.3 Crystal Orientation

XOalign [15] uses the orientation matrix produced by the indexing step in XDS [29] or MOSFLM [30] to calculate the possible multi-axis goniometer settings (χ/φ angles in the case of the PRIGo goniometer) to align crystallographic axes. Alignment of an even-fold symmetry axis along the spindle allows for simultaneous measurements of Bijvoet pairs.

2.6.4 Indexing, Integration, Merging, and Scaling

XDS [29], MOSFLM [30], and HKL-2000 [31] are full suites of programs tackling all aspects of data processing, from indexing to integration, absorption correction, scaling, and merging. Automated processing pipelines such as autoPROC [32], xdsme [15], xia2 [33], or go.com (M. Wang, unpublished) exploit the parallel architecture of XDS and can yield data processing results in a couple of minutes after the end of data collection in an unattended manner. They are therefore available at many beamlines, including beamline X06DA-PXIII at SLS.

The processing of anomalous data has slightly different requirements compared to that of native data. Careful attention is needed at the scaling and absorption correction stages in order to maintain the accuracy of the anomalous measurements of Bijvoet pairs. In XDS, this means that the FRIEDEL'S_LAW=FALSE instruction must be included.

The merging and scaling of SAD data sets collected in various crystal orientations is done with the XDS subroutine XSCALE [29]. If radiation damage is suspected, it might be good to correct individual reflections with the zero-dose extrapolation method [34]. This is done in XSCALE on unmerged data with the keyword CRYSTAL_NAME=a.

2.6.5 Substructure Determination and Phasing

SHELXC/D/E [35], which can be used through the graphical user-interface HKL2MAP [36], are programs for preparing the reflection files (SHELXC), determining the substructure (i.e., locating the anomalous scatterer sites) (SHELXD), and phasing (SHELXE). A non-exhaustive list of automated pipelines for experimental phasing includes autoSHARP [37], PHENIX. AUTOSOL [38], and CRANK2 [39].

3 Methods

3.1 Prior to Heading to the Beamline: Native SAD vs. HA-Derivatization

The decision to phase with native crystals (via native SAD using the anomalous signal from P, S, Cl^-, K^+, or Ca^{2+}) or with crystals containing exogenous heavy elements (via HA-SAD/MAD) is typically made before going to the synchrotron. However, sample properties, as well as beamline limitations, may restrict the available phasing strategies. A major factor is that there is a tradeoff between crystal availability and phasing data quality: native proteins tend to be easier to crystallize and their crystals diffract better than derivatized ones, but native SAD gives the weakest phasing signal and requires the highest data accuracy. For native nucleic acid-only crystals, native SAD can be attempted if the nucleic acid molecule is small (<30 nucleotides) and if the crystal diffracts to nearly atomic resolution [40, 41]. Protein–nucleic acid complexes benefit from the presence of sulfurs in the protein, so that the phasing of large complexes by native SAD is possible [22]. Heavy atom derivatives usually diffract less well than native crystals but almost always give superior phasing data. Similarly, HA-SAD requires fewer parameters and often less collection time than MAD, but at the cost of leaving the SAD phase ambiguity unresolved. The use of covalently bound heavy atoms, as in the case of selenium and bromine, requires incorporation of the heavy atom-labeled monomers into the nucleic acid or protein. By contrast, rapid soaks of heavy atoms into native crystals can be done prior to mounting and cryo-cooling, and if the heavy atoms are incorporated, phasing these derivatives with SAD/MAD can lead to superior phasing data than can be obtained with native SAD.

3.2 Experimental Phasing Data Collection Strategies

A summary flowchart of strategy design prior to data collection is shown in Fig. 2. While the decision whether to pursue native SAD or HA-SAD/MAD is made before going to the synchrotron, most other parameters have to be determined while the crystal is on the diffractometer. We outline the major parameters below.

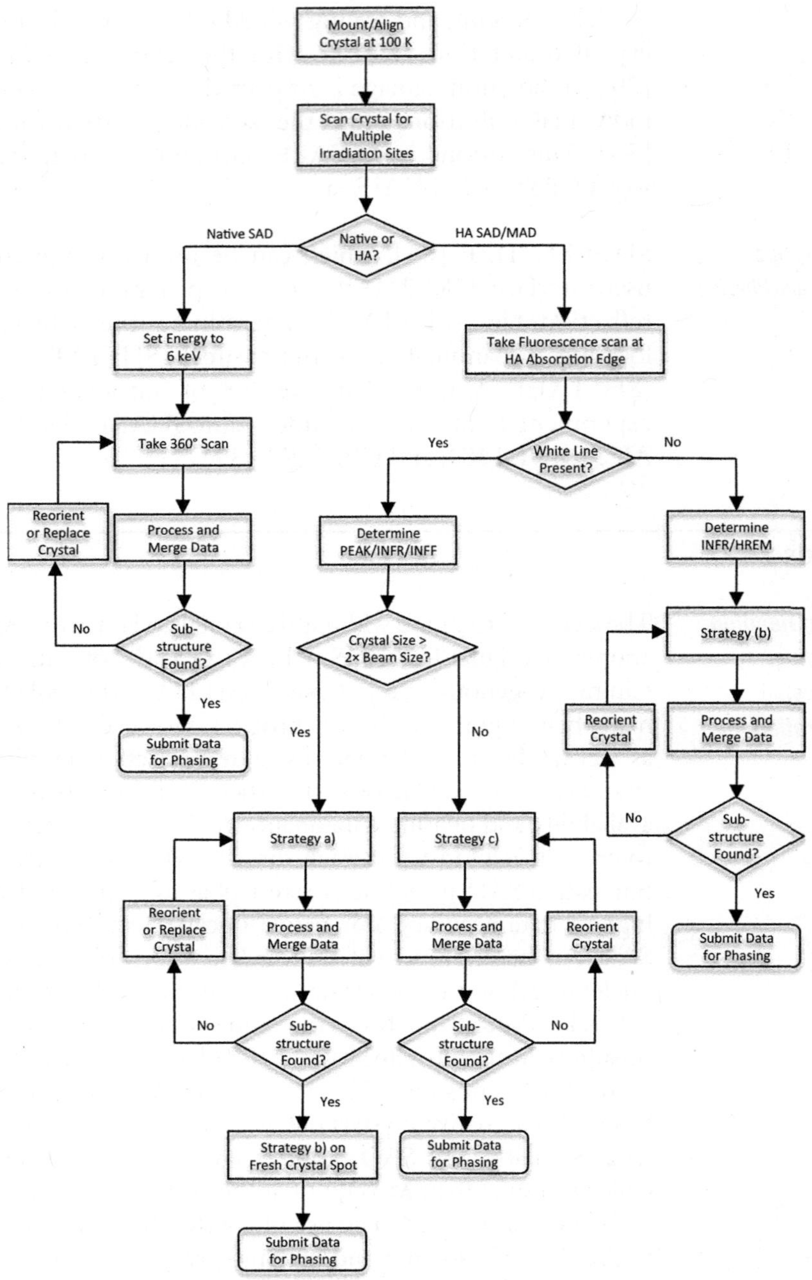

Fig. 2 Flowchart of data collection and strategy design. Strategies (a), (b), and (c) are diagrammed in Fig. 4 and described in Subheading 1.1

3.2.1 Crystal Screening

In macromolecular crystallography, as in most things, the *GIGO* principle holds true: *G*arbage *I*n, *G*arbage *O*ut. Good crystal samples with good diffraction quality and well-separated reflections, low mosaicity and without ice rings will give better data, and thus better electron density maps. Therefore, pre-screening crystals for the above-mentioned qualities is paramount. This can be conveniently done on a home source if beamline time is limited. However, the presence of sample changers at beamline sources has streamlined crystal screening.

Large crystals are always an exciting prospect, but they do not always diffract well, and are not always isomorphous throughout the entire crystal volume. However, large crystals potentially allow for multiple irradiation sites, and thus more data can be collected on a single crystal entity by spreading the total dose over a greater volume. If possible, combining multiple strategies on a single large crystal, e.g., HA-SAD on one area, and MAD on another area, may yield superior phasing data. Testing for isomorphism on a large crystal can be done by performing a raster scan on the entire crystal volume at a low dose.

3.2.2 Strategy for Native SAD

Since the absorption edge of P (for nucleic acids only) and S, Cl^-, K^+, and Ca^{2+} (for protein and protein/nucleic acid complexes) are beyond the reach of most synchrotrons, 6 keV (2.066 Å) is a good compromise between sample and air absorptions, maximum attainable resolution with a flat detector, and anomalous signal strength (P f'', 0.75 e; S f'', 0.95 e) without changes to the experimental setup (e.g., no need for an helium path or a detector 2-θ offset) [22, 42]. Because the anomalous differences measured in a native SAD experiment are so small, high accuracy in the measurements is necessary for substructure solution and subsequent phasing. Thus, data must be collected with high multiplicity to ensure good accuracy. In this respect, merging data sets collected in multiple crystal orientations can effectively reduce systematic instrument errors [22]. For the most difficult cases, it may be advantageous to combine this multi-orientation strategy with the multi-crystal merging approach [21, 43].

3.2.3 Strategy for HA-SAD and MAD

Optimizing the energies for HA-SAD/MAD requires scanning the absorption edge of the heavy atom employed (Fig. 3 shows a typical energy scan at the Ir L-III edge). The peak energy (PEAK) of the absorption edge maximizes f'' and is characterized by a sharp peak called the "white line." A white line is not always present in the fluorescence scan—it is largely element-dependent. In the absence of a white line, one should collect at the lowest energy where f'' is maximized (20 eV above the theoretical absorption edge should be a safe option in most cases)—this is called the high-remote (HREM) energy. HA-SAD data should be collected at the PEAK or HREM energy.

The decision to pursue MAD over SAD usually depends on the quality of the anomalous data from the SAD experiment. Unfortunately, there is no a priori way to determine the phasing quality of SAD data prior to implementing it in phasing programs. SAD data require high completeness and multiplicity; MAD data

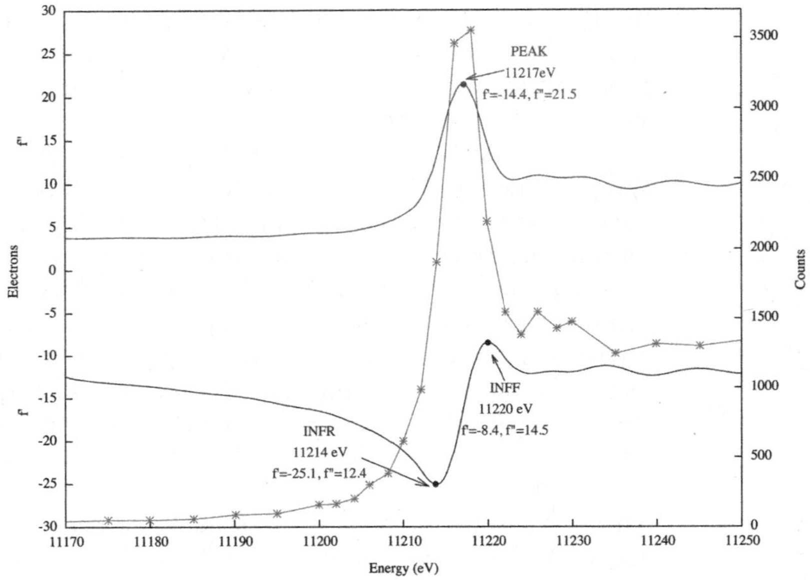

Fig. 3 X-ray fluorescence scan. X-ray fluorescence scans (*green*) and calculated *f′* (*blue*) and *f″* (*red*) curves at the Ir L-III edge. The presence of a *white line* allows for easy determination of the PEAK, INFR, and INFF energies (color figure online)

have less stringent requirements for data completeness [44]. HA-SAD data collection should be done with inverse beam (Strategy (a), Fig. 4) or with an aligned crystal, so that Friedel pairs are collected close together, or simultaneously, in dose terms. The additional time required to invert the sample during the course of an inverse beam experiment is negligible. If one is careful, then collecting additional data for MAD after collecting a SAD data set on the same crystal should be done, especially if the crystal is large (see above). HA-SAD may also benefit from the merging of multiple equivalent crystals [45].

The best MAD strategies maximize both anomalous differences between Friedel pairs (i.e., maximizing f'', see above) and dispersive differences between wavelengths (i.e., maximizing $\Delta f'$). Dispersive differences are maximized by collecting at the inflection points, i.e., at the minimum (rising inflection, INFR) and maximum (falling inflection, INFF or HREM) of f' (Fig. 3) [44, 46]. A MAD experiment with a white line should be done at the PEAK, INFR, and INFF energies. Collecting at the INFR and INFF energies is advantageous due to faster energy switching and to the fact that the two f'' values are usually close, and therefore give rise to near-identical absorption effects. If there is no white line or if the white line is weak, there will be no INFF energy, and a two-wavelength MAD experiment at INFR and HREM energies should be performed instead. The INFR/INFF or INFR/HREM data should be collected by interleaving the energies (Strategy (b), Fig. 4).

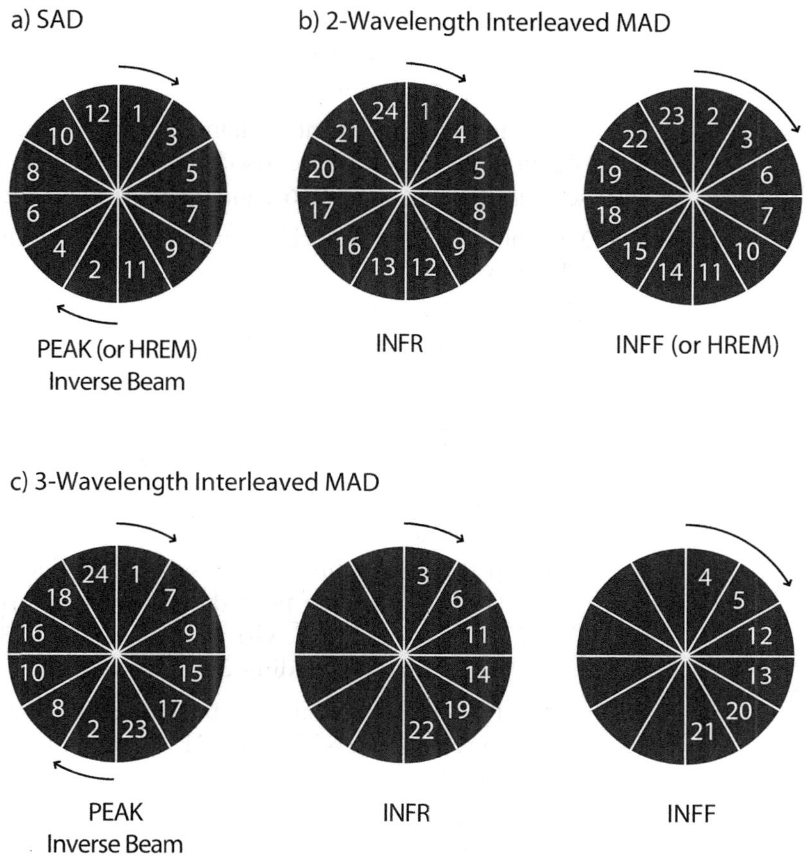

Fig. 4 Diagram of various anomalous data collection strategies. Diagram of (**a**) inverse beam, (**b**) two-wavelength energy interleaving, (**c**) three-wavelength energy interleaving with inverse beam at the first energy. Wedges are 30° and numbered in the order they are collected at each energy. Inverse beam is not necessary if the crystal can be aligned to an even-fold axis or if the crystal symmetry is high (e.g., cubic or hexagonal)

However, if the crystal is small or particularly valuable, a strategy in which all data can be collected simultaneously is advised (Strategy (c), Fig. 4).

3.3 Experimental Phasing Data Collection Procedures

3.3.1 Procedure for Crystal Evaluation and Data Collection Strategy Design

1. Place (manually or by operating the sample changer) a properly mounted (*see* **Note 1**) and cryoprotected crystal on a goniometer with a cryostream maintaining the crystal temperature at 100 K.
2. Center the crystal relative to the X-ray beam.
3. Take at least two test shots: one at $\omega = 0°$, one at $\omega = 90°$.
4. Index the crystal from the two test shots (*see* **Note 2**).
5. Align the crystal to an even-fold symmetry axis, if available (*see* **Note 3**).

6. Evaluate the crystal size and optimal spots for data collection on the crystal (*see* **Note 4**). Make a note if there are more than one well-separated optimal areas.
7. Determine the experimental parameters appropriate for optimal collection: detector distance based on the diffraction (*see* **Note 5**), oscillation angle (*see* **Note 6**), and dose (*see* **Note 7**).
8. Determine the appropriate energy or energies for data collection.
   (a) For native SAD, set the energy to 6 keV. Go to Subheading 3.3.2.
   (b) For heavy atom SAD/MAD, perform a fluorescence scan around the absorption edge of the anomalous scatterer. Select the appropriate X-ray energies using the program CHOOCH [28], which calculates the anomalous scattering factors f' and f'' (*see* **Note 8**).
   - If there is a white line present in the fluorescence scan (*see* Fig. 3), then note the following energies: PEAK, INFR, and INFF. Go to Subheading 3.3.3 for HA-SAD, and Subheading 3.3.4 for MAD.
   - If there is no white line present, then note the following energies: INFR and HREM. Go to Subheading 3.3.3 for HA-SAD, and Subheading 3.3.4 for MAD.

3.3.2 Procedure for Native SAD Data Collection

1. Collect one continuous 360° scan with a total X-ray dose <1 MGy (*see* **Note 7**). If higher dose is required due to the poor resolution of the crystal (>3.0 Å resolution), collect one 360° scan with inverse beam at 30° wedges.
2. Process the data with XDS, paying careful attention to the anomalous contribution to the data. Scale and merge data sets with XSCALE (*see* **Note 9a**).
3. Reorient or replace the crystal (*see* **Note 10**).

Repeat **steps 1–3** until there is sufficient multiplicity in the anomalous data. This can be verified by attempting to find a substructure solution with the program SHELXD.

3.3.3 Procedure for HA-SAD Data Collection

1. At the PEAK or HREM energy, collect one 360° scan with inverse beam at 30° wedges (*see* **Note 11**) (Strategy (a) in Fig. 4) and if possible, a total X-ray dose of <1 MGy (*see* **Note 7**). If the crystal is aligned to an even-fold axis or if the crystal symmetry is high, inverse beam is not necessary.
2. Process the data with XDS, paying careful attention to the anomalous contribution to the data. Scale and merge data with XSCALE (*see* **Note 9b**).

3. Reorient or replace the crystal (*see* **Note 10**).
4. Repeat **steps 1–3** until there is sufficient multiplicity in the anomalous data. This can be verified by finding a substructure solution with the program SHELXD (*see* **Note 12**).
5. If the phasing data for SAD alone is insufficient, continue to Subheading 3.3.4 on a fresh crystal spot.

3.3.4 Procedure for Two-wavelength MAD Data Collection

1. Collect 360° of data each at the INFR and INFF energies, or the INFR and HREM energies, by interleaving the energies with a 30° wedge (Strategy (b) in Fig. 4) and a total X-ray dose of <2 MGy (*see* **Note 7**).
2. Process the data with XDS, paying careful attention to the anomalous contribution to the data. Scale and merge data with XSCALE (*see* **Note 9c**).
3. Reorient the crystal (*see* **Note 10**).
4. Repeat **steps 1–3** until there is sufficient multiplicity in the anomalous data. This can be verified by finding a substructure solution with the program SHELXD.

3.3.5 Procedure for Three-wavelength MAD Data Collection

1. Collect data by interleaving the PEAK energy with inverse beam and the INFR and INFF energies with 30° wedges (Strategy (c) in Fig. 4). If the crystal is aligned to an even-fold axis or if the crystal symmetry is high, inverse beam is not necessary.
2. Process the data with XDS, paying careful attention to the anomalous contribution to the data. Scale and merge data with XSCALE (*see* **Note 9c**).
3. Reorient the crystal (*see* **Note 10**).
4. Repeat **steps 1–3** until there is sufficient multiplicity in the anomalous data. This can be verified by finding a substructure solution with the program SHELXD.

4 Conclusions

Phasing by anomalous dispersion is popular because of its experimental and operational simplicity compared to other methods. Advances in instrumentation have increased the quality achievable for the final data and the efficiency with which they can be collected, but contrary to popular belief, not to the point of dispensing with a thoughtful practitioner. Thus it is still critical to consider all the possible variables in the phasing experiment before proceeding.

5 Notes

1. A properly prepared sample, with attention paid to a clean, single crystal, proper cryoprotection, orientation and mounting, will lead to better data, lower mosaicity, less spurious signal caused by ice rings, etc. Crystals should be mounted on a loop in such a way as to be surrounded by the minimal amount possible of solvent/cryoprotectant. This can be achieved partly by using a loop size slightly bigger than the crystal, and by dabbing the loop to remove excess solvent. If a heavy atom soak was done, it is advisable to do a back-soak prior to mounting it to remove unincorporated heavy atoms, as the latter can increase the diffraction background and give false positives in the fluorescence spectrum.
2. Obtaining the unit cell, space group and orientation matrix is typically performed with two or more images collected in significantly different orientations (90° apart in ω).
3. If possible, set the χ and φ angles that XOalign indicates. If it is not possible to align, skip this step.
4. This can be done by either performing a rastering scan across the crystal, or, if such a feature is not available, by taking test shots at several locations on the crystal and evaluating the quality of the diffraction pattern at the various locations.
5. While for native data it is important to collect to the maximum resolution of the crystal, for phasing, good low resolution diffraction and well-separated reflections are the most critical factors. Thus, reducing the dose and avoiding pushing the resolution may prove advantageous for good phasing.
6. On PILATUS detectors, frames are collected with the fine-slicing method, i.e., at oscillation angles that are typically half the mosaicity of the crystal (as defined by XDS) [27]. An oscillation angle of 0.1° is a safe value to use for most crystals. A data collection speed of 1–2°/s is typical on the PILATUS 2M-F.
7. The key to anomalous data collection is maximizing multiplicity while maintaining radiation damage within tolerable limits. Higher dose does not necessarily lead to better signal-to-noise, especially with modern detectors [47] but it does lead to more radiation damage. The flux of a beam is controlled by the addition of filters. The amount of radiation potentially absorbed by the crystal has to be carefully controlled to minimize radiation damage. While on average a cryo-cooled crystal loses half its diffraction power with a dose of 20 MGy, specific radiation damage affecting anomalous differences occurs at doses <1 MGy [48]. The dose can be calculated with RADDOSE [49] from the exposure time, wedge, flux, beam size, and crystal chemical composition and size, all of which should be easily

obtained during data collection. If possible, the maximum dose should be <1 MGy per 360° of data collected.

8. CHOOCH might not find a suitable PEAK energy in the absence of a white line, or if the white line is weak; thus treat the results of CHOOCH with some caution. If there is a white line and CHOOCH did not determine it correctly, find the PEAK energy directly from the experimental data.
9. Scaling and merging is done with XSCALE, a subroutine of XDS.
   (a) Native SAD data, which consist of multiple 360° scans, are integrated separately and scaled/merged together at the end. Prior to scaling and merging, it is prudent to check for radiation damage during the course of the experiment by comparing statistics for each data set.
   (b) SAD data sets taken at multiple orientations should be processed separately, then scaled and merged together if there are no obvious issues. In inverse beam experiments, the two wedges are also processed separately in XDS.
   (c) For processing of MAD data sets, the integrated data sets for each wavelength should be scaled and merged separately.
10. Starting from a random crystal orientation, we recommend taking 360° scans at incremental χ angles (0°, 5°, 10°, 15°, ...) up to a total dose of about 5 MGy for a single crystal spot. After 5 MGy, we advise to collect on a new crystal and merge equivalent data sets with XSCALE. If it was possible to orient an even-fold axis to collect Bijvoet pairs simultaneously, we also recommend a slight change in the goniometer settings after each 360° scan to minimize systematic errors.
11. Ideally, the sample should be inverted after each frame, but this is unnecessary when collecting the data continuously as is the case on PILATUS detectors. A 30° wedge is a good compromise between data collection time and dose equalization. For samples that are particularly prone to radiation damage, we advise using a smaller wedge size (10°).
12. If the anomalous data are of sufficiently high quality, the SAD data alone can be used for phasing through use of various available phasing programs (*see* Subheading 2.6.4).

References

1. Hendrickson WA (2013) Evolution of diffraction methods for solving crystal structures. Acta Crystallogr A 69:51–59
2. Hendrickson WA (2014) Anomalous diffraction in crystallographic phase evaluation. Q Rev Biophys 47:49–93
3. Hendrickson WA, Horton JR, LeMaster DM (1990) Selenomethionyl proteins produced for analysis by multiwavelength anomalous diffraction (MAD): a vehicle for direct determination of three-dimensional structure. EMBO J 9:1665–1672

4. Doublié S (1997) Preparation of selenomethionyl proteins for phase determination. Methods Enzymol 276:523–530
5. Dauter Z (2010) Carrying out an optimal experiment. Acta Crystallogr D 66:389–392
6. Ennifar E, Walter P, Dumas P (2003) A crystallographic study of the binding of 13 metal ions to two related RNA duplexes. Nucleic Acids Res 31:2671–2682
7. Keel AY, Rambo RP, Batey RT, Kieft J (2007) A general strategy to solve the phase problem in RNA crystallography. Structure 15:761–772
8. Henderson R (1990) Cryoprotection of protein crystals against radiation-damage in electron and X-ray diffraction. Proc R Soc B Biol Sci 241:6–8
9. Hendrickson WA, Teeter MM (1981) Structure of the hydrophobic protein crambin determined directly from the anomalous scattering of sulphur. Nature 290:107–113
10. Kahn R, Fourme R, Bosshard R, Chiadmi M, Risler JL, Dideberg O, Wery JP (1985) Crystal structure study of Opsanus tau parvalbumin by multiwavelength anomalous diffraction. FEBS Lett 179:133–137
11. Hendrickson WA, Pähler A, Smith JL, Satow Y, Merritt EA, Phizackerley RP (1989) Crystal structure of core streptavidin determined from multiwavelength anomalous diffraction of synchrotron radiation. Proc Natl Acad Sci U S A 86:2190–2194
12. Weis WI, Kahn R, Fourme R, Drickamer K, Hendrickson WA (1991) Structure of the calcium-dependent lectin domain from a rat mannose-binding protein determined by MAD phasing. Science 254:1608–1615
13. Hendrickson WA, Ogata C (1997) Phase determination from multiwavelength anomalous diffraction measurements. Methods Enzymol 276:494–523
14. Rupp B (2009) Biomolecular crystallography: principles, practice, and application to structural biology. Garland Science, New York
15. Legrand P. xdsme. http://code.google.com/p/xdsme/
16. Broennimann C, Eikenberry EF, Henrich B, Horisberger R, Huelsen G, Pohl E, Schmitt B, Schulze-Briese C, Suzuki M, Tomizaki T, Toyokawa H, Wagner A (2006) The PILATUS 1M detector. J Synchrotron Radiat 13:120–130
17. Walsh MA, Dementieva I, Evans G, Sanishvili R, Joachimiak A (1999) Taking MAD to the extreme: ultrafast protein structure determination. Acta Crystallogr D 55:1168–1173
18. Rice LM, Earnest TN, Brunger AT (2000) Single-wavelength anomalous diffraction phasing revisited. Acta Crystallogr D 56:1413–1420
19. Clemons WM, Brodersen DE, McCutcheon JP, May JL, Carter AP, Morgan-Warren RJ, Wimberly BT, Ramakrishnan V (2001) Crystal structure of the 30S ribosomal subunit from Thermus thermophilus: purification, crystallization and structure determination. J Mol Biol 310:827–843
20. Voss JE, Vaney M-C, Duquerroy S, Vonrhein C, Girard-Blanc C, Crublet E, Thompson A, Bricogne G, Rey FA (2010) Glycoprotein organization of Chikungunya virus particles revealed by X-ray crystallography. Nature 468:709–712
21. Liu Q, Dahmane T, Zhang Z, Assur Z, Brasch J, Shapiro L, Mancia F, Hendrickson WA (2012) Structures from anomalous diffraction of native biological macromolecules. Science 336:1033–1037
22. Weinert T, Olieric V, Waltersperger S, Panepucci E, Chen L, Zhang H, Zhou D, Rose J, Ebihara A, Kuramitsu S, Li D, Howe N, Schnapp G, Pautsch A, Bargsten K, Prota AE, Surana P, Kottur J, Nair DT, Basolico F, Cecatiello V, Pasqualato S, Boland A, Weichenrieder O, Wang BC, Steinmetz MO, Caffrey M, Wang M (2014) Fast native-SAD phasing for routine macromolecular structure determination. Nat Methods 12(2):131–133
23. Dauter Z (2002) New approaches to high-throughput phasing. Curr Opin Struct Biol 12:674–678
24. Brockhauser S, Ravelli RBG, McCarthy AA (2013) The use of a mini-κ goniometer head in macromolecular crystallography diffraction experiments. Acta Crystallogr D 69:1241–1251
25. Waltersperger S, Olieric V, Pradervand C, Glettig W, Salathe M, Fuchs MR, Curtin A, Wang X, Ebner S, Panepucci E, Weinert T, Schulze-Briese C, Wang M (2015) PRIGo: a new multi-axis goniometer for macromolecular crystallography. J Synchrotron Radiat In Press
26. Alkire RW, Duke NEC, Rotella FJ (2008) Is your cold-stream working for you or against you ? An in-depth look at temperature and sample motion research papers. J Appl Crystallogr 41:1122–1133
27. Mueller M, Wang M, Schulze-Briese C (2012) Optimal fine φ-slicing for single-photon-counting pixel detectors. Acta Crystallogr D 68:42–56
28. Evans G, Pettifer R (2001) CHOOCH: a program for deriving anomalous-scattering factors from X-ray fluorescence spectra. J Appl Crystallogr 34:82–86
29. Kabsch W (2010) XDS. Acta Crystallogr D 66:125–132
30. Leslie A, Powell H (2007) Processing diffraction data with Mosflm. Evol Meth Macromol

Crystallogr NATO Sci II Math Phys Chem 245:41–51
31. Otwinowski Z, Minor W (1997) Processing of X-ray diffraction data. Meth Enzymol 276: 307–326
32. Vonrhein C, Flensburg C, Keller P, Sharff A, Smart O, Paciorek W, Womack T, Bricogne G (2011) Data processing and analysis with the autoPROC toolbox. Acta Crystallogr D 67: 293–302
33. Winter G, Lobley CMC, Prince SM (2013) Decision making in xia2. Acta Crystallogr D 69:1260–1273
34. Diederichs K, McSweeney S, Ravelli RBG (2003) Zero-dose extrapolation as part of macromolecular synchrotron data reduction. Acta Crystallogr D 59:903–909
35. Sheldrick GM (2010) Experimental phasing with SHELXC/D/E: combining chain tracing with density modification. Acta Crystallogr D 66:479–485
36. Pape T, Schneider TR (2004) HKL2MAP: a graphical user interface for macromolecular phasing with SHELX programs. J Appl Crystallogr 37:843–844
37. Vonrhein C, Blanc E, Roversi P, Bricogne G (2007) Automated structure solution with autoSHARP. Methods Mol Biol 364: 215–230
38. Terwilliger TC, Adams PD, Read RJ, McCoy AJ, Moriarty NW, Grosse-Kunstleve RW, Afonine PV, Zwart PH, Hung LW (2009) Decision-making in structure solution using Bayesian estimates of map quality: the PHENIX AutoSol wizard. Acta Crystallogr D 65:582–601
39. Skubák P, Pannu NS (2013) Automatic protein structure solution from weak X-ray data. Nat Commun 4:2777
40. Dauter Z, Adamiak DA (2001) Anomalous signal of phosphorus used for phasing DNA oligomer: importance of data redundancy. Acta Crystallogr D 57:990–995
41. Luo Z, Dauter M, Dauter Z (2014) Phosphates in the Z-DNA dodecamer are flexible, but their P-SAD signal is sufficient for structure solution. Acta Crystallogr D 70:1790–1800
42. Liu Q, Guo Y, Chang Y, Cai Z, Assur Z, Mancia F, Greene MI, Hendrickson WA (2014) Multi-crystal native SAD analysis at 6 keV. Acta Crystallogr D 70:2544–2557
43. Liu Q, Liu Q, Hendrickson WA (2013) Robust structural analysis of native biological macromolecules from multi-crystal anomalous diffraction data. Acta Crystallogr D 69:1314–1332
44. González A (2003) Optimizing data collection for structure determination. Acta Crystallogr D 59:1935–1942
45. Liu Q, Zhang Z, Hendrickson WA (2011) Multi-crystal anomalous diffraction for low-resolution macromolecular phasing. Acta Crystallogr D 67:45–59
46. González A, Pédelacq J, Solà M, Gomis-Rüth FX, Coll M, Samama J, Benini S (1999) Two-wavelength MAD phasing: in search of the optimal choice of wavelengths. Acta Crystallogr D 55:1449–1458
47. Diederichs K (2010) Quantifying instrument errors in macromolecular X-ray data sets. Acta Crystallogr D 66:733–740
48. Holton JM (2009) A beginner's guide to radiation damage. J Synchrotron Radiat 16:133–142
49. Zeldin OB, Gerstel M, Garman EF (2013) RADDOSE-3D: time- and space-resolved modelling of dose in macromolecular crystallography. J Appl Crystallogr 46:1225–1230

Chapter 12

Nucleic Acid Crystallography via Direct Selenium Derivatization: RNAs Modified with Se-Nucleobases

Huiyan Sun, Sibo Jiang, and Zhen Huang

Abstract

Selenium-derivatized RNAs are powerful tools for structure and function studies of RNAs and their protein complexes. By taking the advantage of selenium modifications, researchers can determine novel RNA structures via convenient SAD and MAD phasing. As one of the naturally occurring tRNA modifications, 2-seleno-uridine, which presents almost exclusively at the wobble position of anticodon loop in various bacterial tRNAs (Ching et al., Proc Natl Acad Sci U S A 82:347, 1985; Dunin-Horkawicz et al., Nucleic Acids Res 34:D145–D149, 2006), becomes one of the most promising modifications for crystallographic studies. Our previous studies have demonstrated many unique properties of 2-seleno-uridine, including stability (Sun et al., RNA 19:1309–1314, 2013), minimal structural perturbation (Sun et al., Nucleic Acids Res 40:5171–5179, 2012), and enhanced base-pairing fidelity (Sun et al., Nucleic Acids Res 40:5171–5179, 2012). In this protocol, we present the efficient chemical synthesis of 2-seleno-uridine triphosphate (^{Se}UTP) and the facile transcription and purification of ^{Se}U-containing RNAs (^{Se}U-RNA).

Key words Selenium RNA, 2-Se-uridine triphosphate, SAD and MAD phasing, In vitro transcription

1 Introduction

RNA plays central roles in multiple cellular processes, including genetic information storage, transcription, translation [1], regulation [2] as well as catalysis [3, 4]. The multiple functions of RNAs require highly diversified 3D structures. Thus, elucidating the 3D structures of RNAs will help understanding the RNA functions and the biological processes. X-ray crystallography is one of the most widely used methods for high-resolution structure determination of RNAs and their protein complexes. In addition to crystallization, however, phase problem remains the major challenge in crystallographic studies. Scientists have discovered that large atoms, such as selenium (Se), can serve as anomalous scattering centers for MAD (Multiwavelength anomalous dispersion) and/or SAD (Single-wavelength anomalous dispersion) phasing.

Eric Ennifar (ed.), *Nucleic Acid Crystallography: Methods and Protocols*, Methods in Molecular Biology, vol. 1320,
DOI 10.1007/978-1-4939-2763-0_12, © Springer Science+Business Media New York 2016

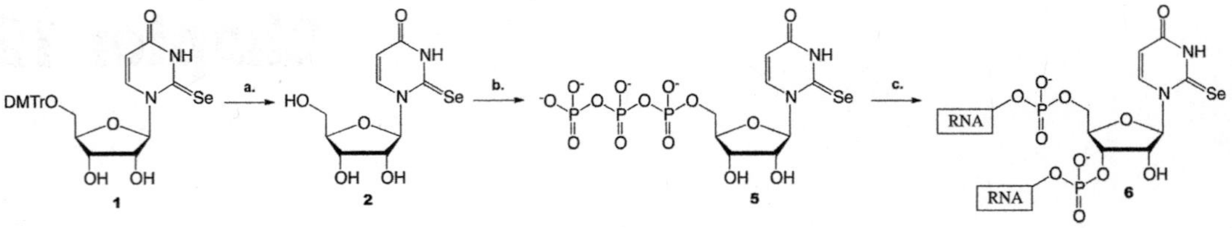

Scheme 1 Chemical synthesis of ^{Se}UTP [1] and transcription of RNA containing ^{Se}U [2]. (**a**) 4 % trifluoroacetic acid; (**b**) $POCl_3$, Me_3PO_4; (tri-*n*-butyl) amine, pyrophosphate, *N,N*-dimethylformamide; the H_2O hydrolysis; (**c**) RNA transcription

Seleno-Met derivatization has been successfully applied for structure determination of novel proteins [5] and nucleic acid–protein complexes [6]. Due to its chemical and biophysical similarity to oxygen and sulfur, selenium atom can be covalently introduced into proteins (yielding Se-Met protein) as well as nucleic acids (yielding Se-NA) by replacing sulfur or oxygen atoms. Huang laboratory has pioneered and developed nucleic acid derivatization and X-ray crystallography with selenium modifications of nucleic acids [7–10]. Among all Se-derivatized nucleic acids, 2-Se-uridine is the only naturally occurring modification. Extensive experimental data show that the 2-seleno-modifiaction has high stability [11] as well as virtually identical structure to the native in RNA structure [12]. One economical and efficient method to incorporate multiple ^{Se}Us into a RNA molecule is through in vitro transcription, since this strategy is well established and favored by most biology laboratories. Here, we present a facile and high-yield synthetic method for 2-selenouridine-triphosphate (^{Se}UTP) and in vitro transcription of ^{Se}U-containing RNAs (Scheme 1).

2 Materials

2.1 ^{Se}UTP Synthesis

1. Round bottom flasks (10 mL).
2. Stir bars.
3. Syringes (1 mL).
4. Needles (23G × 1½ in).
5. Magnetic stir plate.
6. Balloons.
7. Nitrogen gas.
8. Vacuum pump.
9. Rubber septa for 10 mL round bottom flasks.
10. Parafilm.
11. Ice bath.
12. Thin layer chromatography (250 μm).

13. UV lamp.
14. Digital balance.
15. Support stand with extension clamp.
16. Laboratory freezers (−80 °C).
17. High-speed centrifuge tubes (50 mL).
18. Laboratory bench-top centrifuge.
19. High-performance liquid chromatography (HPLC) system.
20. Reverse phase C-18 column (*see* **Note 1**).
21. Lyophilizer.
22. 1-(5′-*O*-4,4′-dimethoxytrityl-β-D-ribofuranosyl)-2-selenouridine.
23. Trifluoroacetic acid.
24. Dichloromethane.
25. Methanol.
26. Trimethyl phosphate.
27. Proton-sponge (Sigma-Aldrich, Saint Louis, MI, USA).
28. Phosphorus oxychloride ($POCl_3$).
29. Tributylammonium pyrophosphate.
30. Tributylamine (TBA).
31. Dimethylformamide (DMF).
32. Triethylammonium bicarbonate (1 M).
33. 3 M NaCl.
34. Ethanol, ≥99.8 %.
35. Buffer A: 10 mM triethylamine-acetic acid (TEAAc) buffer solution in deionized, pH 7.4.
36. Buffer B: 10 mM TEAAc buffer solution in 50 % acetonitrile and deionized water (v/v), pH 7.4.

2.2 ^{Se}U-RNA Transcription

1. T7 RNA polymerase (*see* **Note 2**).
2. 10× Reaction buffer for transcription: 400 mM Tris–HCl (pH 7.5), 60 mM $MgCl_2$, 100 mM NaCl, 20 mM spermidine.
3. 100 mM ATP, CTP, GTP, and UTP solutions. UTP is substituted with ^{Se}UTP in ^{Se}U-RNA transcription experiment.
4. 100 mM DTT.
5. 100 mM $MgCl_2$.
6. RNase-free water.
7. Centrifuge tubes, 0.5 mL, sterile.
8. Pipettes (adjustable-volume).
9. Sterile pipette tips.

10. Transcription template. Linearized plasmid templates are applied in self-cleaving mutant and wild-type hammerhead-ribozyme transcription (50 ng/μL) and a double-stranded DNA template (55 nucleotides) is applied in non-self-cleaving hammerhead-ribozyme transcription (1 μM).
11. A heating block set at 37 °C.

2.3 Polyacrylamide Gel Electrophoresis for Se-RNA Purification

1. Acrylamide/bis-acrylamide (19:1), 40 % (w/v) solution. This solution is commercially available at many major scientific supply companies.
2. 5× TBE (Tris-Borate-EDTA) solution. This solution is commercially available at many major scientific supply companies.
3. 10 % (w/v) ammonium persulfate solution (APS) in water. Store at 4 °C.
4. *N,N,N,N*-tetramethyl-ethylenediamine (TEMED). Store at 4 °C.
5. Necessary polyacrylamide gel electrophoresis (PAGE) apparatus.
6. Gel loading-dye (pH 8.0): 0.25 % (w/v) bromophenol blue, 0.25 % (w/v) xylene cyanol FF, 0.2 M EDTA, 50 % glycerol, water.
7. Preparative high-speed centrifuge tubes (50 mL) sterile and preparative centrifuge.
8. UV lamp.
9. 3 M sodium chloride (NaCl) solution.
10. Ethanol, ≥99.8 %.
11. Laboratory freezers (−80 °C) and refrigerator (4 °C).

3 Methods

3.1 Selenouridine Nucleoside Synthesis and Purification

The synthesis of 2-selenouridine nucleoside was performed under argon. All organic solvents in this experiment should be redistilled, and all solid reagents should be dried under reduced pressure prior to use.

1. Weigh out 1-(5′-*O*-4,4′-dimethoxytrityl-β-D-ribofuranosyl)-2-selenouridine (compound **1**, Scheme 1, 0.4 mmol) and place in a 10 mL round bottom flask together with a stir bar.
2. Seal the round bottle flask with septa and seal the septa with parafilm.
3. Attach the flask to a high vacuum to dry the sample through a clean needle for 1 h.
4. Attach a balloon to a syringe and then inflate the balloon with argon.

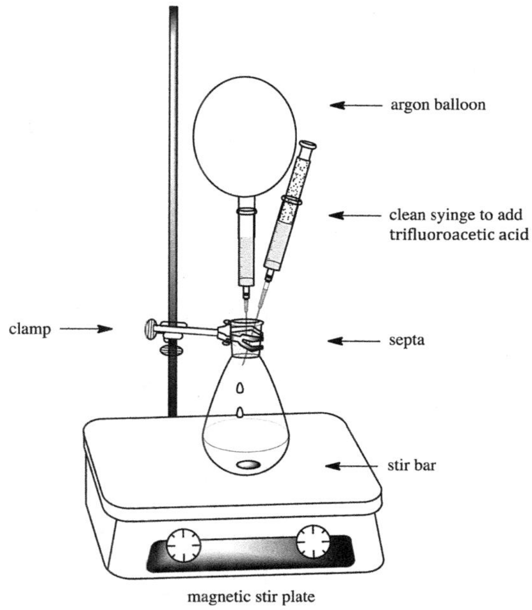

Fig. 1 Illustration of apparatus for selenouridine nucleoside synthesis

5. Remove the round bottom flask from the high vacuum and quickly insert the balloon to it through a clean needle to fill the whole flask with argon.
6. Securely place the round bottom flask above a magnetic stir plate, add 3 mL of dichloromethane into the reaction through a clean syringe and stir vigorously.
7. Inject trifluoroacetic acid dropwisely into the flask via a clean syringe (an immediate orange color should be observed once the acid is added) until the reaction pH reach 4 (Fig. 1).
8. Monitor the reaction for 1 h and check it with TLC plates using 10 % methanol in dichloromethane as eluent and use uridine sample as a reference to the 2-Se-uridine product (Fig. 2).
9. After the starting material is completely converted to 2-Se-uridine product, slowly add methanol into the reaction drop by drop to quench the dimethoxytrityl group until the orange color entirely disappears (*see* **Note 3**).
10. The reaction solution is transferred into a high-speed centrifuge tubes (15 mL) and another 3 to 4 mL of dichloromethane is added into the same centrifuge tube, shake gently to dissolve impurity.
11. Centrifuge the suspension and the supernatant is removed.

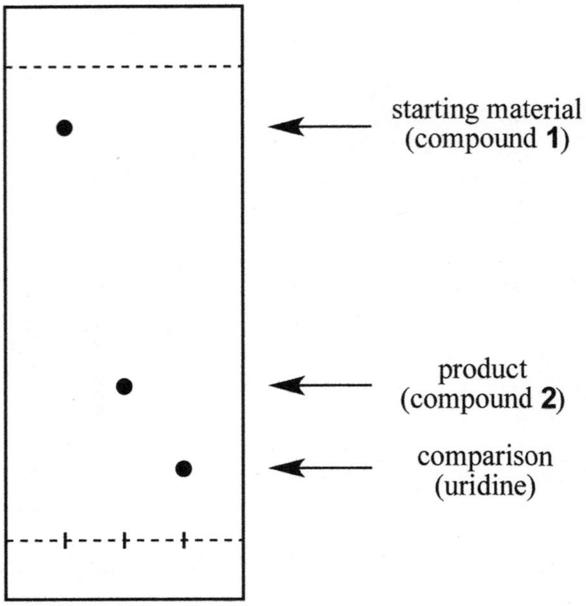

Fig. 2 Illustration of TLC analysis of selenouridine product, starting material and uridine (comparison)

12. Re-dissolve the pellet with a minimal amount of methanol and add another 5 mL of dichloromethane to re-precipitate the product and remove the by-products.
13. Repeat procedure **12** two times to obtain a pure 2-Se-uridine product.

3.2 Selenouridine Triphosphate Synthesis and Purification

The synthesis of 2-selenouridine triphosphate was performed under argon. All organic solvents in this experiment should be anhydrous, and all solid reagents should be dried under reduced pressure prior to use.

1. Weigh out 2-selenouridine nucleoside (compound **2**, Scheme 1, 0.35 mmol), proton sponge (0.7 mmol) and tributylammonium pyrophosphate (0.53 mmol) and place each compound into separate round bottom flasks (10 mL).
2. Place a stir bar into each flask and cap with a dried septum. Use parafilm to wrap around the septa and make sure the flask is sealed properly.
3. Attach the flasks to high vacuum via a needle and dry the compounds for 2 h.
4. Remove the 2-selenouridine nucleoside flask (Flask 1) from the vacuum hose first and quickly insert argon balloon into the flask through a needle to fill the flask with argon gas.

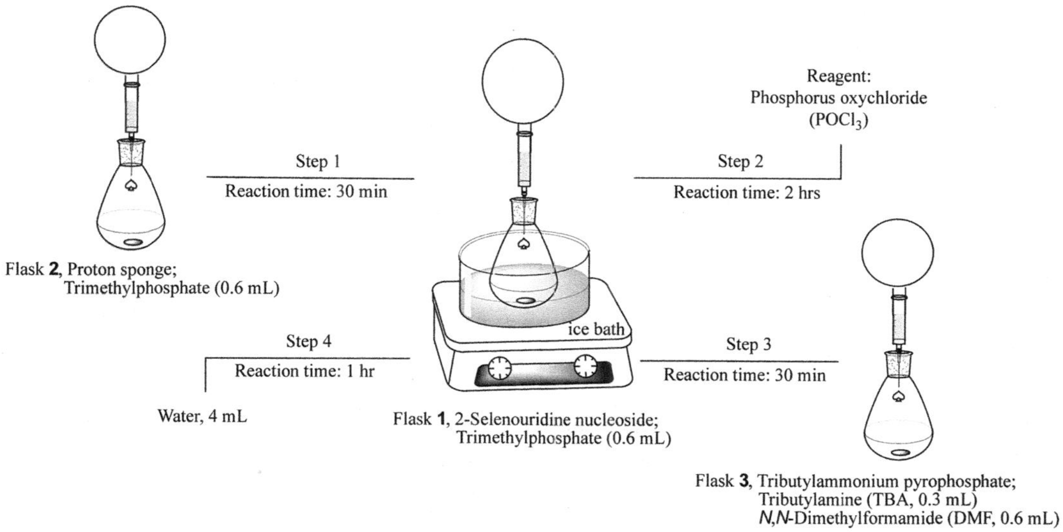

Fig. 3 Illustration of the synthesis of 2-selenouridine triphosphate, includes four steps, reagents and reaction time

5. Add 0.6 mL of trimethylphosphate into Flask 1 (Fig. 3) to dissolve the 2-selenouridine nucleoside (compound 2) and stir vigorously.
6. Prepare an ice bath and place it on the top of a magnetic stir plate.
7. Place Flask 1 in the ice bath securely (*see* Fig. 3 for the reaction setup) and make sure the septa area of the flask is above the ice.
8. Remove the proton sponge flask (Flask 2) from the vacuum hose next and quickly insert argon balloon into it through a needle to fill the flask with argon gas.
9. Add 0.6 mL of trimethylphosphate into Flask 2 (Fig. 3) to completely dissolve the proton sponge.
10. Use a clean syringe to transfer the dissolved proton sponge from Flask 2 into Flask 1 and stir the mixture vigorously for 30 min on ice bath (**step 1**, Fig. 3).
11. Add phosphorus oxychloride ($POCl_3$, 0.7 mmol, 0.065 mL) into Flask 1 reaction mixture dropwisely via a clean syringe and stir vigorously for 30 min in ice bath (**step 2**, Fig. 3).
12. After 30 min of stirring of the reaction mixture, check the formation of intermediate **3** with TLC plates by using 15 % methanol in dichloromethane as eluent. The TLC plate set up is shown in Fig. 4.
13. Remove the tributylammonium pyrophosphate flask (Flask 3) from the vacuum hose and quickly insert argon balloon into it through a needle to fill the flask with argon gas.

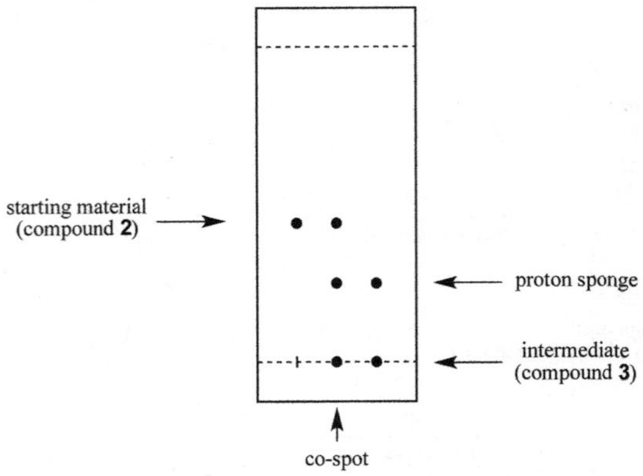

Fig. 4 Illustration of TLC analysis of intermediate phosphorodichloridate **3**, starting material and proton sponge

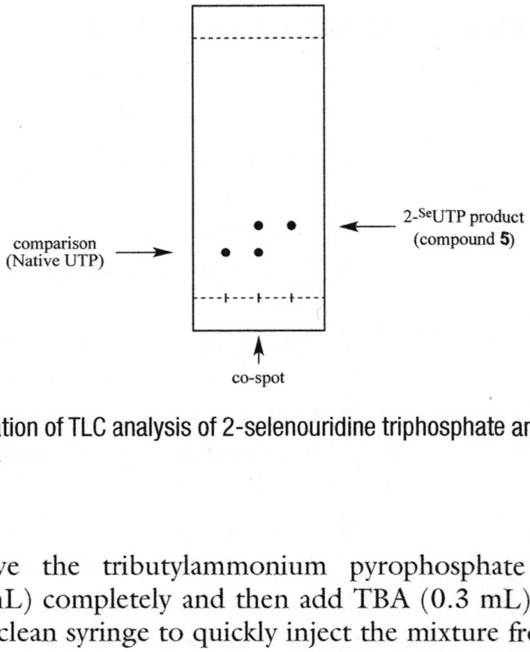

Fig. 5 Illustration of TLC analysis of 2-selenouridine triphosphate and native UTP (comparison)

14. Dissolve the tributylammonium pyrophosphate in DMF (0.6 mL) completely and then add TBA (0.3 mL), mix well. Use a clean syringe to quickly inject the mixture from Flask 3 into the reaction (Flask 1) and stir vigorously in ice bath for 15 min (**step 3**, Fig. 3).
15. Add water (4 mL) to the reaction mixture (Flask 1) to hydrolyze compound **4** to final product 2-selenouridine triphosphate (compound **5**, **step 4**) and let the reaction continuously stir for another 3 h at room temperature.
16. Check the formation of final product 2-selenouridine triphosphate(compound**5**)byTLCplate,usingisopropanol:ammonium hydroxide:water (5:3:2) as eluent and use uridine triphosphate (native UTP) sample as a comparison to the 2-Se-uridine triphosphate product. The TLC plate set up is shown in Fig. 5.

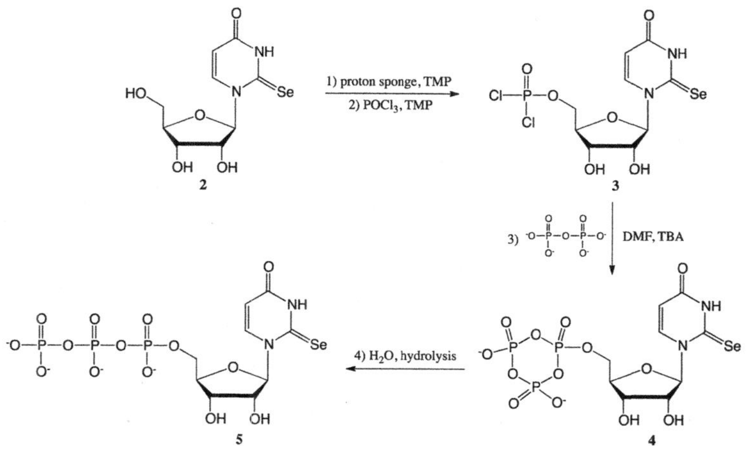

Scheme 2 Chemical synthesis of ^{Se}UTP [1]. (1) Proton sponge, Me_3PO_4; (2) $POCl_3$, Me_3PO_4; (3) tributylammonium pyrophosphate, TBA, DMF; (4) H_2O hydrolysis

17. After the reaction, open the reaction flask (Flask 1) and transfer the reaction mixture into a clean centrifuge tube (50 mL).
18. Add 0.68 mL 3 M NaCl and 20 mL ethanol into the crude ^{Se}UTP, cap the centrifuge, use parafilm to seal and shake the tube to mix completely.
19. Place the tube into −80 °C freezer for half an hour and use centrifuge (10,000 ×*g*) to collect ^{Se}UTP sample (yellow pellet).
20. Remove the supernatant from the tube and let it air dry. Use water (~200 μL) to re-dissolve the ^{Se}UTP yellow pellet.
21. The final concentration of ^{Se}UTP is measured and adjusted via UV–Vis to 100 mM for RNA transcription.
22. To analyze the purity of the synthesized ^{Se}UTP (compound 5, Scheme 2), we use RP-HPLC and C18 column. The HPLC analysis was performed with a gradient from 100 % buffer A (20 mM triethylammonium acetate in water) to 40 % buffer B (20 mM triethylammonium acetate in 50 % acetonitrile and 50 % water) in 15 min. The ^{Se}UTP HPLC analysis compared with native UTP is shown in Fig. 6.

3.3 2-Se-RNA Transcription

1. Add all the reaction components into a centrifuge tube (0.5 mL, sterile) and mix well, *see* Table 1. The total volume of the reaction is 100 μL.
2. Place the reaction tube in the heating block (37 °C) and incubate for 3 h.
3. Heat inactivate T7 RNA polymerase at 70 °C for 10 min.
4. Transfer the transcribed ^{Se}U-RNA to a spin column to isolate transcribed RNA from unreacted NTPs.
5. Use a clean sterile tube to store the crude ^{Se}U-RNA.

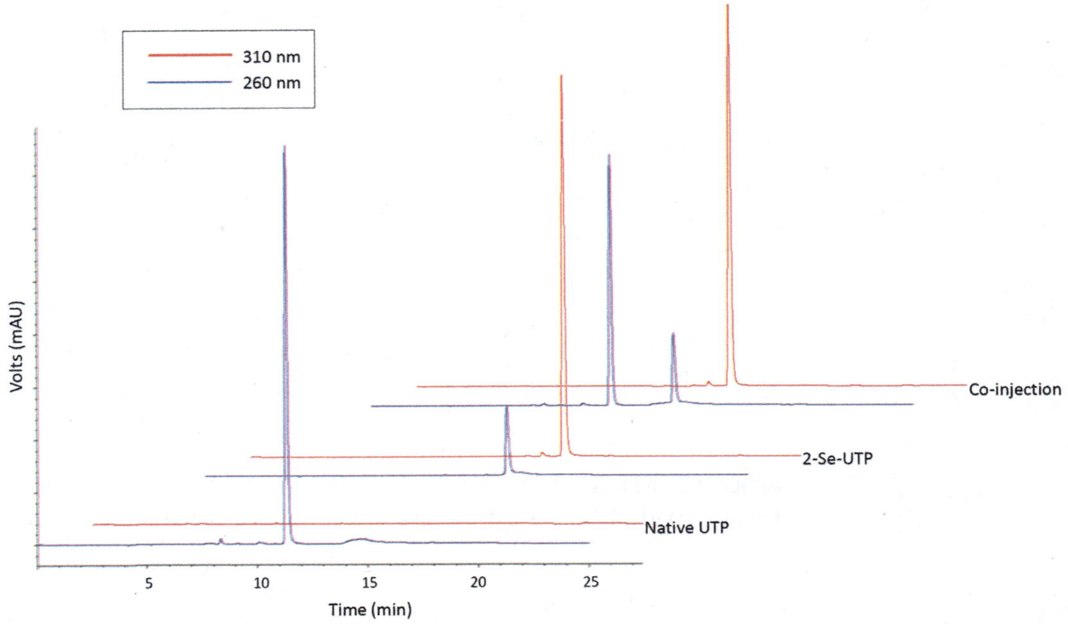

Fig. 6 HPLC profiles: native UTP monitored at 260 and 307 nm (retention time: 11.2 min); 2-^{Se}UTP monitored at 260 and 307 nm (retention time: 14.1 min); coinjection of both native UTP and ^{Se}UTP monitored at 260 and 307 nm (retention time: 11.2 min and 14.1 min)

Table 1
Reaction components of the RNA transcription

10× Reaction buffer	10 μL
100 mM Dithiothreitol (DTT)	10 μL
100 mM ATP	5 μL
100 mM CTP	5 μL
100 mM GTP	5 μL
100 mM 2-^{se}UTP	10 μL
Inorganic pyrophosphatase (0.1 U/μL)	4 μL
100 mM $MgCl_2$	5 μL
DNA template (200 ng/μL)	10 μL
T7 RNA polymerase	10 μL
Water (RNase-free)	26 μL

3.4 Gel Purification of 2-Se-RNA

1. Mix the crude ^{Se}U-RNA with the gel loading-dye (1:1, v/v) and load the sample on a preparative gel.
2. Let the dyes run to proper positions and remove the gel from gel electrophoresis apparatus and place it on a flat surface.

3. Remove the glass plate from one side of the PAGE gel and place a layer of plastic wrap on top of it.
4. Place a preparative or analytical TLC plate on top of the plastic wrap (the silica-coated side faces the plastic wrap) and flip the entire glass plates over.
5. Carefully remove the remaining glass plate from the PAGE gel with a spatula.
6. Use a UV lamp to visualize the gel and use a clean knife to carefully cut off the RNA band.
7. Transfer the cutted gel to a clean tube and grind the gel into fine pieces.
8. Add RNase-free water into the tube (2× gel volume) and soak the gel by rotating the tube for 4 h at room temperature. Collect the liquid and soak the gel again.
9. Centrifuge the collected and combined suspension at 1,000 × *g* for 10 min and transfer the supernatant to another tube.
10. Filter the supernatant with a 0.2 μm nylon syringe filter to remove fine gel particles.
11. RNA was precipitated by NaCl/ethanol precipitation (following the procedure in Subheading 3.2, **steps 17–20**) to obtain a purified 2-Se-RNA. If the RNA volume is too big, lyophilize it (to reduce the volume) first before NaCl/ethanol precipitation.
12. Redissolve the ^{Se}U-RNA pellet with RNase-free water and quantify the RNA by UV–Vis absorption at 260 or 307 nm.
13. Store the ^{Se}U-RNA sample in −80 °C freezer. For longer-term storage, the ^{Se}U-RNA can be stored after lyophilization.

4 Notes

1. We use Ultimate XB-C18 250 mm × 4.6 mm, 5 μm reverse phase columns.
2. We use commercially available AmpliScribe™ T7-Flash™ enzyme solution and this enzyme is included in the AmpliScribe™ T7-Flash™ Transcription Kit (Epicentre, Madison, WI, USA).
3. A white suspension is observed, indicating the formation of 2-Se-uridine product.

References

1. Watson JD (1963) Involvement of RNA in the synthesis of proteins. Science 140:17–26
2. Serganov A, Patel DJ (2007) Ribozymes, riboswitches and beyond: regulation of gene expression without proteins. Nat Rev Genet 8:776–790
3. Schmeing TM, Ramakrishnan V (2009) What recent ribosome structures have revealed about the mechanism of translation. Nature 461: 1234–1242
4. Ponting CP, Oliver PL, Reik W (2009) Evolution and functions of long noncoding RNAs. Cell 136:629–641
5. Hendrickson WA, Horton JR, LeMaster DM (1990) Selenomethionyl proteins produced for

analysis by multiwavelength anomalous diffraction (MAD): a vehicle for direct determination of three-dimensional structure. EMBO J 9:1665

6. Ferre-D'Amare AR, Zhou K, Doudna JA (1998) Crystal structure of a hepatitis delta virus ribozyme. Nature 395:567–574
7. Carrasco N, Ginsburg D, Du Q, Huang Z (2001) Synthesis of selenium-derivatized nucleosides and oligonucleotides for X-ray crystallography. Nucleosides Nucleotides Nucleic Acids 20:1723–1734
8. Caton-Williams J, Huang Z (2008) Biochemistry of selenium-derivatized naturally occurring and unnatural nucleic acids. Chem Biodivers 5:396–407
9. Sheng J, Huang Z (2010) Selenium derivatization of nucleic acids for X-Ray crystal-structure and function studies. Chem Biodivers 7:753–785
10. Lin L, Sheng J, Huang Z (2011) Nucleic acid X-ray crystallography via direct selenium derivatization. Chem Soc Rev 40:4591–4602
11. Sun H, Jiang S, Caton-Williams J, Liu H, Huang Z (2013) 2-Selenouridine triphosphate synthesis and Se-RNA transcription. RNA 19:1309–1314
12. Sun H, Sheng J, Hassan AEA, Jiang S, Gan J, Huang Z (2012) Novel RNA base pair with higher specificity using single selenium atom. Nucleic Acids Res 40:5171–5179

Chapter 13

Practical Radiation Damage-Induced Phasing

Chloe Zubieta and Max H. Nanao

Abstract

Although crystallographers typically seek to mitigate radiation damage in macromolecular crystals, in some cases, radiation damage to specific atoms can be used to determine phases de novo. This process is called radiation damage-induced phasing or "RIP." Here, we provide a general overview of the method and a practical set of data collection and processing strategies for phasing macromolecular structures using RIP.

Key words Radiation damage, Phasing, Heavy atoms, Synchrotron, RIP, RIPAS

1 Introduction

1.1 RIP Background

High brilliance macromolecular crystallography (MX) synchrotron beamlines are critical resources in the portfolio of structural biology. Early in their development, however, it was realized that radiation damage was a significant issue in data collected from these beamlines and warranted deeper investigation [1]. Since that time, significant resources have been invested in understanding and mitigating the effects of X-ray-induced radiation damage. Cryo-cooling [2–6], advanced data collection strategies [7–10], radiation damage scavengers [11–19], optimized dose rate [20, 21], and optimized photon energies [22–25] have all been studied with the common aim of reducing the amount of radiation damage accumulated during the course of collecting MX diffraction data. The motivations for this considerable body of work include the loss of diffractive power of the crystal as a function of dose and the potential complication to the interpretation of diffraction data [26] (for crystals with very radiation sensitive groups for example).

Radiation damage in MX can be broadly divided into two types: general and specific, although the two are related. General damage typically refers to crystal scale changes and includes, for example, changes to the Wilson B-factor, unit cell dimensions and mosaicity versus dose. This type of damage affects the overall packing and orientation of the macromolecules in the crystal.

Eric Ennifar (ed.), *Nucleic Acid Crystallography: Methods and Protocols*, Methods in Molecular Biology, vol. 1320,
DOI 10.1007/978-1-4939-2763-0_13, © Springer Science+Business Media New York 2016

Specific radiation damage, on the other hand, refers to localized changes in the crystal. Examples of specific radiation damage include disruption of disulphide bonds, reduction of the occupancy of metal atoms, debromination of brominated nucleic acids, and decarboxylation of glutamate and aspartate side chains [1, 27]. While no benefit or use has of yet been identified for general damage to crystals, in some cases specific radiation damage can be a benefit instead of an impediment. More than a decade ago it was shown that not only can specific radiation damage be used to determine phases in macromolecular crystals, but also to determine the "substructure" of radiation damage de novo (i.e., without external phases or knowledge about the radiation damage substructure) [28, 29]. This method of phasing was named "RIP" (Radiation damage-Induced Phasing) and the core concept is that the specific changes induced by radiation damage can be exploited in a manner analogous to the single isomorphous replacement (SIR) method. In RIP, the changes to structure factors come from radiation damage, rather than the addition of a heavy atom. Experimentally, radiation damage can be used in a variety of ways (*see* Subheading 1.3), but in its initial incarnation, a RIP experiment was described as follows and shown graphically in Fig. 1:

1. A minimal, low-dose dataset is collected.
2. The same position on the crystal is subjected to a high-dose "burn."
3. A second minimal low-dose dataset is collected.

The first dataset is typically called the "before" dataset and can be thought of as analogous to the derivative dataset in SIR. The second dataset is typically called the "after" dataset and is analogous to the native dataset in SIR. Because radiation damage results in a reduction in the total number of electrons, the after dataset will have fewer electrons than the before dataset. For this reason the after dataset is treated as the native and the before dataset as the "heavy atom" derivative dataset in keeping with SIR conventions. Recently, a modification to the data collection protocol has been described in which one large dataset is collected and the definition of the "before" and "after" datasets is done at the analysis stage [30]. If a favorable balance between specific and general radiation damage can be achieved, RIP offers a number of advantages over other phasing methods. Compared to sulfur SAD, RIP has the advantage of being possible at short wavelengths, where absorbance issues are less pronounced, as well as producing much larger percent differences to structure factors. Compared to SIR, MIR and selenomethionine MAD, RIP has the advantage of not requiring any chemical modification of the protein. Since data are collected from the same position from a single crystal in RIP, RIP is less likely to suffer from non-isomorphism between the two datasets, which can be an advantage over isomorphous replacement methods. Finally, RIP does not require the use of a tunable energy beamline.

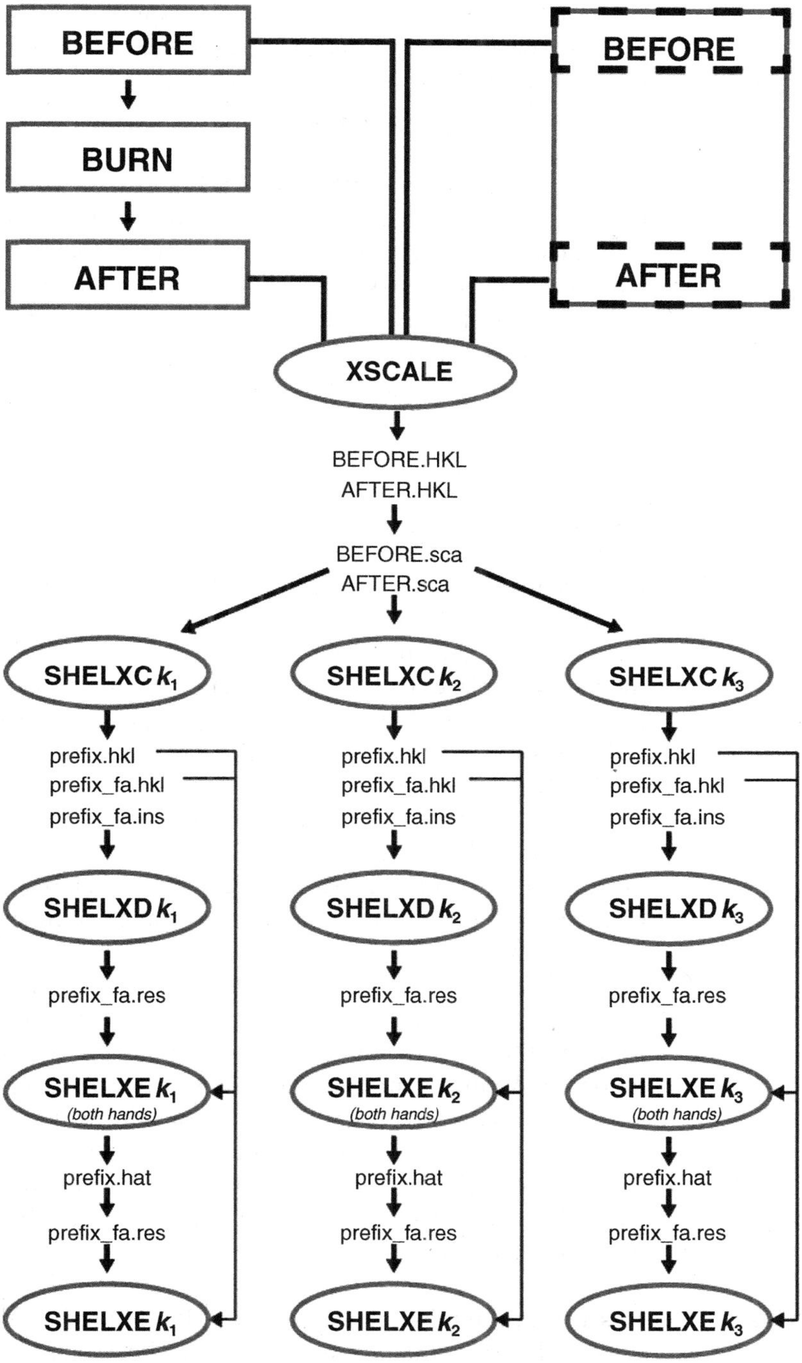

Fig. 1 Workflow for RIP data collection, data preparation, substructure determination, and phasing. Datasets are represented as *grey boxes*, sub-datasets are represented as *black dashes*, and computer programs are shown as *grey ovals*. The *upper left* corner represents the classic RIP experiment, and the *upper right* represents a segmented RIP experiment. Example filenames are listed below the programs that created them. In this example, only three values of *k* are shown, but a more typical value is 20

1.2 RIP Particularities (Compared to S/MIR and S/MAD)

While RIP is similar to SIR in many respects, after its discovery, two significant RIP-specific challenges were identified. It was discovered that RIP substructures are different from SIR substructures, and that conventional scaling methods are often inadequate for RIP.

RIP substructures differ from SIR substructures in a number of practically important ways. First, RIP substructures contain both positive and negative peaks, whereas SIR substructures contain only positive peaks. This is due to the fact that radiation can induce not just a reduction in occupancy at specific sites in the crystal, but also movements and re-arrangements. The most well characterized of such movements is the breaking of disulfide bonds followed by the movement of the Sγ atoms to new positions [1, 27, 31]. In this situation, the substructure will contain positive peaks on the starting positions of Sγ atoms and negative peaks on the new position(s) of the Sγ atoms. In addition to the presence of negative peaks, RIP substructures typically are comprised of many low occupancy sites, compared to the relatively few, high occupancy sites in SIR. These two differences can make substructure solution more difficult in RIP than in anomalous dispersion or isomorphous replacement. The most effective means of dealing with this problem is to bootstrap up from an initial substructure that contains only positive sites, as will be discussed in more detail in Subheadings 3.3 and 3.4.

In examining early RIP data in both model phased RIP difference maps and RIP difference Patterson maps, a large number of negative peaks, and many short, negative vectors, respectively were discovered. This suggested that conventional scaling methods were over-estimating the contribution of the "after" dataset. It is still not fully understood why conventional heavy atom scaling algorithms are not as effective for RIP, but this has introduced the need for a modification to scaling protocols compared to other phasing methods. An inelegant, but effective, solution to this problem is to scale the two datasets together using conventional means and then down-weight the after dataset by a small percentage. The percentage of downweighting is varied empirically, as the optimum downweighting varies from crystal to crystal. This process is described in Subheading 3.2.

1.3 RIP Use Cases

It is a common misconception that RIP is limited to X-ray damage to crystals that contain disulfide bonds. The reality is that any kind of specific radiation damage can be used. This includes damage to metal sites (including all of the commonly used elements for S/MIR or S/MAD), and even damage induced by ultraviolet light [32–35]. RIP is not limited to de novo phasing, but can also be used in structure validation and as an aid to model building. While the technique has varied uses and applications, in this review

Table 1
Use cases for specific radiation damage

	Required RIP signal	Reference
Substructure determination		
RIP substructure determination	High	[29, 30, 36]
Cross-validation of anomalous and RIP substructures	Low	[37]
Phasing		
Phasing with substructure entirely from RIP	High	[29, 30, 36]
Phasing with substructure from external sources (e.g., molecular replacement)	Medium	[28, 38]
Phase combination with other sources	Low	[37, 39]
Model building		
Aid in chain tracing by using damage as waypoints	Low	
Location of atoms in radiation sensitive ligands	Low	[40]
Validate molecular replacement solution	Low	[41]

we will focus on basic substructure determination and phasing. It is important to remember, however, that there are many other uses for specific radiation damage, as summarized in Table 1.

2 Materials

Here, we describe a basic RIP workflow using XDS/XSCALE [42], SHELX [43], and CCP4 [44]. It is amenable to parallelization, which is critical for trying multiple scale factors for the "after" dataset. It should however be noted that in challenging cases, once the substructure has been determined, SHARP [45] should be used in preference to this high throughput workflow because of its more rigorous treatment of radiation damage. Furthermore, SHARP can calculate the correct overall scale between the before and after datasets using the heavy atom substructure.

3 Methods

3.1 RIP Data Collection

The basic RIP data collection described in Subheading 1.1 and shown schematically in Fig. 1 has been shown to be effective in many cases. In collecting the before and after datasets, dose is likely to be the most critical parameter. One should aim for the smallest dose possible that yields acceptable data for the "before" and "after" datasets. In this respect, the strategy used for the collection of these two datasets can borrow from strategies developed for MAD that, for example, reduce dose at the expense of resolution and redundancy (*see* **Note 1**). The dose required for the "burn," however can be difficult to predict. This process can be aided by using RADDOSE [46–48], a flux-calibrated beamline [49] and/or an online microspectrophotometer [50] to monitor radiation damage (*see* **Note 2**). Even with these tools, burn dose determination typically requires some trial and error. An alternative approach is to collect a very large number of images and to create pseudo "before" and "after" datasets [30] (*see* **Note 3**).

3.2 RIP Data Preparation

The same care should be taken in indexing and scaling the before and after datasets together that one normally uses for MAD or S/MIR data treatment [51]. When using XDS, it is easy to ensure consistent indexing between the before and after datasets using the REFERENCE_DATA_SET keyword. When performing a RIP with anomalous scattering (RIPAS) treatment, it is also important to set the FRIEDEL'S_LAW keyword to be "False" to keep Friedel mates separate. Although not strictly necessary because SHELXC scales the before and after datasets, a scaling step using XSCALE (Fig. 1) can be useful. XSCALE can be helpful in identifying any obvious problems (e.g., if the scale factor is very different from 1.0 or there is a very high temperature factor between the two datasets). Once scaled together, data are converted to SCALEPACK format. It is recommended that data be left in unmerged format (MERGED = FALSE). SHELXC is then used to produce a HKL file of Fa's, an input file for SHELXD, and the native HKL file that will be used later by SHELXE for phasing and density improvement. A "prefix" must be specified which is the root filename of all the files that SHELXC will create. As an example, we will simply use "prefix" (Fig. 1). SHELXC has been modified to specifically handle RIP data, using the "RIP" and "DSCA" keywords [36]. The RIP keyword is used to specify the after dataset and the DSCA keyword is used to alter the weight of the after dataset to account for the changes in scattering power due to radiation damage. In order to include anomalous differences, one can alternatively use the RIPA keyword.

As mentioned earlier, conventional scaling algorithms tend to overweight the contribution of the "after" dataset, so at this point in the process, multiple down-weightings are applied to the "after" dataset. This is done with the DSCA keyword. A typical RIP analysis might contain 20 such down-weights (k) and vary from 0.9 to 1.0, although this range can and should be optimized (*see* **Note 4**). Optimization can be done by obtaining scale factors from SHARP, or by examining the relationship between SHELXD statistics like CC(all)/CC(weak) as a function of k. When phasing with disulfides, one should specify the number of disulfides to search for using the DSUL keyword. Before the highly parallelized version of

SHELXD was released, care would be taken at this stage to set the resolution limits of substructure determination based on RIP difference statistics. Specifically, the <d'/sig(d)> statistic was inspected in the SHELXC output and the high resolution limit was set to the resolution at which this value dropped below 1.3. Recently, however, we have found that the success of RIP substructure determination can sometimes be improved by including data beyond a value of 1.3 or even using the full resolution of the data, but using a large number of trials (20,000–50,000). This number of trials would have previously been impractical, but on a multi-core machine with the parallelized version of SHELXD, run times can be reasonable (e.g., 20,000 trials with the example trypsin dataset in space group P212121, 45483 reflections, maximum resolution 1.3 Å, takes eighteen minutes on a dual Intel Xeon X5675 6 core (12 core total) 3.07 GHz machine). An alternative to SHELXC is to use XPREP to prepare SHELXD input files and Fa's, but the after dataset will then need to be down-weighted by some other method. An advantage of using XPREP is that it produces many other useful diagnostics and plots, including facile generation of origin-removed Patterson maps. In either case, at this stage, one should have a set of three files which will be used for SHELXD: an input file (.ins file), a HKL file of the Fa's, and an HKL file of the before file. For example: prefix_fa.ins, prefix_fa.hkl and prefix.hkl (Fig. 1). Once input files for SHELXD and SHELXE have been produced, substructure determination can commence with SHELXD.

3.3 Substructure Determination

SHELXD can be started at this stage, using the prefix_fa files: prefix_fa.ins and prefix_fa.hkl (Fig. 1). Once the program has completed (*see* **Note 5**), many of the same metrics of substructure solution success can be used for RIP compared to S/MIR or S/MAD. However, there are some important differences. First, the CC(all)/CC(weak) ratio is typically greatly reduced compared to other methods. Second, the typical criterion of correct substructures having a rapid drop-off in site occupancy is rarely met for RIP, since RIP substructures contain a large number of sites without a large internal discrimination in peak height. The presence of a cluster of high CC(all)/CC(weak) solutions that are separated from the rest of the solutions is generally quite predictive of correct solutions, but in our experience, the relationship is less strong in RIP than in other methods. As an example, in Fig. 2, one can see that there are plots with almost no distinct cluster, but with a reasonably correct substructures. One metric which has been consistently predictive of RIP success is a non-flat relationship between k and the values of CC(all)/CC(weak). If a peak can be seen just below $k=1$, this is usually a good indication that phasing and density improvement should be pursued. An example can be seen in Fig. 2. On the left hand panel, one can see that as k is varied, the

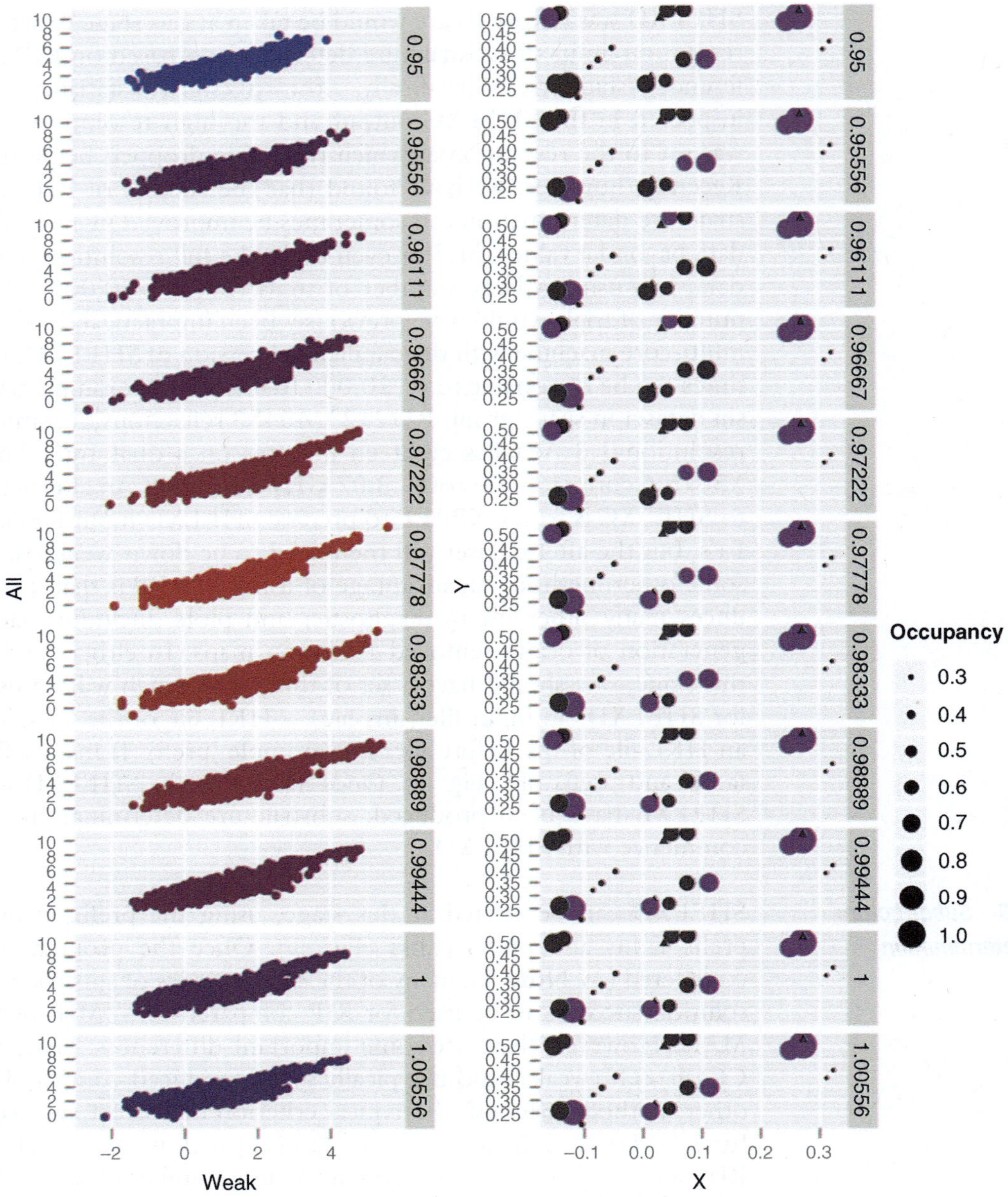

Fig. 2 Correspondence of SHELXD CC(all)/CC(weak) with substructure correctness versus *k*. At left are plots of SHELXD CC(all)/CC(weak) for different values of k. The color is scaled from *blue* to *red* according to the lowest to highest CC(all)/CC(weak), i.e., the trials from the *k* value with the best CC(all)/CC(weak) are shown in *red*. At right is a graphical representation of the correctness of substructure solutions. For this representation, a model phased RIP difference map is created and peak searched (for peaks >8σ) with ANODE [53] to produce a "gold standard" substructure. This substructure is projected into the XY plane, and the strength (i.e., RIP peak height) of individual damage sites is shown by the diameter of individual *circles*—the stronger RIP sites are shown as *circles* or *triangles*. Positive sites are represented as *circles* and negative sites are represented as *triangles*. The correctness of SHELXD sites is then compared to this standard, and individual *circles/triangles* are colored *purple* when the site has been correctly identified. Data are from a RIP experiment on trypsin [30]

CCs reach a maximum at $k = 0.9778$. When the substructure with the best CC(all)/CC(weak) from this run is compared with the known RIP substructure for similarity, and correctly identified sites are colored *purple*, one can see that this value of k also produces the most correct substructure. Despite these potentially useful metrics, we generally recommend treating statistics at this stage cautiously, and to not abandon phasing even if the metrics seem unpromising.

3.4 Phasing, Substructure Improvement and Phase Improvement

The next step is to use SHELXE for substructure improvement, phase calculation and phase improvement (Fig. 1, bottom rows). If SHELXD produces a substructure with the incorrect hand, a map with the mirror image of the correct features is produced. Therefore, one must run SHELXE in both hands. SHELXE handles the substructure inversion (and space group change when applicable) with the "-i" flag. All that is needed is to run SHELXE with and without the "-i" flag. This can easily be done in parallel. Note that statistics on the resultant map, such as the Pseudo-Free correlation coefficient and contrast will be the same for the correct and incorrect hands. The most critical parameter for RIP using SHELXE is the solvent content. Although reasonable estimates of solvent content can be obtained by calculation of the Matthews coefficient, it is usually worthwhile to try several solvent content values. It is at this stage that one begins the iterative process of substructure update. Recall that the substructure from SHELXD contains only positive sites in the RIP substructure. Negative sites can be identified by the difference Fourier analysis built into SHELXE with the "-b" option. This option causes SHELXE to produce a new substructure file with extension ".hat" at the completion of the run that contains both positive and negative sites, with occupancies scaled to RIP difference peak height. This ".hat" file can be copied to prefix_fa.res (*see* **Note 6** and Fig. 1). This process must be performed several times in difficult cases to fully elaborate the RIP substructure. An example of the improvements to the phase error that are possible is shown in Fig. 3. In this figure, we compare the RIP substructure at each round of improvement by this process in the same manner as in Fig. 1. One difference is that in order to highlight the improvement to minor sites, we have created a reference substructure that includes RIP peaks down to 6σ (compared to 8σ before). One can see that in the initial substructure, none of the negative sites (triangles) are identified, and indeed neither are some of the strong positive sites. As the process continues, more and more of the sites (both positive and negative) are identified, the phase errors diminish, and the maps improve and converge.

One optional but important parameter is the inclusion of cycles of automatic chain tracing. In our experience, this has been the difference between solving and not solving several RIP structures.

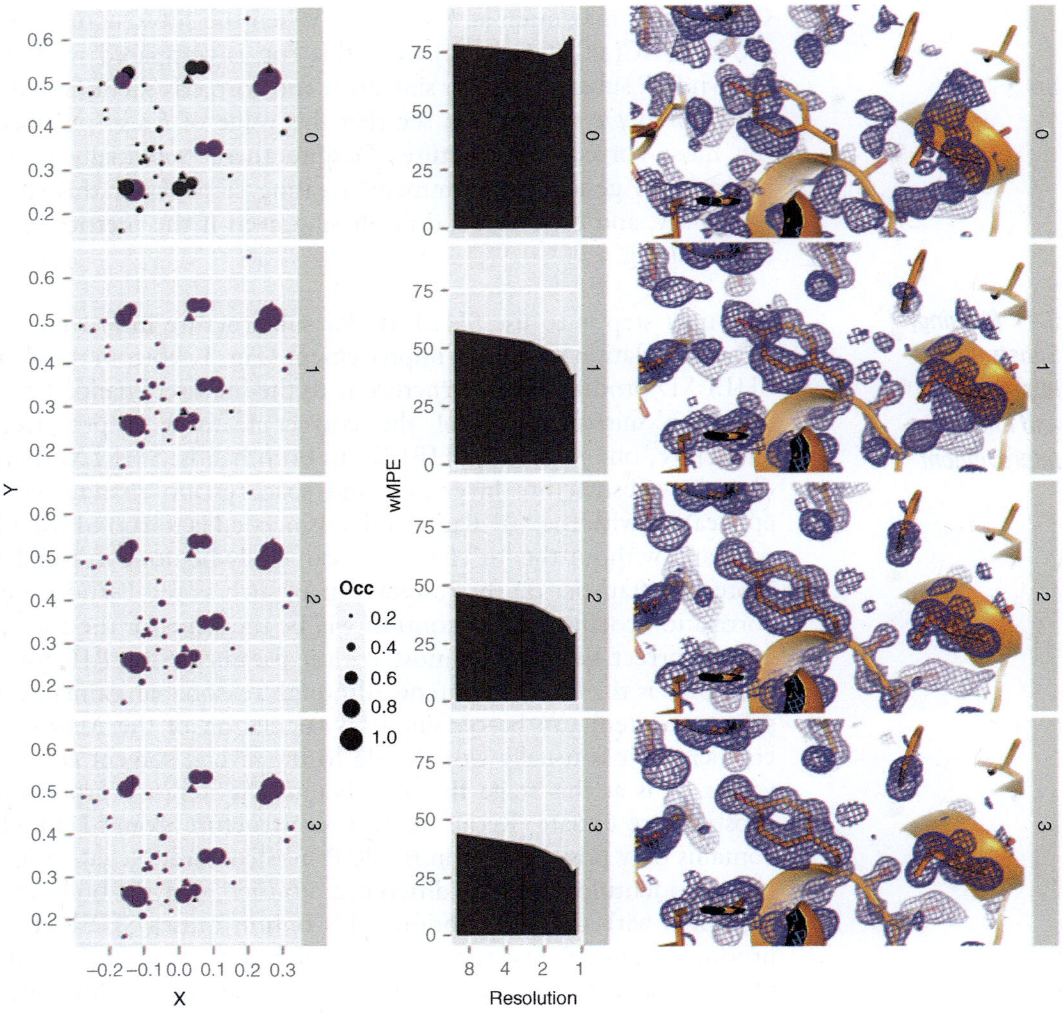

Fig. 3 Phase and map improvement can be improved by iterative improvement of the RIP substructure. RIP substructures are improved by using the RIP peak positions obtained by SHELXE difference Fourier analysis. The iteration number is indicated at the right hand of each panel. Iteration 0 is the initial substructure from SHELXD, and subsequent iterations [1–3] are from difference Fourier analysis. The left hand column shows the improvement to the RIP substructures at each stage of the substructure improvement. As in Fig. 2, the substructure coordinates are projected onto the XY plane, and correctly identified sites are colored *purple*. The middle column shows the improvement to the weighted mean phase error compared to the refined trypsin structure as the substructure is improved. The right hand column shows the concomitant improvement to the electron density maps at each stage of the phasing. Maps are contoured at 1σ

The only drawback of including this option ("-a") is the computational time. This option also makes it easy to resolve which substructure hand is correct, since automatic building will only be successful in electron density maps from the correct substructure hand. Finally, there are numerous newer features of SHELXE that have thus far not been systematically tested in the context of RIP, but could also provide a needed boost in difficult cases.

These include the use of the "Free lunch algorithm" ("-e") and the substructure optimization option ("-z").

At the end of the SHELXE run(s), log files are inspected to perform a triage of the multiple SHELXC/D/E threads for all values of *k*. In the absence of this, the number of maps that one must manually inspect can rapidly become overwhelming. The parameters which appear to be the most predictive of the correctness of the resultant electron density map are the correlation coefficient of the partially automatically built model when chain tracing is enabled ("partial CC") and the Pseudo-free correlation coefficient ("PFCC"). Any phase sets with partial CCs above ~20 % and PFCCs above 65 % should be inspected manually. Phase sets with partial CCs above 25 % and PFCCs above 70 % are almost always solved. Once phases have been obtained, RIP does not differ from other phasing methods. Cycles of manual or automatic model improvement are performed, followed by refinement using standard programs such as PHENIX, Refmac, or CNS. It should be noted that non-canonical bond lengths such as those induced by radiation damage to disulfide bonds can be difficult to properly model and refine with standard bond length libraries [52].

The workflow described here is specific to macromolecular crystals with radiation sensitive groups. As with any other phasing technique, RIP cannot be expected to be universally applicable. However, by following the guidelines described above for the careful collection and processing of crystallographic data, one can dramatically improve the success rate of obtaining interpretable phases by RIP.

4 Notes

1. As in S/MIR and S/MAD, it is inadvisable to attempt to get high resolution in any of the datasets.
2. In planning the dose for all datasets, one must be careful to consider the size of the X-ray beam in comparison to the crystal. Similarly, the profile of the beam is also important: is it a Gaussian? A "Top hat"? These parameters will have a large impact on the accuracy of the dose calculation
3. As an approximate target for the burn, we generally aim for the crystal to have absorbed 2 MGy by the end of the burn. When collecting in "segmented" mode, one can simply plan for the data collection to achieve 2 MGy at some point in the data collection.
4. There are two strategies for preventing filenames from colliding with each other when using multiple values of *k*. The first is to run each set of SHELXC/D/E for a given *k* in its own directory. The second is to include *k* in the root filename (the prefix). For example protein_0.9888.

5. It is possible to begin the next step (SHELXE) without waiting for SHELXD to complete, since SHELXD continuously creates .res files with the best solution.
6. In preparing the .hat file for use in the next round of SHELXE one must remove the "END" line that divides the positive from the negative sites in the ".hat" file before proceeding, or only the positive sites will be carried forward.

References

1. Ravelli RB, McSweeney SM (2000) The "fingerprint" that X-rays can leave on structures. Structure 8:315–328
2. Garman EF, Owen RL (2006) Cryocooling and radiation damage in macromolecular crystallography. Acta Cryst D 62:32–47
3. Teng TY (1990) Mounting of crystals for macromolecular crystallography in a free-standing thin film. J Appl Crystallogr 23:387–391
4. Hope H (1988) Cryocrystallography of biological macromolecules: a generally applicable method. Acta Cryst B 44:22–26
5. Cosier J, Glazer AM (1986) A nitrogen-gas-stream cryostat for general X-ray diffraction studies. J Appl Crystallogr 19:105–107
6. Garman EF, Schneider TR (1997) Macromolecular cryocrystallography. J Appl Crystallogr 30:211–237
7. Popov AN, Bourenkov GP (2003) Choice of data-collection parameters based on statistic modelling. Acta Cryst D 59:1145–1153
8. Bourenkov GP, Popov AN (2006) A quantitative approach to data-collection strategies. Acta Cryst D 62:58–64
9. Bourenkov GP, Popov AN (2010) Optimization of data collection taking radiation damage into account. Acta Cryst D 66:409–419
10. Dauter Z (2010) Carrying out an optimal experiment. Acta Cryst D 66:389–392
11. Murray J, Garman E (2002) Investigation of possible free-radical scavengers and metrics for radiation damage in protein cryocrystallography presented at the "Second International Workshop on Radiation Damage to Crystalline Biological Samples" held at Advanced Photon Source, Chicago, USA, in December 2001. J Synchrotron Radiat 9:347–354
12. Holton JM (2006) XANES measurements of the rate of radiation damage to selenomethionine side chains. J Synchrotron Radiat 14: 51–72
13. Kauffmann B, Weiss MS, Lamzin VS, Schmidt A (2006) How to avoid premature decay of your macromolecular crystal: a quick soak for long life. Structure 14:1099–1105
14. Southworth-Davies RJ, Garman EF (2006) Radioprotectant screening for cryocrystallography. J Synchrotron Radiat 14:73–83
15. Barker AI, Southworth-Davies RJ, Paithankar KS, Carmichael I, Garman EF (2009) Room-temperature scavengers for macromolecular crystallography: increased lifetimes and modified dose dependence of the intensity decay. J Synchrotron Radiat 16:205–216
16. Macedo S, Pechlaner M, Schmid W, Weik M, Sato K, Dennison C, Djinović-Carugo K (2009) Can soaked-in scavengers protect metalloprotein active sites from reduction during data collection? J Synchrotron Radiat 16:191–204
17. Nowak E, Brzuszkiewicz A, Dauter M, Dauter Z, Rosenbaum G (2009) To scavenge or not to scavenge: that is the question. Acta Cryst D 65:1004–1006
18. Kmetko J, Warkentin M, Englich U, Thorne RE (2011) Can radiation damage to protein crystals be reduced using small-molecule compounds? Acta Cryst D 67:881–893
19. Allan EG, Kander MC, Carmichael I, Garman EF (2013) To scavenge or not to scavenge, that is STILL the question. J Synchrotron Radiat 20:23–36
20. Owen RL, Yorke BA, Gowdy JA, Pearson AR (2011) Revealing low-dose radiation damage using single-crystal spectroscopy. J Synchrotron Radiat 18:367–373
21. Southworth-Davies RJ, Medina MA, Carmichael I, Garman EF (2007) Observation of decreased radiation damage at higher dose rates in room temperature protein crystallography. Structure 15:1531–1541
22. Shimizu N, Hirata K, Hasegawa K, Ueno G, Yamamoto M (2006) Dose dependence of radiation damage for protein crystals studied at various X-ray energies. J Synchrotron Radiat 14:4–10
23. Weiss MS, Panjikar S, Mueller-Dieckmann C, Tucker PA (2005) On the influence of the incident photon energy on the radiation damage in crystalline biological samples. J Synchrotron Radiat 12:304–309

24. Homer C, Cooper L, Gonzalez A (2011) Energy dependence of site-specific radiation damage in protein crystals. J Synchrotron Radiat 18:338–345
25. Fourme R, Honkimäki V, Girard E, Medjoubi K, Dhaussy A-C, Kahn R (2012) Reduction of radiation damage and other benefits of short wavelengths for macromolecular crystallography data collection. J Appl Crystallogr 45: 652–661
26. Garman EF (2010) Radiation damage in macromolecular crystallography: what is it and why should we care? Acta Cryst D 66:339–351
27. Burmeister WP (2000) Structural changes in a cryo-cooled protein crystal owing to radiation damage. Acta Cryst D 56:328–341
28. Evans G, Polentarutti M, Djinovic-Carugo K, Bricogne G (2003) SAD phasing with triiodide, softer X-rays and some help from radiation damage. Acta Cryst D 59:1429–1434
29. Ravelli RBG, Leiros H-KS, Pan B, Caffrey M, McSweeney S (2003) Specific radiation damage can be used to solve macromolecular crystal structures. Structure 11:217–224
30. De Sanctis D, Nanao MH (2012) Segmenting data sets for RIP. Acta Cryst D 68:1152–1162
31. Leiros H-K, McSweeney SM, Smalås AO (2001) Atomic resolution structures of trypsin provide insight into structural radiation damage. Acta Cryst D 57:488–497
32. Nanao MH, Ravelli RBG (2006) Phasing macromolecular structures with UV-induced structural changes. Structure 14:791–800
33. Schönfeld DL, Ravelli RBG, Mueller U, Skerra A (2008) The 1.8-Å crystal structure of α1-acid glycoprotein (Orosomucoid) solved by UV RIP reveals the broad drug-binding activity of this human plasma lipocalin. J Mol Biol 384: 393–405
34. De Sanctis D, Tucker PA, Panjikar S (2011) Additional phase information from UV damage of selenomethionine labelled proteins. J Synchrotron Radiat 18:374–380
35. Panjikar S, Mayerhofer H, Tucker PA, Mueller-Dieckmann J, de Sanctis D (2011) Single isomorphous replacement phasing of selenomethionine-containing proteins using UV-induced radiation damage. Acta Cryst D 67:32–44
36. Nanao MH, Sheldrick GM, Ravelli RBG (2005) Improving radiation-damage substructures for RIP. Acta Cryst D 61:1227–1237
37. Ravelli RBG, Nanao MH, Lovering A, White S, McSweeney S (2005) Phasing in the presence of radiation damage. J Synchrotron Radiat 12:276–284
38. Schiltz M, Dumas P, Ennifar E, Flensburg C, Paciorek W, Vonrhein C, Bricogne G (2004) Phasing in the presence of severe site-specific radiation damage through dose-dependent modelling of heavy atoms. Acta Cryst D 60: 1024–1031
39. Rudiño-Piñera E, Ravelli RBG, Sheldrick GM, Nanao MH, Korostelev VV, Werner JM, Schwarz-Linek U, Potts JR, Garman EF (2007) The solution and crystal structures of a module pair from the Staphylococcus aureus-binding site of human fibronectin – a tale with a twist. J Mol Biol 368:833–844
40. Ravelli RBG, Gigant B, Curmi PA, Jourdain I, Lachkar S, Sobel A, Knossow M (2004) Insight into tubulin regulation from a complex with colchicine and a stathmin-like domain. Nature 428:198–202
41. Grininger M, Ravelli RBG, Heider U, Zeth K (2004) Expression, crystallization and crystallographic analysis of DegS, a stress sensor of the bacterial periplasm. Acta Cryst D 60:1429–1431
42. Kabsch W (2010) XDS. Acta Cryst D 66: 125–132
43. Sheldrick GM (2010) Experimental phasing with SHELXC/D/E: combining chain tracing with density modification. Acta Cryst D 66: 479–485
44. Winn MD, Ballard CC, Cowtan KD, Dodson EJ, Emsley P, Evans PR, Keegan RM, Krissinel EB, Leslie AG, McCoy A, McNicholas SJ, Murshudov GN, Pannu NS, Potterton EA, Powell HR, Read RJ, Vagin A, Wilson KS (2011) Overview of the CCP4 suite and current developments. Acta Cryst D 67:235–242
45. De La Fortelle E, Bricogne G (1997) Maximum-likelihood heavy-atom parameter refinement for multiple isomorphous replacement and multiwavelength anomalous diffraction methods. Methods Enzymol 276: 472–494
46. Paithankar KS, Garman EF (2010) Know your dose: RADDOSE. Acta Cryst D 66:381–388
47. Paithankar KS, Owen RL, Garman EF (2009) Absorbed dose calculations for macromolecular crystals: improvements to RADDOSE. J Synchrotron Radiat 16:152–162
48. Zeldin OB, Gerstel M, Garman EF (2013) RADDOSE-3D: time- and space-resolved modelling of dose in macromolecular crystallography. J Appl Crystallogr 46:1225–1230
49. Owen RL, Holton JM, Schulze-Briese C, Garman EF (2009) Determination of X-ray flux using silicon pin diodes. J Synchrotron Radiat 16:143–151

50. McGeehan J, Ravelli RBG, Murray JW, Owen RL, Cipriani F, McSweeney S, Weik M, Garman EF (2009) Colouring cryo-cooled crystals: online microspectrophotometry. J Synchrotron Radiat 16:163–172
51. McCoy AJ, Read RJ (2010) Experimental phasing: best practice and pitfalls. Acta Cryst D 66:458–469
52. Carpentier P, Royant A, Weik M, Bourgeois D (2010) Raman-assisted crystallography suggests a mechanism of X-ray-induced disulfide radical formation and reparation. Structure 18:1410–1419
53. Thorn A, Sheldrick GM (2011) ANODE: anomalous and heavy-atom density calculation. J Appl Crystallogr 44:1285–1287

Chapter 14

Soaking Hexammine Cations into RNA Crystals to Obtain Derivatives for Phasing Diffraction Data

Robert T. Batey and Jeffrey S. Kieft

Abstract

Solving a novel RNA structure by x-ray crystallography requires a means to obtain initial phase estimates. This is a challenge because many of the tools available for solving protein structures are not available for RNA. We have developed a reliable means to use hexammine cations to address this challenge. The process involves engineering the RNA to introduce a reliable hexammine binding site into the structure, then soaking crystals of these RNAs with an iridium (III) or cobalt (III) compound in a "directed soaking" strategy. Diffraction data obtained from these crystals then can be used in SAD or MAD phasing. In many cases, suitable derivatives can be obtained by soaking the hexammine into RNA crystals that have not been engineered. Considerations for using this method and example protocols are presented.

Key words RNA, Hexammine, Structure solution, Heavy atoms, Iridium

1 Introduction

Obtaining phase information is a necessary step to determine a macromolecular structure by x-ray crystallography. For proteins, this step in the process has become increasingly less challenging as the number of solved protein structures has grown. This is because phasing can often be readily accomplished using solved structures of homologous proteins to obtain initial phase estimates by the molecular replacement (MR) method and powerful programs such as PHASER [1, 2]. For applications that require experimental phases, expressing the protein as a selenomethionine derivative is a robust method that now is used for over half of protein structures solved [3, 4]. Therefore, only the most challenging problems—or an unusual situation—require different phasing approaches.

For RNA crystallography, the ability to use MR or biosynthetic incorporation of a heavy atom is significantly limited. With respect to MR, RNA crystallographers are less likely to use this approach, in part because the number and diversity of RNA structures available in the database is more limited than for proteins. Although

Eric Ennifar (ed.), *Nucleic Acid Crystallography: Methods and Protocols*, Methods in Molecular Biology, vol. 1320,
DOI 10.1007/978-1-4939-2763-0_14, © Springer Science+Business Media New York 2016

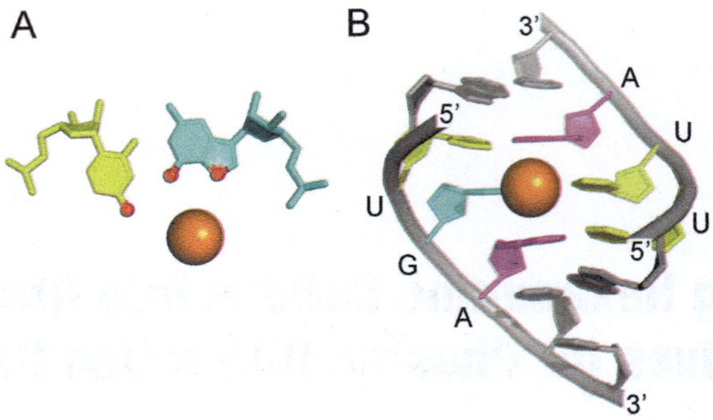

Fig. 1 Binding of hexammines to the cation binding G-U wobble motif. (**a**) Idealized G-U wobble pair. The small balls on the nucleobases indicate functional groups in the major groove with a partial negative charge form a surface for cation binding. The *large sphere* denotes the location of the central metal atom of a bound hexammine complex. For simplicity, the amines are not shown. (**b**) Example of cation binding to a single G-U wobble motif within an RNA helix. The G-U pair and flanking sequences are labeled. The *orange sphere* is the central atom of the iridium (III) hexammine complex, the amines have been removed for simplicity

molecular replacement-based methods using small A-form RNA helices have been developed [5–7], solving most RNA structures requires obtaining experimental phases using diffraction data from derivatized crystals. Synthetic approaches to the direct incorporation of selenium, bromine, or other anomalous scattering atoms are typically used on small oligonucleotides [8–18]; for larger RNAs with complex tertiary architecture these approaches are less useful.

An alternative to the above approaches is a throwback to the earliest days of macromolecular crystallography: soaking. In this method, a crystal is derivatized by soaking the crystal in a heavy atom-containing solution (generally a cation to take advantage of the polyanionic nature of RNA), hoping that heavy atoms will bind to one or more specific locations within the RNA structure. Although often successful, a disadvantage to this method is that in most cases one cannot predict with any certainty if suitable specific binding sites exist, and hence is a trial-and-error method that can be time-consuming. The fickle and unpredictable nature of this process gave rise to the well-earned nickname of "soak-and-pray" [9, 19, 20].

To address the inherently uncertain and unpredictable nature of soaking heavy atoms into RNA crystals, we developed a highly general "directed soaking" method. This approach involves engineering of one or more reliable non-structure-perturbing cation binding sites into the RNA sequence of interest and then soaking hexammine cations suitable for phasing into crystals containing this RNA [21]. The method described here is based on the observation that G-U wobble base-pairs in A-form RNA helices create a binding site for many cations [22–28], including hexammine complexes known to be very useful for phasing RNA diffraction data (Fig. 1) (examples: [27, 29–39]). Indeed, hexammine complexes have proven their worth in phasing diffraction data from crystals of complexes as large as the ribosome (examples: [40–42]). However, not all G-U wobble pairs work equally well to localize a hexammine cation because the identity and orientation of the base-pairs that flank the G-U pair matter. The reason for this effect has been analyzed elsewhere and versions appropriate for this method have been experimentally identified [21]. Here, we recommend specific versions (see below). Overall, the method comprises mutating the RNA sequence of interest to include a G-U wobble motif that robustly binds a hexammine cation, crystallizing this RNA and then soaking hexammine complexes into the crystals to obtain the derivative (*see* **Note 1**). We have had success with both cobalt (III) and iridium (III) hexammine, but even cesium has been used to phase when localized at this motif [43]. Diffraction data collected from these crystals at appropriate x-ray energies is then used for multiple- or single-wavelength anomalous diffraction (MAD or SAD) phasing.

Choosing the right version of the G-U wobble motif to engineer into the RNA sequence is important, as not all versions of the motif bind hexammine cations with the same characteristics. We undertook a systematic study of motifs containing single G-U wobble pairs and tandem G-U wobbles to find those that bound hexammines with the lowest B-factor (best localized) and with the highest anomalous signal [21]. We found that the arrangement of both partially negatively and positively charged functional groups in the major groove affected cation binding. Based on these findings, we empirically identified both single and double G-U wobble motifs that could robustly and reliably bind a hexammine cation and could therefore be used to introduce a binding site for directed soaking (Fig. 2a, b).

This method has advantages and disadvantages. A strength is that it eliminates the need to try many different cations or to generate chemically modified RNAs (*see* **Note 2**). The method has been used to successfully derivatize crystals grown at diverse ionic strengths, including very high monovalent concentrations that would be thought to be disruptive to electrostatic interaction between the hexammine and RNA [35]. The method requires mutation of the RNA but because the mutations involve only a few nucleotides, the G-U cation binding module can be introduced into parts of RNA that are nonessential for function and generally do not perturb the structure. Thus, perturbation of local RNA structure by the mutation(s) is not a significant concern. In some cases, like the SAM-I riboswitch, the site is already present in the natural sequence and thus no modification is required [27]. Finally, no crystal soaking protocol can be truly general. Each crystal is different and thus the amount of soaking time, number of steps, concentration of heavy atom, etc. must be optimized. Here, we present two protocols that were successful for obtaining derivatives of RNA

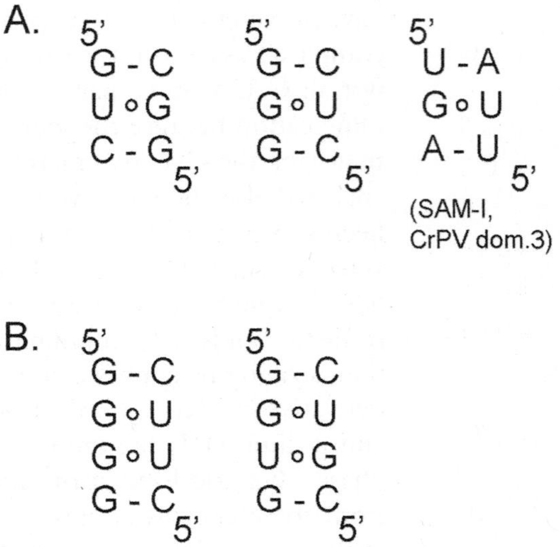

Fig. 2 Empirically determined versions of the G-U wobble motif that bind hexammines suitable for phasing [21]. (**a**) Single G-U motif versions that can robustly bind a hexammine cation. The rightmost was used to solve two different RNA structures: domain 3 from the Cricket paralysis virus internal ribosome entry site [35], and the *S*-adenosylmethionine riboswitch type I [27]. (**b**) Two tandem G-U motifs that robustly bind a hexammine cation

crystals grown in either high-salt or low-salt conditions; these protocols form the basis for protocols tailored to each crystal form. Finally, it should be noted that because of the polyanionic nature of RNA, that there are typically at least one or two hexammine binding sites in a typical RNA. A number of recent RNA structures were solved using hexammines without engineering specific binding sites into the sequence (examples: [33, 36–38, 44–47]). Together, these considerations now make iridium (III) hexammine, with or without prior engineering of the RNA, overwhelmingly the current method of choice for obtaining phase information.

2 Materials

2.1 Synthesis of Iridium (III) Hexammine Chloride

1. Iridium (III) chloride ($IrCl_3$).
2. Ammonium hydroxide.
3. Teflon tape.
4. Sintered-glass filters.
5. Rotary evaporator.
6. Concentrated hydrochloric acid (HCl).
7. Absolute ethanol.
8. Ultrapure water.
9. Heavy-walled Ace pressure tube (Sigma-Aldrich, Saint-Louis, MO, USA).

2.2 Generation of Hexammine Acetate from Hexammine Chloride

1. Cobalt (III) hexammine chloride or iridium (III) hexammine chloride.
2. Lead acetate.
3. Ultrapure water.

2.3 Soaking of Hexammine into Crystals (Examples)

1. Low-salt crystallization solution A: 50 mM Sodium 2-(N-morpholino)ethanesulfonic acid (MES) pH 5.6, 100 mM Magnesium acetate, 15 % 2-methyl-2,4-pentanediol (MPD).
2. High-salt crystallization solution B: 1.4 M Lithium sulfate, 40 mM Magnesium acetate, 50 mM (4-(2-hydroxyethyl)-1-piperazineethanesulfonic acid) HEPES-NaOH pH 7.5.
3. Stabilizing solution: 2.0 M Lithium sulfate, 40 mM Magnesium acetate, 50 mM HEPES-NaOH pH 7.5, 0.5 mM spermidine-HCl.
4. Derivative solution: 3.0 M Lithium acetate, 40 mM Magnesium acetate, 0.5 mM spermidine-HCl, 50 HEPES-NaOH pH 7.5, 200 mM hexammine acetate.

3 Methods

3.1 Design of the RNA/ Introduction of the G-U Wobble Motif

1. The directed soaking method described here can be employed in two ways (Fig. 3). In the first, the cation-binding G-U wobble motif is included in the design of the library of RNAs used in initial crystal screens. The advantage is that if diffracting crystals are obtained from this library, no additional RNAs need to be made to generate crystals suitable for directed soaking. In the second method, libraries of RNAs that do not contain the motif are screened to identify diffracting crystals. Once crystals are found, new versions of the crystallizing RNA are made that contain the motif in one or more places and this RNA is used to grow crystals that then can be used in directed soaking. An advantage to this second method is that if the wild type and motif-containing RNAs crystallize under similar conditions, it provides evidence that introduction of the motif does not alter the structure of the RNA (*see* **Note 3**).
2. In either of the two manifestations described above, an important decision is where to place the cation-binding G-U wobble motif. The goal is to place the motif where it will not affect crystallization, structure, or function. In every RNA there are likely many suitable locations and in most cases several should be tried. Guiding principles are:
   (a) Place the motif into existing Watson–Crick paired helical elements.
   (b) Iridium hexammine bound with high occupancy to one motif is likely enough to provide sufficient phasing power

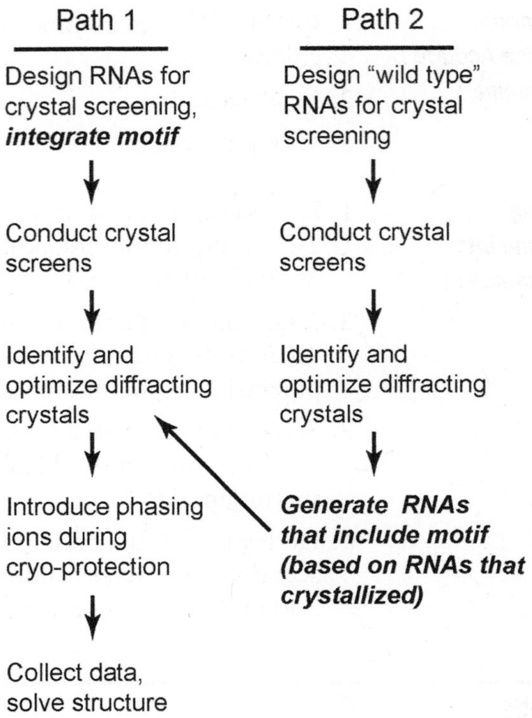

Fig. 3 Different ways to integrate the use of the cation binding G-U motif and hexammines into an overall RNA crystallization and structure-determination strategy

when combined with weaker endogenous sites for RNAs below 150 nucleotides. However, when possible, include multiple sites in different helices, especially for larger RNAs.

(c) Place the motif at least 1–2 base-pairs from the end of helices. The G-U wobble can introduce slight local changes in helical conformation that could affect end-to-end helical stacking often important for crystallization.

(d) Place the motif in parts of the RNA not essential for function or for formation of the global fold (Fig. 4a). This information must be gleaned from other studies (phylogeny, mutagenesis coupled with biophysical, biochemical, and functional studies, etc.).

(e) Attempt to find locations where the motif can be introduced with a minimum of mutation. A location in the sequence where a point substitution converts the wild type sequence to a cation binding motif is ideal (Fig. 4b).

(f) RNA is often systematically engineered to enhance crystallizability by altering the lengths of helical elements that are not critical for function [48–50]; if this is done, the motif can be included within the engineered helix or helices.

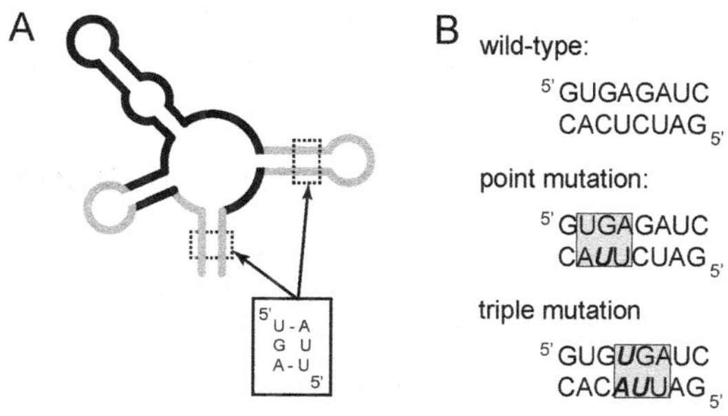

Fig. 4 Examples of how to engineer RNA to include the cation binding motif. (**a**) A stylized RNA secondary structure is shown. *Black* portions are conserved regions known to be essential for structure and function and *gray regions* are variable that can be altered without effect. Two good locations for a cation binding G-U wobble motif (a successfully used version is boxed) are shown with *hatched boxes*. (**b**) Examples of ways an RNA helix could be mutated to introduce a good hexammine cation binding site. In one, a point mutation is used to convert a stretch of Watson–Crick base-pairs to the motif. In the other, three sites are mutated to give the same motif. Both methods work, but the first could be simpler to accomplish with standard mutagenesis methods

3. RNA containing the motif is transcribed using T7 RNA polymerase or chemically synthesized, purified, and crystallized using methods beyond the scope of this protocol (*see* **Note 4**).

3.2 Synthesis of Iridium (III) Hexamine

Unfortunately, at the time of this writing, iridium (III) hexammine is not commercially available. This protocol is an adaptation of that presented by Galsbøl et al. [51]. In particular, the glassware and methods used in this protocol are revised to use readily available materials and ease of implementation for any biochemistry laboratory.

1. In a heavy-walled Ace pressure tube, add 2 g of $IrCl_3$ to 35 mL (fill tube almost to top) of ammonium hydroxide. Seal the tube well with Teflon tape, particularly around the O-ring seal (*see* **Note 5**). Screw the top on tight and set halfway into a silicone oil bath set at 150 °C. Incubate for 4 days. If volume decreases due to a leak in the seal, remove, cool down and refill with fresh ammonium hydroxide.
2. Over the course of the first few days, most of the solid should dissolve and the solution transition from a dark brown to an orange/yellow solution and finally to a nearly colorless solution. These changes reflect the formation of colored tri-, tetra-, and penta-ammine compounds prior to the hexa-ammine compound, which is colorless.

3. When synthesis is complete, remove the tube from the silicon oil bath and allow to completely cool. Then incubate on slushy ice.
4. Pass solution through a sintered glass filter to remove solid material.
5. Remove solvent by rotary evaporation, with the water bath set at 50 °C to heat the solution while evaporating.
6. Resuspend solid in 5 mL of water and place in a polypropylene 50 mL conical tube. Note that the solid material will not completely go into solution; this is normal.
7. Add 2 mL concentrated HCl to resuspension. A large amount of white precipitate should appear. Centrifuge at 5,000 × *g* for 5 min to pellet the precipitate.
8. Remove supernatant (usually light yellow in color) and wash pellet with 5 mL of 2:1 (v/v) water:concentrated HCl by vigorously vortexing. Centrifuge to pellet the precipitate and remove supernatant.
9. Repeat **step** 7 two times.
10. After last spin, remove supernatant and wash the pellet three times with 5 mL of absolute ethanol.
11. Air dry and resuspend in ultrapure H_2O. Add sufficient water (total of ~15 mL) to bring most of the white solid into solution. Spin to pellet the insoluble material and transfer the supernatant to fresh microfuge tubes.
12. Collect an absorbance spectrum of the material. There should be an obvious peak at 251 nm. Calculate the concentration of iridium (III) hexammine chloride using the literature value for the extinction coefficient of 92 $M^{-1}cm^{-1}$ at 251 nm [51].
13. Store solution at −20 °C.

3.3 Introducing the Hexammine into Crystals by Soaking: Low Ionic Strength Example

The following protocol is based on a successful example for soaking hexammine cations into crystals grown under relatively low ionic strength conditions [34]. Several parameters must be optimized for each crystal form (*see* **Notes 6** and 7).

1. Grow crystals of the RNA of interest of suitable size and quality for diffraction data collection. For this example, the crystals were grown in low-salt crystallization condition A in a sitting-drop vapor diffusion experiment at 30 °C.
2. Unseal the well containing the crystals and add solution matching the conditions of the drop plus 1 mM hexammine chloride and any cryoprotecting agents (soaking solution) to the crystallization drop. Add enough to double the size of the drop (e.g., add 1 μL soaking solution to a 1 μL drop).

3. Reseal the well and allow the drop and crystals to equilibrate for 5 min.
4. Unseal the well containing the crystals and add 20 μL of the soaking solution.
5. Reseal the well and allow the drop and crystals to equilibrate for 5 min.
6. Unseal the well containing the crystals and remove 15 μL of the well volume, then add 15 μL of the soaking solution.
7. Reseal the well and allow the drop and crystals to equilibrate for 10 min.
8. Harvest the crystals with a cryo-loop and flash-cool in liquid nitrogen.

3.4 Generating Cobalt (III) Hexammine Acetate from the Chloride Salt (See Note 8)

1. Dissolve 0.668 g of cobalt (III) hexammine chloride in 25 mL ultrapure water to generate a solution of 100 mM hexammine.
2. Dissolve 1.42 g lead acetate in 5 mL ultrapure water.
3. Combine the lead acetate and cobalt (III) hexammine solutions in a 50 mL conical tube. Lead chloride will immediately precipitate. Allow 30 min at room temperature for complete precipitation.
4. Centrifuge the conical tube in a clinical centrifuge to pellet the precipitate. Collect the supernatant which contains the hexammine acetate salt and transfer to a clean large plastic weigh boat.
5. Allow the weight boat and solution to sit on the bench until the water has completely evaporated. Dissolve the resultant crystals in 5 mL of ultrapure water, resulting in a solution of cobalt (III) hexammine acetate at ~500 mM.
6. Store at −20 °C.

3.5 Generating Iridium (III) Hexammine Acetate from the Chloride Salt (See Note 8)

This is based on the same procedure used to produce cobalt (III) hexammine acetate. However, because iridium (III) hexammine is not commercially sold and thus not available in large amounts, we adapted the protocol slightly to use smaller amounts of starting material.

1. Dissolve 0.0535 g lead acetate in 0.5 mL of 188 mM iridium (III) hexammine chloride in a micro-centrifuge tube (188 mM was the concentration of our stock of iridium (III) hexammine chloride solution; adjust the amount of lead acetate for other concentrations).
2. Lead chloride will immediately precipitate. Allow 30 min at room temperature for complete precipitation.

3. Centrifuge tube in a micro-centrifuge at maximum velocity to pellet the precipitate. Collect the supernatant which contains the hexammine acetate salt and transfer to a small clean plastic weigh boat.
4. Allow the weigh boat and solution to sit on the bench until the water has completely evaporated. The resultant crystals can then be stored at −20 °C until needed. Before use, dissolve in ultrapure water to the desired concentration.

3.6 Introducing the Hexammine by Soaking: High Salt Example

Crystals grown under higher ionic strength (high salt) present a challenge because the high concentration of cation can compete for the hexammine binding sites, precluding specific binding. This mandates using much higher concentrations of hexammine in the soaking solution; however, because the hexammine chloride salts are only soluble to ~100 mM under these conditions, a suitable concentration may not be achievable. In addition, hexammine sulfates are relatively insoluble and thus crystals grown under high sulfate conditions have the additional complication that the hexammine cations precipitate when added to the crystallization drop. The following is a protocol that was used to overcome these challenges by replacing the sulfate in the crystallization drop with acetate and using hexammine acetate salts [35]. Again, several parameters must be optimized (*see* **Note 6**).

1. Grow crystals of the RNA of interest of suitable size and quality for diffraction data collection. For this example, the crystals were grown against a well containing high-salt crystallization solution B in a hanging-drop vapor diffusion experiment at 30 °C.
2. To stabilize the crystals, unseal the well, remove the well solution, and replace with stabilizing solution.
3. Reseal the well and allow the drop to equilibrate for 24 h at 30 °C.
4. Unseal the well and transfer the crystals to a second soaking tray containing 50 μL of the well solution. This is most readily done using a cryo-loop.
5. Cover/seal the soaking tray and allow the drop and crystals to equilibrate for 10 min.
6. Uncover the soaking tray and add 50 μL of soaking a solution made with 10 % derivative solution + 90 % stabilizing solution.
7. Reseal the well and allow the drop and crystals to equilibrate for 10 min.
8. Unseal the well and remove 50 μL of the solution, then immediately add 50 μL of a solution made with 20 % derivative solution + 80 % stabilizing solution.

9. Reseal the well and allow the drop and crystals to equilibrate for 10 min.
10. Repeat **steps 8** and **9**, each time increasing the percentage of derivative solution by 10 % until the crystals are in essentially a 100 % derivative solution (ten incremental soaks).
11. Unseal the well and remove most of the solution surrounding the crystals, then add 100 μL of 100 % derivative solution. Repeat.
12. Reseal the well and allow the drop and crystals to equilibrate for 1 h.
13. Harvest the crystals with a cryo-loop and flash-cool in liquid nitrogen.

4 Notes

1. Even if the G-U motif is not included in the RNA, soaking with hexammine cations can often lead to a derivative suitable for phasing. There are endogenous hexammine sites in many RNA structures that are robust enough for this purpose. However, inclusion of the engineered cation binding G-U motif ensures that a strong high-occupancy site is present. This, in turn, makes weaker endogenous sites easier to locate and use in phasing.
2. The directed soaking method we describe here was originally designed to work with hexammine cations, but other cations are also likely to be useful. This is because the G-U motif is a fairly general cation binding site. Indeed, there is one example of cesium being used in conjunction with the motif for successful phasing [43]. We have not, however, made an exhaustive survey of various cations.
3. If crystals of the RNA of interest have been obtained using RNA that does not contain the G-U motif for cation binding, the same or very similar crystallization conditions are likely to be successful in crystallizing the RNAs that contain the motif. Inclusion of the motif, in our experience, does not significantly alter the conditions under which an RNA crystallizes or the quality of the crystals.
4. Protocols for crystallizing various RNAs are beyond the scope of this chapter. Discussions of the many strategies and protocols used to crystallize RNAs can be found in this book and other useful reviews [50, 52, 53].
5. A difficulty commonly encountered with the Ace pressure tube is that the FETFE O-ring seal is not compatible with ammonium hydroxide at high temperature and rapidly degrades.

This is resolved by wrapping the O-ring in Teflon tape prior to sealing the tube. An alternative and more expensive option is a Teflon-lined hydrothermal synthesis reactor (Changshu Deschem Chemical Glassware Equipment Factory, Changshu, China). The advantage to the use of this vessel is that it is simpler to use and the seal is completely lined with inert materials.

6. Parameters that may vary for different crystal forms include:
   (a) Concentration of the hexammine cation.
   (b) Time of soaking.
   (c) Number of "steps" in the procedure; i.e., is the cation concentration increased in steps and if so, how many?
   (d) Concentration and identity of any included cryoprotectants.
7. Here, we present procedures for soaking cations into the crystal after they have grown. It is also possible to co-crystallize the hexammine with the RNA [27]. If the initial crystals were grown without hexammine, then including them in the drop during crystallization may mandate some re-optimization of the crystallization conditions to account for the increased ionic strength association with the trivalent hexammine cation.
8. The acetate salts of hexammines are more soluble than the chloride salts, and the sulfate salts are insoluble. For most applications, the chloride salts are acceptable as the hexammine concentration used is well within the range of solubility. However, for some applications higher concentrations of hexammines may be needed. The example presented here is one where the crystal was grown under very high salt conditions and the hexammine concentration must be high to compete for the cation binding sites [35].

Acknowledgements

The authors thank current and former members of our labs for thoughtful discussions and technical assistance and David Costantino for critical reading of this manuscript. R.T.B. is supported by NIH grants GM073850 and GM083953. J.S.K. is supported by NIH grants GM097333 and GM081346. J.S.K. is an Early Career Scientist of the Howard Hughes Medical Institute.

References

1. McCoy AJ, Grosse-Kunstleve RW, Adams PD, Winn MD, Storoni LC, Read RJ (2007) Phaser crystallographic software. J Appl Crystallogr 40:658–674
2. McCoy AJ (2007) Solving structures of protein complexes by molecular replacement with Phaser. Acta Crystallogr D Biol Crystallogr 63:32–41

3. Doublie S (1997) Preparation of selenomethionyl proteins for phase determination. Methods Enzymol 276:523–530
4. Hendrickson WA, Horton JR, LeMaster DM (1990) Selenomethionyl proteins produced for analysis by multiwavelength anomalous diffraction (MAD): a vehicle for direct determination of three-dimensional structure. EMBO J 9: 1665–1672
5. Robertson MP, Scott WG (2007) The structural basis of ribozyme-catalyzed RNA assembly. Science 315:1549–1553
6. Robertson MP, Scott WG (2008) A general method for phasing novel complex RNA crystal structures without heavy-atom derivatives. Acta Crystallogr D Biol Crystallogr 64: 738–744
7. Robertson MP, Chi YI, Scott WG (2010) Solving novel RNA structures using only secondary structural fragments. Methods 52:168–172
8. Brandt G, Carrasco N, Huang Z (2006) Efficient substrate cleavage catalyzed by hammerhead ribozymes derivatized with selenium for X-ray crystallography. Biochemistry 45: 8972–8977
9. Golden BL (2000) Heavy atom derivatives of RNA. Methods Enzymol 317:124–132
10. Jiang J, Sheng J, Carrasco N, Huang Z (2007) Selenium derivatization of nucleic acids for crystallography. Nucleic Acids Res 35: 477–485
11. Carrasco N, Buzin Y, Tyson E, Halpert E, Huang Z (2004) Selenium derivatization and crystallization of DNA and RNA oligonucleotides for X-ray crystallography using multiple anomalous dispersion. Nucleic Acids Res 32: 1638–1646
12. Hobartner C, Micura R (2004) Chemical synthesis of selenium-modified oligoribonucleotides and their enzymatic ligation leading to an U6 SnRNA stem-loop segment. J Am Chem Soc 126:1141–1149
13. Hobartner C, Rieder R, Kreutz C, Puffer B, Lang K, Polonskaia A, Serganov A, Micura R (2005) Syntheses of RNAs with up to 100 nucleotides containing site-specific 2'-methylseleno labels for use in X-ray crystallography. J Am Chem Soc 127:12035–12045
14. Salon J, Sheng J, Jiang J, Chen G, Caton-Williams J, Huang Z (2007) Oxygen replacement with selenium at the thymidine 4-position for the Se base pairing and crystal structure studies. J Am Chem Soc 129:4862–4863
15. Sheng J, Jiang J, Salon J, Huang Z (2007) Synthesis of a 2'-Se-thymidine phosphoramidite and its incorporation into oligonucleotides for crystal structure study. Org Lett 9: 749–752
16. Baugh C, Grate D, Wilson C (2000) 2.8 A crystal structure of the malachite green aptamer. J Mol Biol 301:117–128
17. Kieft JS, Zhou K, Grech A, Jubin R, Doudna JA (2002) Crystal structure of an RNA tertiary domain essential to HCV IRES-mediated translation initiation. Nat Struct Biol 9: 370–374
18. Martick M, Scott WG (2006) Tertiary contacts distant from the active site prime a ribozyme for catalysis. Cell 126:309–320
19. Golden BL, Gooding AR, Podell ER, Cech TR (1996) X-ray crystallography of large RNAs: heavy-atom derivatives by RNA engineering. RNA 2:1295–1305
20. Wedekind JE, McKay DB (2000) Purification, crystallization, and X-ray diffraction analysis of small ribozymes. Methods Enzymol 317: 149–168
21. Keel AY, Rambo RP, Batey RT, Kieft JS (2007) A general strategy to solve the phase problem in RNA crystallography. Structure 15: 761–772
22. Masquida B, Westhof E (2000) On the wobble GoU and related pairs. RNA 6:9–15
23. Varani G, McClain WH (2000) The G x U wobble base pair. A fundamental building block of RNA structure crucial to RNA function in diverse biological systems. EMBO Rep 1:18–23
24. Kieft JS (1997). Structure and thermodynamics of a metal ion binding site in the RNA major groove: cobalt (III) hexammine as a probe. [Thesis]. Type, University of California, Berkeley, USA
25. Cate JH, Doudna JA (1996) Metal-binding sites in the major groove of a large ribozyme domain. Structure 4:1221–1229
26. Colmenarejo G, Tinoco I Jr (1999) Structure and thermodynamics of metal binding in the P5 helix of a group I intron ribozyme. J Mol Biol 290:119–135
27. Montange RK, Batey RT (2006) Structure of the S-adenosylmethionine riboswitch regulatory mRNA element. Nature 441:1172–1175
28. Stefan LR, Zhang R, Levitan AG, Hendrix DK, Brenner SE, Holbrook SR (2006) MeRNA: a database of metal ion binding sites in RNA structures. Nucleic Acids Res 34:D131–D134
29. Batey RT, Gilbert SD, Montange RK (2004) Structure of a natural guanine-responsive riboswitch complexed with the metabolite hypoxanthine. Nature 432:411–415

30. Cate JH, Gooding AR, Podell E, Zhou K, Golden BL, Kundrot CE, Chech TR, Doudna JA (1996) Crystal structure of a group I ribozyme domain: principles of RNA packing. Science 273:1678–1685
31. Cochrane JC, Lipchock SV, Strobel SA (2007) Structural investigation of the GlmS ribozyme bound to its catalytic cofactor. Chem Biol 14: 97–105
32. Kazantsev AV, Krivenko AA, Harrington DJ, Holbrook SR, Adams PD, Pace NR (2005) Crystal structure of a bacterial ribonuclease P RNA. Proc Natl Acad Sci U S A 102: 13392–13397
33. Toor N, Keating KS, Taylor SD, Pyle AM (2008) Crystal structure of a self-spliced group II intron. Science 320:77–82
34. Pfingsten JS, Costantino DA, Kieft JS (2006) Structural basis for ribosome recruitment and manipulation by a viral IRES RNA. Science 314:1450–1454
35. Costantino DA, Pfingsten JS, Rambo RP, Kieft JS (2008) tRNA-mRNA mimicry drives translation initiation from a viral IRES. Nat Struct Mol Biol 15:57–64
36. Peselis A, Serganov A (2012) Structural insights into ligand binding and gene expression control by an adenosylcobalamin riboswitch. Nat Struct Mol Biol 19:1182–1184
37. Smith KD, Lipchock SV, Ames TD, Wang J, Breaker RR, Strobel SA (2009) Structural basis of ligand binding by a c-di-GMP riboswitch. Nat Struct Mol Biol 16:1218–1223
38. Garst AD, Heroux A, Rambo RP, Batey RT (2008) Crystal structure of the lysine riboswitch regulatory mRNA element. J Biol Chem 283:22347–22351
39. Huang L, Serganov A, Patel DJ (2010) Structural insights into ligand recognition by a sensing domain of the cooperative glycine riboswitch. Mol Cell 40:774–786
40. Ban N, Nissen P, Hansen J, Moore PB, Steitz TA (2000) The complete atomic structure of the large ribosomal subunit at 2.4 A resolution. Science 289:905–920
41. Cate JH, Yusupov MM, Yusupova GZ, Earnest TN, Noller HF (1999) X-ray crystal structures of 70S ribosome functional complexes. Science 285:2095–2104
42. Clemons WM Jr, May JL, Wimberly BT, McCutcheon JP, Capel MS, Ramakrishnan V (1999) Structure of a bacterial 30S ribosomal subunit at 5.5 A resolution. Nature 400: 833–840
43. Gilbert SD, Rambo RP, Van Tyne D, Batey RT (2008) Structure of the SAM-II riboswitch bound to S-adenosylmethionine. Nat Struct Mol Biol 15:177–182
44. Edwards AL, Reyes FE, Heroux A, Batey RT (2010) Structural basis for recognition of S-adenosylhomocysteine by riboswitches. RNA 16:2144–2155
45. Ren A, Rajashankar KR, Patel DJ (2012) Fluoride ion encapsulation by Mg2+ ions and phosphates in a fluoride riboswitch. Nature 486:85–89
46. Serganov A, Huang L, Patel DJ (2009) Coenzyme recognition and gene regulation by a flavin mononucleotide riboswitch. Nature 458:233–237
47. Johnson JE Jr, Reyes FE, Polaski JT, Batey RT (2012) B12 cofactors directly stabilize an mRNA regulatory switch. Nature 492: 133–137
48. Kieft JS, Costantino DA, Filbin ME, Hammond J, Pfingsten JS (2007) Structural methods for studying IRES function. Methods Enzymol 430:333–371
49. Edwards AL, Garst AD, Batey RT (2009) Determining structures of RNA aptamers and riboswitches by X-ray crystallography. Methods Mol Biol 535:135–163
50. Reyes FE, Garst AD, Batey RT (2009) Strategies in RNA crystallography. Methods Enzymol 469:119–139
51. Galsbøl FH, Simonsen K (1990) The preparation, separation and characterization of some ammine complexes or Iridium(III). Acta Chem Scand 44:796–801
52. Golden BL (2007) Preparation and crystallization of RNA. Methods Mol Biol 363: 239–257
53. Golden BL, Kundrot CE (2003) RNA crystallization. J Struct Biol 142:98–107

Chapter 15

Using Molecular Replacement Phasing to Study the Structure and Function of RNA

Marco Marcia

Abstract

In recent years a wide variety of RNA molecules regulating fundamental cellular processes has been discovered. Therefore, RNA structure determination is experiencing a boost and many more RNA structures are likely to be determined in the years to come. The broader availability of experimentally determined RNA structures implies that molecular replacement (MR) will be used more and more frequently as a method for phasing future crystallographic structures. In this report we describe various aspects relative to RNA structure determination by MR. First, we describe how to select and create MR search models for nucleic acids. Second, we describe how to perform MR searches on RNA using available crystallographic software. Finally, we describe how to refine and interpret the successful MR solutions. These protocols are applicable to determine novel RNA structures as well as to establish structural-functional relationships on existing RNA structures.

Key words Nucleic acid sequence homology, De novo structure design, Long noncoding RNA, RNA structure, Homology modeling

1 Introduction

Since the determination of the three-dimensional structure of $tRNA^{Phe}$ [1], the rate of RNA structure determination has experienced a continuous growth. In the past decade the number of new RNA structures deposited in public databases has increased exponentially and in 2013 the 1000th RNA structure has been determined. About 55 % of these structures were determined by X-ray crystallography. The total number of RNA structures still represents only about 1 % of the total number of available protein structures, but the trend shows that RNA crystallography is rapidly expanding. The increasing success in RNA crystallography is also related to the fact that a wide variety of new RNA molecules is currently being discovered, enlarging the pool of attractive RNA targets for structural studies. Therefore, it can be expected that many more RNA structures will be determined in the future [2–4].

Eric Ennifar (ed.), *Nucleic Acid Crystallography: Methods and Protocols*, Methods in Molecular Biology, vol. 1320,
DOI 10.1007/978-1-4939-2763-0_15, © Springer Science+Business Media New York 2016

The crystallographic phase problem in RNA crystallography can be overcome using the same strategies as for protein crystallography [5], i.e., using isomorphous replacement (SIR, MIR, SIRAS, MIRAS), anomalous dispersion (SAD, MAD), or molecular replacement (MR). MR is the most widely used method. Statistics show that about half of the available RNA structures have been solved by MR [6]. In the future, considering the constant increase in the number of RNA structures—and thus of potential MR search models—it can be expected that the use of MR will be even wider [7].

General methods to solve macromolecular structures by MR have been described in previous chapters in this series, with particular focus on protein structure determination [8]. Although the theoretical principles are identical, MR strategies applied to RNA structure determination present practical differences in respect to those used for solving protein structures [6]. Some software is currently unsuitable to process RNA structures, structural homology in RNAs follows different principles than in proteins, and RNA MR search models should be selected according to different principles than protein models [6]. Most importantly, MR approaches to RNA structure determination has broad applications that extend beyond the typical applications of MR approaches to protein structure determination. MR has been used to determine novel RNA structures [9], to map ion-binding sites in RNA molecules [10, 11], to visualize RNAs in different structural conformations [11], to reveal structural differences across series of RNA mutants [12], to solve structures of RNA multisubunit complexes [13, 14], and to study ligand binding to RNA [15]. Specifically in our lab, we have used MR to determine structures of the group II intron, a highly structured ribozyme capable of self-splicing and retrotransposition [16]. We could identify 74 site-bound ions and 5 different catalytic conformations of this large ribozyme [11, 17]. Our experimental approach constitutes an illustrative example of how MR can support studies on RNAs.

This report describes the details of our experimental protocols, and extends beyond our specific applications to provide a broader overview in support of studies on other RNA molecules besides the group II intron.

2 Materials

Most modern Unix-like operating computers are suitable for performing the calculations required to phase and interpret RNA structures by MR. Most applications can also be run on Mac or—in fewer instances—on Windows operating computers. In our lab we perform crystallographic calculations on a 64-bit computer operating Ubuntu 11.10 and equipped with an Intel core i7-2600

CPU, with 8 3.40-GHz processors, and with an 8-GB RAM module. Accessibility to all software packages is managed by the SBGrid consortium (http://www.sbgrid.org/) (*see* **Note 1**).

3 Methods

RNA structure determination by MR results from the successful accomplishment of sequential steps including (1) obtaining high quality diffraction data, (2) selecting and preparing a suitable MR search model, (3) performing the MR search, and (4) interpreting and refining the output solution (Fig. 1).

3.1 Preparation of High-Quality Diffraction Data Sets for the Target

The first essential piece of data to input into a MR search is a high quality dataset containing the structure factors of the target. To obtain such dataset, the steps described below should be considered. For detailed protocols of each experimental step, references to previous chapters of this series are indicated.

1. Prepare a highly pure and homogeneous sample of the target RNA. Methods to prepare pure and homogeneous RNA samples for crystallization were described by Ferre-D'Amare

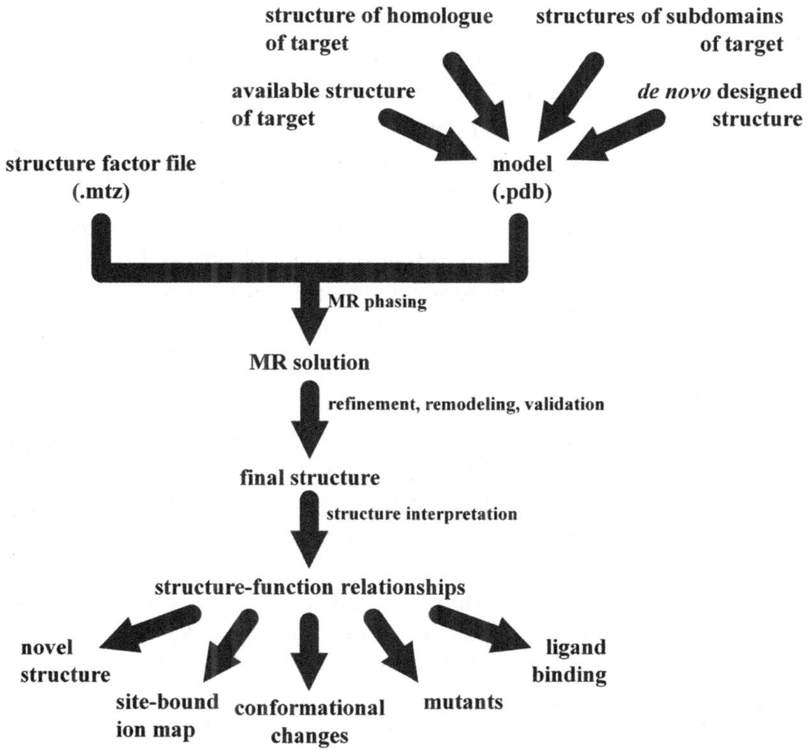

Fig 1 Flowchart of a typical MR approach for RNA structure determination

and Doudna [18]. Specifically, in our lab we purify the group II intron following a non-denaturing purification method [19], which has been observed to yield higher conformational homogeneity in large RNA molecules [20].

2. Crystallize the target RNA. Methods describing RNA crystallization were previously described by Doudna and Golden [21, 22]. Specifically, crystals of the group II intron were grown by hanging drop vapor diffusion using 2 μL sample drops and 300 μL crystallization solution at 30 °C, using EasyXtal 15-Well Tool plates with a white O-ring (*see* **Note 2**).
3. Harvest the crystals of the target RNA. A detailed protocol for crystal harvesting has been described by Garman and Owen [23]. Specifically, crystals of the group II intron were harvested after growing as elongated rods reaching dimensions of (20–50) × (50–100) × (100–400) μm^3 [11]. Crystals were flash-frozen under a 100 K cryostream at an in-house X-ray generator (MicroMax 007 HF source equipped with the AFC-11 4-axis partial χ goniometer and the Saturn 944^+ CCD detector), stored in liquid nitrogen and shipped to synchrotron facilities in dry-shippers.
4. Collect diffraction data of the target RNA. Synchrotron sources are nowadays most suitable for data acquisition, providing high beam intensity, rapid measuring time, high detection sensitivity, and sophisticated automation. Procedures for semi-automated data acquisition at synchrotron sources were described extensively in Garman and Sweet [24] and in Arzt et al. [25]. Specifically, diffraction datasets of the group II intron crystals were collected at beamlines 24ID-C/E at the Advanced Photon Source (APS, Argonne, USA, http://necat.chem.cornell.edu/) using a 50 % attenuated, 30–70 μm-wide X-ray beam. Helical data collection [26] along the longest crystal dimension was the most successful acquisition mode, guaranteeing minimal radiation damage and allowing for the acquisition of highly complete (>95 %) and redundant (>3-fold) datasets.
5. Process the collected diffraction data. Indexing, integration and scaling can be performed manually using available software, automatically using data processing pipelines, or in situ using pipelines developed and accessible at the various synchrotron beamlines.

   For manual processing four major software packages are available: XDS (X-ray Detector Software [27], http://xds.mpimf-heidelberg.mpg.de/html_doc/XDS.html), Mosflm ([28], http://www.mrc-lmb.cam.ac.uk/harry/mosflm/), D*Trek ([29], http://pdfs.mbc-als.org/dtrek_overview99.pdf), and HKL2000 ([30], http://www.hkl-xray.com/sites/default/

files/HKL2000manual/index.htm). Detailed instructions on the use of each package are reported in the corresponding references and in the indicated online manuals. A typical script used for running XDS [27] to solve structures of the group II intron [11] is reported in Table 1.

For processing through automated pipelines, xia2 is recommended ([31], http://www.ccp4.ac.uk/newsletters/newsletter48/articles/Xia2/manual.pdf).

Finally, pipelines developed at various beamlines worldwide include: RAPD (The Rapid Automated Processing of Data) at APS, Argonne, USA (https://rapd.nec.aps.anl.gov/rapd/); ISPyB (Information System for Protein CrystallographY Beamlines) at ESRF, Grenoble, France [32]; *AutoProcess* at the Canadian Light Source, Saskatchewan, Canada (http://cmcf.lightsource.ca/user-guide/autoprocess/); XDSME (XDS Made Easier) at SOLEIL, Gif-sur-Yvette, France (http://code.google.com/p/xdsme/); go.com at SLS, Villigen, Switzerland (http://www.psi.ch/sls/pxii/computing); autoxds at SSRL, Menlo Park, USA (http://smb.slac.stanford.edu/facilities/software/xds/#autoxds_script); and *fast_dp* at Diamond Light Source, Didcot, UK (http://diamond.ac.uk/mx-home/Common/Common-Manual/Data-Analysis/Automated-Software-Pipeline.html).

3.2 Preparation of the MR Search Model

After obtaining a high quality dataset for the target RNA, the choice of an appropriate MR search model is the most critical step in determining the success of the structure determination procedure [33]. The structure of such model should closely resemble the structure of the target (RMSD < 2.5 Å) [33]. Since in most cases the structural similarity cannot be determined a priori, a trial and error process needs to be followed. The guidelines below indicate how to prepare MR search models starting from (1) available structures of the target, (2) available structures of homologues of the target, (3) available structures of subdomains of the target, or (4) de novo designed, idealized segments of RNA [6].

3.2.1 Using an Available Structure of the Target

1. Download the coordinates of the model structure from public databases, i.e., the Protein Data Bank (PDB, www.pdb.org) or the Nucleic Acid Database (NDB, http://ndbserver.rutgers.edu/), and save it in the local working directory.
2. Prepare the coordinate file of the model by removing solvent molecules and ligands, using a text editor.
3. If multiple copies of the target molecule are present in the asymmetric unit of the structure, or if multiple structures of the target molecule are available, select the one that shows higher occupancy of all residues, lowest B-factors, smallest

Table 1
XDS scripts

```
!=================== XDS.INP ===========================
!
JOB=ALL !XYCORR INIT COLSPOT IDXREF DEFPIX INTEGRATE CORRECT
!
!============ DETECTOR PARAMETERS ====================
!
DETECTOR=ADSC
MINIMUM_VALID_PIXEL_VALUE=1 OVERLOAD=65535
DIRECTION_OF_DETECTOR_X-AXIS=1.0 0.0 0.0
DIRECTION_OF_DETECTOR_Y-AXIS=0.0 1.0 0.0
TRUSTED_REGION=0.0 1.00 !Trust to edge of detector
NX=3072 NY=3072 QX=0.10259 QY=0.10259
!
!=========== PROCESSING CONTROL PARAMETERS ============
!
MAXIMUM_NUMBER_OF_PROCESSORS=8
!
!============== GEOMETRICAL PARAMETERS ================
!
ORGX=1521.19309874 ORGY=1536.04737304 !Beam center in pixels
DETECTOR_DISTANCE=400.0 !(mm)
ROTATION_AXIS= 1.0 0.0 0.0
OSCILLATION_RANGE=0.5!degrees(>0)
X-RAY_WAVELENGTH=0.97949!Angstroem
INCIDENT_BEAM_DIRECTION=0.0 0.0 1.0
FRACTION_OF_POLARIZATION=0.95
POLARIZATION_PLANE_NORMAL=0.0 1.0 0.0
!
!========= PARAMETERS CONTROLLING REFINEMENTS =========
!
REFINE(IDXREF)=BEAM AXIS ORIENTATION CELL !DISTANCE
REFINE(INTEGRATE)=BEAM ORIENTATION CELL !DISTANCE AXIS
REFINE(CORRECT)=DISTANCE BEAM ORIENTATION CELL AXIS
FRIEDEL'S_LAW=TRUE
STRICT_ABSORPTION_CORRECTION=TRUE
!
!========= CRITERIA FOR ACCEPTING REFLECTIONS =========
!
INCLUDE_RESOLUTION_RANGE=50.0 2.65 !Angstroems; Used by
!DEFPIX,INTEGRATE,CORRECT
!
!============ SELECTION OF DATA IMAGES ================
!
NAME_TEMPLATE_OF_DATA_FRAMES=../crystal1_???.img
MAXIMUM_NUMBER_OF_PROCESSORS=8
MAXIMUM_NUMBER_OF_JOBS=14
BACKGROUND_RANGE=1 5
DATA_RANGE=1 140
SPACE_GROUP_NUMBER=19  !0 for unknown crystals; cell constants are ignored.
UNIT_CELL_CONSTANTS=89.95    95.58   226.40  90.000  90.000  90.000
```

(continued)

Table 1
(continued)

```
!====================== XSCALE.INP ==========================
!
!===================== DATA FILES ===========================
!
OUTPUT_FILE=crystal1.ahkl
INPUT_FILE= ./XDS_ASCII.HKL
INCLUDE_RESOLUTION_RANGE=50 2.65
!
!====================== PARAMETERS ==========================
!
!MAXIMUM_NUMBER_OF_PROCESSORS=16
RESOLUTION_SHELLS= 12 10 8 5 3.5 3.2 3.0 2.9 2.8 2.7 2.65
SPACE_GROUP_NUMBER=19
UNIT_CELL_CONSTANTS=89.95    95.58   226.40  90.000  90.000  90.000

!======================= XDSCONV.INP ==============================
!
!======================= INPUT FILE ==============================
!
INPUT_FILE=crystal1.ahkl    XDS_ASCII !UNIQUE OLDHKL ANOMAL NORMAL
!
!====================== OUTPUT FILE ==============================
!
OUTPUT_FILE=crystal1.ccp4asc CCP4 !CCP4_F CCP4_I SHELX XtalView
!
!==================== OTHER PARAMETERS ==========================
SPACE_GROUP_NUMBER=19  !if omitted taken from header of input file
UNIT_CELL_CONSTANTS=89.95    95.58   226.40  90.000  90.000  90.000
FRIEDEL'S_LAW=TRUE
INCLUDE_RESOLUTION_RANGE=50 2.65

!================= F2MTZ.INP =======================
!
TITLE XDS to MTZ
FILE crystal1.ccp4asc
SYMMETRY   19
CELL      89.95    95.58   226.40  90.000  90.000  90.000
LABOUT  H K L FP SIGFP
CTYPOUT H H H  F   Q
END

!============== XDSMTZ.COM =================
!================ F2MTZ ===================
!
f2mtz HKLOUT temp.mtz <F2MTZ.INP
!
!================= CAD ====================
cad HKLIN1 temp.mtz HKLOUT crystal1.mtz<<EOF
LABIN FILE 1 ALL
END
EOF
```

Scaled intensities were converted to amplitudes using XDSCONV and the file format was converted to .mtz in f2mtz in CCP4 [35] for further processing number of non-modeled residues. Avoid selecting molecules that may be distorted due to crystal packing forces [34].

4. If the available structure was obtained by NMR, the whole NMR ensemble can be used as a model, i.e., in Phaser [35, 36].

3.2.2 Using the Structure of a Homologue of the Target

Structural homology in RNAs follows different principles than structural homology in proteins, primarily because the structure of RNA helical motifs is maintained when the nucleotides in the partner strands co-vary [6]. Consequently, RNA molecules may possess similar tertiary structure despite possessing very low sequence identity. Therefore, identifying RNA homologues is more successful using structure-similarity alignment algorithms, such as CMfinder [37], Infernal [38, 39], or LocARNA [40], rather than sequence-similarity algorithms used for proteins [6]. After identifying a homologue of known structure for the target RNA, such structure can be prepared to be used as a MR search model. The following considerations should be taken into account.

1. Inspect the alignment of target and model to identify regions of discrepancy. Such discrepancies can be represented by stems or loops forming tertiary interactions that may be present in the model but not in the target, or vice versa [41].
2. Inspect the structure of the model visually, to identify regions involved in crystal contacts and surface regions that may be flexible (B factor higher than the average) or distorted by crystallographic packing forces.
3. Based on the above considerations, prune the selected model if necessary. First, aim to reduce the discrepancy between model and target. For instance, in a text editor remove from the model structure the coordinates of loops or stems that are not present in the target. Second, remove regions of high flexibility (high B factors) or subject to unnatural constraints due to crystal packing.
4. If initial MR searches are unsuccessful, prune the coordinates further by creating minimal search models, following an approach that resembles the use of polyalanine [42], polyserine [43, 44], or polyglycine [45, 46] models in protein structure determination. The identity of the individual nucleotides is less important than that of individual amino acids in proteins [6]. The contribution provided by the nucleobase moieties of the model to the structure factor is minimal in respect to the contribution provided by the sugar–phosphate backbone [6]. Therefore, such minimal RNA models can be composed, that is, of the phosphate groups or of the sugar–phosphate backbone of the initial model. Such selections can be created using text editors. One alternative convenient method is to use the software PyMOL [47] and the script reported in Table 2.

Table 2
PyMOL script used to create minimal search models of RNA coordinates

```
###
### Load the model
###
load .\model.pdb
###
### Create a selection of phosphate groups only
###
select p, model and chain A and name P+OP1+OP2+O3'+O5'
cmd.create(None,"p",zoom=0)
cmd.delete("p")
set_name obj01, phosphates
###
### Create a selection of sugar-phosphate backbone
###
select back, model and chain A and name
P+OP1+OP2+O3'+O5'+C1'+C2'+C3'+C4'+C5'+O2'+O4'
cmd.create(None,"back",zoom=0)
cmd.delete("back")
set_name obj02, backbone
###
### Create a selection of nucleobases only
###
select base, model and chain A and name
N1+C2+O2+N3+C4+O4+C5+C6+N4+N6+N7+C8+N9+N2+O6
cmd.create(None,"base",zoom=0)
cmd.delete("base")
set_name obj03, bases
###
### Save the files
###
save .\model_P.pdb, phosphates
save .\model_backbone.pdb, backbone
save .\model_bases.pdb, bases
save .\P-bb-bases.pse,format=pse
```

3.2.3 Using Structures of Domains of (Homologues of) the Target

Multi-domain MR searches are often used in protein crystallography, when rigid-body motions amongst different domains of the target molecule hamper structure determination (*see* **Note 3**). Similar multi-domain searches are also possible for RNA. However, the prediction of the domain composition of RNA molecules is performed differently than for proteins. In RNA, structural domains can be defined by calculating the "helical diagram" [48]—often referred to as secondary structure—of the molecule computationally or experimentally. Computational predictions can be performed with various online accessible software, e.g., mFOLD [49] or ViennaRNA [50]. By contrast, experimental helical diagrams—which are generally more accurate—can be obtained by various probing methods, i.e., SHAPE (selective 2′-hydroxyl acylation analyzed by primer extension [51]), in-line probing [52], hydroxyl radical footprinting [53], enzymatic probing [54], chemical probing [55], and other higher-throughput methods [56]. The references indicated above correspond to previous chapters in this series and report detailed protocols on how to perform such experiments.

After defining the secondary structure of the target, the following considerations should be taken into account in order to prepare models of the individual domains to input in the MR search (*see* **Note 4**).

1. Identify domains of the target based on the position of multi-way junctions derived from the secondary structure maps. The domains can be defined in a hierarchical manner. For instance, the group II intron possesses a 6-way junction connecting its six primary domains (D1–6). Additionally, within D1 a 5-way junction delimits five subdomains (D1i–ii, D1a, D1b, D1c, D1d). Finally, each subdomain can be further subdivided into smaller modules by internal loops or two-way junctions (i.e., the coordination loop separating D1d1 and D1d2, the D1i-loop separating D1i and D1ii, the EBS2 junction separating modules of D1d2 in class IIA/IIB introns) [16].
2. If structures of the domains identified in the previous step are available, these can serve as MR search models.
3. Prune the coordinates of such domains following similar considerations described above (*see* Subheading 3.2.2). Exclude inter-domain junction regions or any poorly ordered segment, which may possess high flexibility.
4. If necessary, create minimal models for each domain to be used in the MR search, following the guidelines above. Create poly-phosphate or sugar–phosphate backbone models by removing the coordinates of the nucleobases in PyMOL.
5. Keep the number of search models to a minimum. For instance, coaxially stacked helices can generally be considered as single search ensembles. Alternatively, if necessary, split into individual modules the coordinates of subdomains, whose helical arrangement is broken by internal loops or 2-way junctions.

3.2.4 Using De Novo Designed Models

Helical diagrams may reveal the presence of domains for which no experimental structure is available (*see* **Note 5**). Alternatively, the use of available structures may result in an unsuccessful MR search. In these cases, MR can still be attempted using de novo designed models [6]. Such approach was used successfully to determine the structure of the 70-nucleotide-long L1-ligase ribozyme [9, 57–59]. The experimental approach was described comprehensively in Robertson et al. [57]. Briefly, the authors followed the steps indicated below.

1. Draw ideal RNA motifs mimicking individual helical stems of the target. Such idealized helices can be drawn in the software Coot [60] using the "Ideal RNA/DNA…" option on the "Other Modelling Tools" dialog box in the main tab "Calculate". For a given sequence, choices are available to build DNA or RNA, A or B form helices, and single or double strands. Note that RNA helices are generally of the A form [59]. Coaxially stacked helices may be modeled as single stems, in order to reduce the number of input search models. If appropriate and according to the secondary structure map of the target, the helical stems designed in Coot may be capped by simple loop structures, i.e., GNRA tetraloops. Coordinates for such loops can be extracted from available RNA structures downloaded from public databases. At this stage, the sum of all selected helical segments may represent only a small portion of the target molecule (50 % of the total scattering or less). Such "undersampling" is not necessarily problematic, and may actually enhance the success of the MR search [57].
2. Obtain a starting MR solution in Phaser [36] (*see* Subheading 3.2.6). Inspect and edit it in Coot [60] (*see* Subheading 3.2.9). Refine this initial solution using Refmac5 [61] (*see* Subheading 3.2.10).
3. Reiterate the steps above to include new idealized helical motifs until no more helical fragments can be added.
4. Discard the model, convert the calculated phase probability distributions to Hendrickson–Lattman coefficients within CCP4 [35] and import them into CNS [62] or Phenix [63] to generate a solvent-flattened electron density map.
5. Use the resulting map for manual model building without considering information from the previously used idealized helical models. Examine the resulting structure against composite omit maps and generate refined phases by simulated annealing (*see* Subheading 3.2.11).

3.2.5 Structure Phasing

Various software packages are available to perform MR searches, e.g., Phaser [36, 64], AMoRe [65], BALBES [66], MolRep [67, 68], Auto-Rickshaw [69], Bias Removal Server [70], CaspR [71], BRUTEPTF [72], MR pipeline [73], EPMR [74], Queen of Spades [75], SOMoRe [76], COMO [77]. With few exceptions (e.g., BALBES), most packages are suitable to perform MR searches on nucleic acids. In our lab, we routinely use Phaser [36] via the CCP4 [35] or Phenix [63] interfaces or MolRep [67, 68] *via* the CCP4 interface [35] or via command scripts. For technical aspects relative to the use of specific software packages we refer to previous issues of this and other book series (indicated by the references above), where such packages are described. Here, we report details of our experimental protocol to use Phaser and MolRep.

3.2.6 Phaser

We use Phaser for automated MR searches using structure factors and models prepared as described above (*see* Subheadings 3.1 and 3.2). We follow the steps below.

1. User-friendly graphical interfaces (GUI) for Phaser can be found in the CCP4 GUI under "Program List" → "Phaser MR" or in the Phenix GUI under "Molecular Replacement" → "Phaser-MR". The following instructions apply to the CCP4 GUI.
2. In the title section, input a title for the search (optional) and select "Phaser mode" → "automated search". Options are also available to perform separate rotation or translation searches or to perform an analysis of the cell content and packing only.
3. In the "Define data" folder, input the structure factor file for the target RNA, collected as described above (*see* Subheading 3.1). Run Phaser using the space group from the structure factor file (the .mtz file), or, alternatively, use the option "mtz space group and enantiomorph" if a bias remains in the definition of the space group after data integration and scaling.
4. In the "Define ensembles" folder, introduce the coordinates of the search models selected as described above (*see* Subheading 3.2). For multi-domain searches, introduce the coordinates of each domain/subdomain as an individual ensemble. The estimated similarity of the ensemble to the target structure can be indicated as a percentage of sequence identity, or as an estimate value of root mean square deviation (RMSD) between the structures. The correlation between sequence identity and structural RMSD was established for proteins by Chothia and Lesk [78], but it has not been determined so far for RNA. Therefore, a range of values for the parameter "Similarity of PDB to target structure" should be tested.
5. In the "Define composition of the asymmetric unit" folder, the total scattering should be defined describing the composition of the crystals of the target. We generally define this parameter inputting the total molecular weight of the nucleic acid(s) composing the asymmetric unit. Alternatively, the sequence can be input.
6. In the "Search parameters" folder, define the list of ensembles to be used in the search. In multi-domain searches, the order in which the ensembles are searched may affect the success of the search [33, 79]. The domains of closest structural similarity to the target and the largest domains should be generally searched for first.
7. Finally, in the "Additional parameters" folder, adjust the value for the "packing criterion" option. The 5 % default value can be increased if MR searches fail due to high number of "clashes".

Table 3
Phaser input and output files used for solving structures of the group II intron

```
#=== PHASER COMMAND SCRIPT GENERATED BY CCP4I ===
#
TITLE MR
MODE MR_AUTO
ROOT "~/phaser"
#================= DEFINE DATA =================
HKLIN "./crystal1.mtz"
LABIN  F=FP SIGF=SIGFP
RESOLUTION HIGH 3.0 LOW 50.0
SGALTERNATIVE SELECT HAND
#=============== DEFINE ENSEMBLES ===============
ENSEMBLE ensemble1 &
    PDB "./model.pdb" IDENT 95.0
#============== DEFINE COMPOSITION ==============
COMPOSITION BY ASU
COMPOSITION NUCLEIC MW 125000 NUMBER 1
#============== SEARCH PARAMETERS ==============
SEARCH ENSEMBLE ensemble1 NUMBER 1
#=========== ADDITIONAL PARAMETERS ==============
#=============== OUTPUT CONTROL ================
TOPFILES 1
#============== EXPERT PARAMETERS ===============

#========================= PHASER.SOL ================================
#   MR
SPACEGROUP P 21 21 21
SOLU SET RFZ=19.4 TFZ=64.4 PAK=8 LLG=11272 TFZ==55.4 LLG=12116 TFZ==57.4
SOLU 6DIM ENSE ensemble1 EULER 200.932 0.572 339.173 FRAC 0.53105 1.00377 -
1.49528 BFAC 0.00000
```

8. At the end of the MR search, inspect the solution files (.sol) reported by Phaser. A typical successful solution results in a TFZ score of higher than 8 [80]. Typical Phaser input and .sol files can be found in Table 3.

3.2.7 MolRep

We use MolRep to solve RNA structures for which automated MR searches (i.e., in Phaser, see above) are unsuccessful, but for which weak experimental phases could be obtained from heavy-atom derivatized crystals. Such weak experimental phases can be obtained using protocols described in previous issues of this series [81–83]. To perform phased MR in MolRep, we follow the steps described below.

1. Open MolRep in the CCP4 GUI ("Program List" → "Molrep"). Alternatively, run MolRep from manual script files. A typical Molrep input script can be found in Table 4.
2. In the "Title" folder, input a title for the MolRep search (optional) and do a "search for model in the map" performing

Table 4
MolRep input file for phased molecular replacement

```
#================== MOLREP =====================
#
molrep <<stop
#
#============ Output documentation ============
#
_DOC  Y
#
#== Structure factor with experimental phases ==
#
_FILE_F crystal1_poorphases.mtz
_F  F
_SIGF  SIGF
_PH    PHIB
_END    <--- end of MTZ block
#
#================= Model =======================
#
_FILE_M model.pdb
#
#============ Other parameters =================
#
_NP     200
_NMON 1
_RESMAX 3.4
_END
stop
```

"rotation and translation function". Input the structure factor and phase file obtained by experimental phasing as a .mtz file.

3. Select "use experimental phases from input mtz file".
4. Input the coordinates of the search model in the field "Model in".
5. Set the number of expected molecules composing the asymmetric unit ("_NMON") and the highest resolution edge to include in the calculations ("_RESMAX") (*see* **Note 6**).
6. After a partial model has been positioned successfully by MolRep, such model can be used to improve the initial experimental phases. Using the newly determined phases, MolRep can be run iteratively to position other subdomains of the target.

3.2.8 Structure Refinement, Validation and Interpretation

Structure refinement and validation are important steps to assess and confirm a MR phasing solution [84]. Additionally, it is generally only upon refinement that important information on the target RNA can be obtained, i.e., information on ligands, ions, catalytic states, conformational changes, thus fulfilling the scope of the MR experiment.

Previous protocols in this series described macromolecular refinement procedures in detail [85]. Here, we specifically report procedures that we routinely use in our lab for refining macromolecular structures using Refmac5 [61] and Phenix [63].

3.2.9 Structure Remodeling in Coot

1. After obtaining a MR solution, inspect it visually in Coot [60]. In the σ_A-weighted $F_o - F_c$ electron density map, localize regions that display differences between the model and the phased electron density obtained by MR for the target.
2. Inspect the σ_A-weighted $2F_o - F_c$ electron density map and correct for clashes, nucleotides or chains not residing in regions of strong electron density, frame shifts and differences in the sequence. If the $2F_o - F_c$ electron density map is sufficiently well defined, complete missing portions of the model structure—if any—by adding the corresponding residues. Any such remodeling actions can be performed manually in Coot or semiautomatically using the convenient Coot plugin RCrane [86, 87].
3. Launch RCrane in Coot from the main tab "Extensions" → "Launch RCrane". After launching the program, "RCrane" becomes available as one of the main tabs. RCrane provides options for building new chains in direction 5′–3′ or 3′–5′, for extending existing chains, and for optimizing the geometry of existing chains (the latter task can be performed even in the absence of an electron density map) [86, 87].
4. For manual remodeling, delete residues or modify the identity of nucleobases in Coot using "Calculate" → "Model/fit/refine" → "Delete" or "Calculate" → "Model/fit/refine" → "Simple mutate", respectively. Differently from RCrane, the latter options do not result in map fitting or in geometrical regularization.

3.2.10 Structure Refinement in Refmac5

1. Open Refmac5 in the CCP4i GUI, selecting "Program list" → "Refmac5". Alternatively, prepare an input script similar to the one reported in Table 5.
2. In the Title folder, select "restrained refinement" using "no prior phase information". Alternatively, in the first round of refinement, select "TLS & restrained refinement" to model anisotropic displacements of the atoms. TLS refinement [88] is particularly helpful if the input structure factors are limited to medium-to-low resolution. In this same folder, input the structure factor of the target (*see* Subheading 3.1), the coordinates of the manually remodeled MR solution (see points 1 and 2 above), and—if appropriate—a TLS input file which can

Table 5
Refmac input file

```
#================= REFMAC5 =========================
refmac5
XYZIN "crystal1_phaser.pdb" XYZOUT "crystal1_refmac1.pdb"
HKLIN "crystal1.mtz" HKLOUT "crystal1_refmac1.mtz"
TLSIN "crystal1_tlsin.tls" TLSOUT "crystal1_tlsout.tls"
LIBOUT "crystal1_lib.cif"
make check NONE
make -
    hydrogen YES -
    hout NO -
    peptide NO -
    cispeptide YES -
    ssbridge YES -
    symmetry YES -
    sugar YES -
    connectivity NO -
    link NO
refi -
    type REST -
    resi MLKF -
    meth CGMAT -
    bref ISOT
refi tlsc 10
ncyc 10
scal -
    type SIMP -
    LSSC -
    ANISO -
    EXPE
solvent YES
weight AUTO
monitor MEDIUM -
    torsion 10.0 -
    distance 10.0 -
    angle 10.0 -
    plane 10.0 -
    chiral 10.0 -
    bfactor 10.0 -
    bsphere 10.0 -
    rbond 10.0 -
    ncsr 10.0
labin  FP=FP SIGFP=SIGFP FREE=FreeR_flag
labout  FC=FC FWT=FWT PHIC=PHIC PHWT=PHWT DELFWT=DELFWT PHDELWT=PHDELWT
FOM=FOM
temp set 20.0
PNAME unknown
DNAME unknown091011
RSIZE 80
EXTERNAL WEIGHT SCALE 10.0
EXTERNAL USE MAIN
EXTERNAL DMAX 4.2
END
```

be created within the Refmac5 GUI using the "Create TLSin" button. Finally, define the desired names for the output files.

3. In the "TLS Parameters" folder, set "initial B factors to 20" and perform "10 cycles" of TLS refinement (only in the first round of refinement).
4. In the "Refinement Parameter" folder, select to run "10 cycles" using "automatic weighting" and "experimental sigmas to weight the X-ray terms". Refine "isotropic" or "anisotropic" temperature factors depending on the resolution of the diffraction data (isotropic temperature factor refinement should be used if resolution is worse than 2 Å). Exclude FreeR_flagged data (typically 5 % of the reflections).
5. In the "Scaling" folder, select the "Babinet scaling" option to take into account the effects of bulk solvent (*see* **Note 7**).

3.2.11 Structure Validation via Calculation of Simulated-Annealing Omit Maps in Phenix

Simulated-annealing refinement and calculation of omit maps are common procedures after phasing a macromolecular structure by MR. In general, these procedures can help resolving model bias, which is a typical matter of concern in MR searches [84]. In addition, simulated-annealing omit maps may help revealing conformational changes in the target or it may support the identification of functionally and structurally important solvent molecules [11]. The following steps explain how to calculate simulated-annealing omit maps in Phenix [63].

1. Open the Phenix GUI. Open the omit map input interface by selecting "Maps" → "Autobuild – create omit map".
2. In the "Configure" tab, select the input and output parameters.
3. In the "Input" folder, select the input data files. These should be the structure factor of the target (*see* Subheading 3.1) and the refined target coordinates as obtained by MR, remodeling and refinement (*see* Subheadings 3.4.1 and 3.4.2). As "Omit map type", select "simulated annealing". As "Omit region", select either "composite" or "omit selection". In the "Composite" mode, Phenix sequentially selects boxes of the unit cell and calculates omit maps for each box. At the end of the calculation, Phenix assembles a complete omit map of the whole structure. Such an option is suitable to validate the overall MR solution, particularly when this has been obtained using de novo designed artificial models. In the "Omit selection" mode, Phenix requires that one selects a specific region of the structure to be omitted in the calculation. Such selection can be done using the "Omit map options" and inputting the desired selection under "Omit selection". This latter operation mode is faster and it is thus particularly useful to confirm the identity and position of precise portions of the model. For instance, we used it to confirm a conformational change and

the position of catalytic solvent molecules in the active site of the group II intron [11].

4. In the "Output" folder, select the title and the output directory (optional).

3.2.12 Structure Interpretation: Identification and Modeling of Solvent Molecules

Proper identification of solvent molecules is more fundamentally important for RNAs than for proteins, because solvent molecules are essential for RNA folding, structural stabilization and function [89–92]. However, interpreting solvent molecule binding sites in RNA may be difficult, especially at the low resolution (>3 Å) typically obtained for large RNA molecules [93]. For identifying and modeling solvent molecules in RNA structures, MR strategies can be very valuable [11]. For instance, for a target of known structure, multiple new structures may be determinable by MR in the presence of different combinations of metal ions or ligands. The comparison of such structures could then lead to the identification of the metal-ion and ligand binding sites in the target. We successfully applied such an approach to identify physiologically relevant site-bound ions in the structure of the group II intron [11]. We followed the steps below:

1. Crystallize a target of known structure in the presence of physiological ions (typically, potassium and magnesium) and of heavy ions that mimic physiological ions. Potassium can be effectively replaced by thallium and rubidium (*see* **Note 8**), which possess similar ionic properties [94]. Cesium may also be used as a monovalent ion, but it should be considered that cesium has a significantly larger radius than potassium. Additionally, cesium sites possess analogous properties to cobalt hexammine, which in turns is similar to hexahydrated magnesium [94, 95]. Magnesium can be effectively replaced by manganese, cadmium or zinc [96, 97]. Hexahydrated magnesium binding sites can also be replaced by cobalt-, iridium- or osmium-hexamine [94, 95, 98, 99]. Strontium or barium may also be used as divalent ions, but it should be considered that their ionic radius is more similar to that of potassium than to that of magnesium, so that such ions may also occupy potassium binding sites [11].
2. Collect the diffraction data of the structures containing heavy atoms at tunable synchrotron beamlines, using suitable X-ray energy for observing anomalous scattering of the heavy ions. Anomalous scattering tables are available at http://skuld.bmsc.washington.edu/scatter/AS_periodic.html [100].
3. Solve all structures by MR, using the known target structure as a search model (*see* Subheading 3.2.1). If the ions or ligand induce structural changes that prevent obtaining a straightforward

MR solution, the search model can be subdivided into multiple individual ensembles (*see* Subheading 3.2.3).

4. Calculate anomalous difference Fourier electron density maps for the structures solved in the presence of heavy atoms. For these calculations, use the software SFall [101] and FFT [102] in CCP4. In SFall, select "generate structure factors and phases from coordinates," and "append F_C/PHI_C to all columns from existing .mtz file". Input the refined coordinates of each given structure solved by MR and the corresponding structure factors. Successively, run FFT "to generate anomalous map" and "output map in CCP4 format to cover all atoms in PDB file". Input the .mtz file generated by SFall and the refined coordinates of each structure solved by MR. The output of FFT is the desired anomalous difference Fourier electron density maps. Model the corresponding heavy atoms at sites where such maps shows peaks significantly higher than noise (i.e., $>3\sigma$).
5. In structures solved with physiological ions, model potassium at all sites that are systematically replaced by heavy analogues of potassium, and magnesium at sites that show binding of heavy analogues of magnesium, as indicated in points 1 and 4 above.
6. Additionally, when modeling physiological ions, respect the following geometrical constraints. Potassium forms bonds with distances of 2.8–3.5 Å from ligand atoms of the RNA nucleotides [103–105]. Magnesium adopts an octahedral coordination geometry with bond lengths of 2.0–2.2 Å [94, 104–106]. Note that magnesium possesses the same number of electrons and therefore the same scattering power as water or sodium.
7. If peaks of $2F_o - F_c$ electron density remain unoccupied after modeling the RNA molecule and the ions, consider the possibilities of modeling other solvent molecules to these sites, compatibly with the composition of the crystallization solutions. Typical components of RNA sample buffers are polyamines and anions. Polyamines can generally be recognized by structural sites that display elongated electron density peaks and that position the nitrogen atoms of the ligand at suitable distance to engage in ionic interactions with the RNA phosphates [107, 108]. Anions or zwitterions, such as Cl^-, SO_4^{2-} [109], or sulfonate buffer molecules [110] show weak anomalous scattering Fourier electron density peaks at low X-ray energy, and can be recognized from the distribution of the electrostatic potential at their binding sites. Additionally, these molecules generally position their sulfur or chlorine atoms at approximately 3.8 Å from neighboring nitrogen or oxygen atoms.

8. Finally, after modeling all ions, adjust their occupancy by comparing their solvent B-factors with the B-factors of surrounding nucleotides. B-factor refinement can also be performed in Phenix.

4 Notes

1. For running SBGrid, all Linux machines need installations of a recent version (6.17+) of csh/tcsh [111]. Other details for subscription or hardware requirements can be found at http://www.sbgrid.org/.
2. EasyXtal 15-Well Tool plates with a black O-ring (Qiagen, Venlo, The Netherlands) or other commercial plates that we tested do not provide the same isolation and are thus not recommendable for crystallization at 30 °C. Instead, they are suitable for experiments run at lower temperatures.
3. For multi-domain searches in Phaser, it needs to be considered that the structure of individual RNA domains obtained in isolation may be different from that of the same domain in the context of the full molecule [19, 112, 113]. Therefore, the correct assessment of the secondary structure maps of model and target is important in order to select the MR search domains appropriately.
4. Currently, multi-domain MR searches have significant chances of success using a maximum of 6–8 helical fragments, each not shorter than 12–15 bp [6].
5. For large RNA molecules, the use of ideal helices as MR search models may be challenging. In these cases, homology models or de novo predicted tertiary models of domains or subdomains of the target may serve as more appropriate MR search models [6, 114].
6. Excluding higher resolution shells from phased MR calculations in MolRep may sometimes be beneficial, possibly because these shells contain data of lower accuracy than the lower resolution shells.
7. In Refmac5, changing from simple scaling to Babinet scaling excludes mask bulk solvent artifacts. It requires that the option "calculate contribution from the solvent region" is unchecked. Solvent mask parameters can also be optimized by running Refmac5 with the keyword "solvent optimize", which writes out R and R_{free} for different combinations of VDW probe, ion probe, and shrinkage sizes [61].
8. Thallium chloride is insoluble in water. Therefore, anions different from chloride should be chosen in all buffers when

using thallium. Acetate is generally an efficient substitute. Additionally, because of its similarity to potassium and because it can be readily absorbed through the skin, thallium is highly toxic and should be handled with extreme care.

Acknowledgments

I thank Prof. Anna Pyle and all members of the Pyle lab for constructive discussion and critical reading of the manuscript. All group II intron diffraction data were collected at beamlines 24-ID-C and E, NE-CAT, APS, Argonne (IL), USA, which are supported by a grant from the National Institute of General Medical Sciences (P41 GM103403) from the National Institutes of Health. Use of the Advanced Photon Source, an Office of Science User Facility operated for the US Department of Energy (DOE) Office of Science by Argonne National Laboratory, was supported by the US DOE under Contract No. DE-AC02-06CH11357. Specifically, at NE-CAT I thank Dr. Kanagalaghatta Rajashankar for precious discussion and for his help with phased molecular replacement experiments. This project was supported by the National Institute of Health (RO1GM50313).

References

1. Robertus JD, Ladner JE, Finch JT et al (1974) Structure of yeast phenylalanine tRNA at 3 Å resolution. Nature 250:546–551
2. Cruz JA, Westhof E (2009) The dynamic landscapes of RNA architecture. Cell 136:604–609
3. Novikova IV, Hennelly SP, Sanbonmatsu KY (2012) Sizing up long non-coding RNAs: do lncRNAs have secondary and tertiary structure? Bioarchitecture 2:189–199
4. Westhof E, Romby P (2010) The RNA structurome: high-throughput probing. Nat Methods 7:965–967
5. Muirhead H, Perutz MF (1963) Structure of haemoglobin. a three-dimensional fourier synthesis of reduced human haemoglobin at 5.5 A resolution. Nature 199:633–638
6. Marcia M, Humphris-Narayanan E, Keating KS et al (2013) Solving nucleic acid structures by molecular replacement: examples from group II intron studies. Acta Crystallogr D 69:2174–2185
7. Rossmann MG (2001) Molecular replacement–historical background. Acta Crystallogr D 57:1360–1366
8. Toth EA (2007) Molecular replacement. Methods Mol Biol 364:121–148
9. Robertson MP, Scott WG (2007) The structural basis of ribozyme-catalyzed RNA assembly. Science 315:1549–1553
10. Kazantsev AV, Krivenko AA, Pace NR (2009) Mapping metal-binding sites in the catalytic domain of bacterial RNase P RNA. RNA 15:266–276
11. Marcia M, Pyle AM (2012) Visualizing group II intron catalysis through the stages of splicing. Cell 151:497–507
12. Lau MW, Ferre-D'Amare AR (2013) An in vitro evolved glmS ribozyme has the wild-type fold but loses coenzyme dependence. Nat Chem Biol 9:805–810
13. Grigg JC, Ke A (2013) Structural determinants for geometry and information decoding of tRNA by T Box Leader RNA. Structure 21:2025–2032
14. Polikanov YS, Blaha GM, Steitz TA (2012) How hibernation factors RMF, HPF, and YfiA turn off protein synthesis. Science 336:915–918
15. Stoddard CD, Widmann J, Trausch JJ et al (2013) Nucleotides adjacent to the ligand-binding pocket are linked to activity tuning in the purine riboswitch. J Mol Biol 425:1596–1611

16. Pyle AM (2010) The tertiary structure of group II introns: implications for biological function and evolution. Crit Rev Biochem Mol Biol 45:215–232
17. Marcia M, Somarowthu S, Pyle AM (2013) Now on display: a gallery of group II intron structures at different stages of catalysis. Mob DNA 4:14
18. Ferre-d'Amare AR, Doudna JA (1997) Establishing suitability of RNA preparations for crystallization. Determination of polydispersity. Methods Mol Biol 74:371–377
19. Toor N, Keating KS, Taylor SD et al (2008) Crystal structure of a self-spliced group II intron. Science 320:77–82
20. Pereira MJ, Behera V, Walter NG (2010) Nondenaturing purification of co-transcriptionally folded RNA avoids common folding heterogeneity. PLoS One 5:e12953
21. Doudna JA (1997) Preparation of homogeneous ribozyme RNA for crystallization. Methods Mol Biol 74:365–370
22. Golden BL (2007) Preparation and crystallization of RNA. Methods Mol Biol 363:239–257
23. Garman E, Owen RL (2007) Cryocrystallography of macromolecules: practice and optimization. Methods Mol Biol 364:1–18
24. Garman E, Sweet RM (2007) X-ray data collection from macromolecular crystals. Methods Mol Biol 364:63–94
25. Arzt S, Beteva A, Cipriani F et al (2005) Automation of macromolecular crystallography beamlines. Prog Biophys Mol Biol 89:124–152
26. Flot D, Mairs T, Giraud T et al (2010) The ID23-2 structural biology microfocus beamline at the ESRF. J Synchrotron Radiat 17:107–118
27. Kabsch W (1993) Automatic processing of rotation diffraction data from crystals of initially unknown symmetry and cell constants. J Appl Crystallogr 26:795–800
28. Leslie AW, Powell H (2007) Processing diffraction data with MOSFLM. In: Sussman JL, Read RJ (eds) Evolving methods for macromolecular crystallography. Springer, Amsterdam, pp 41–51
29. Pflugrath JW (1999) The finer things in X-ray diffraction data collection. Acta Crystallogr D 55:1718–1725
30. Otwinowski Z, Minor W (1997) Processing of X-ray diffraction data collected in oscillation mode. Methods Enzymol 276:307–326
31. Winter G, Lobley CM, Prince SM (2013) Decision making in xia2. Acta Crystallogr D 69:1260–1273
32. Delageniere S, Brenchereau P, Launer L et al (2011) ISPyB: an information management system for synchrotron macromolecular crystallography. Bioinformatics 27:3186–3192
33. Evans P, McCoy A (2008) An introduction to molecular replacement. Acta Crystallogr D 64:1–10
34. Costanzi S (2012) Homology modeling of class a G protein-coupled receptors. Methods Mol Biol 857:259–279
35. Collaborative computational project number 4 (1994) The CCP4 suite: programs for protein crystallography. Acta Crystallogr D 50:760–763
36. McCoy AJ, Grosse-Kunstleve RW, Adams PD et al (2007) Phaser crystallographic software. J Appl Crystallogr 40:658–674
37. Yao Z, Weinberg Z, Ruzzo WL (2006) CMfinder – a covariance model based RNA motif finding algorithm. Bioinformatics 22:445–452
38. Nawrocki EP (2014) Annotating functional RNAs in genomes using Infernal. Methods Mol Biol 1097:163–197
39. Nawrocki EP, Eddy SR (2013) Infernal 1.1: 100-fold faster RNA homology searches. Bioinformatics 29:2933–2935
40. Will S, Reiche K, Hofacker IL et al (2007) Inferring noncoding RNA families and classes by means of genome-scale structure-based clustering. PLoS Comput Biol 3:e65
41. Somarowthu S, Legiewicz M, Keating KS et al (2013) Visualizing the ai5gamma group IIB intron. Nucleic Acids Res 42:1947–1958
42. DeLano WL, Brunger AT (1995) The direct rotation function: Patterson correlation search applied to molecular replacement. Acta Crystallogr D 51:740–748
43. Minor DL, Lin YF, Mobley BC et al (2000) The polar T1 interface is linked to conformational changes that open the voltage-gated potassium channel. Cell 102:657–670
44. Storici P, Capitani G, De Biase D et al (1999) Crystal structure of GABA-aminotransferase, a target for antiepileptic drug therapy. Biochemistry 38:8628–8634
45. Fabiane SM, Sohi MK, Wan T et al (1998) Crystal structure of the zinc-dependent beta-lactamase from Bacillus cereus at 1.9 A resolution: binuclear active site with features of a mononuclear enzyme. Biochemistry 37:12404–12411
46. Hausrath AC, Gruber G, Matthews BW et al (1999) Structural features of the gamma subunit of the Escherichia coli F(1) ATPase revealed by a 4.4-A resolution map obtained by x-ray crystallography. Proc Natl Acad Sci U S A 96:13697–13702

47. Schrodinger LLC is a software company. Maybe the website can be mentioned: www.schrodinger.com
48. Moore PB (1999) Ribosomes and the RNA world. In: Gesteland RF, Cech TR, Atkins JF (eds) The RNA world. Cold Spring Harbor Laboratory Press, Cold Spring Harbor, NY, pp 119–135
49. Zuker M (2003) Mfold web server for nucleic acid folding and hybridization prediction. Nucleic Acids Res 31:3406–3415
50. Hofacker IL, Fekete M, Stadler PF (2002) Secondary structure prediction for aligned RNA sequences. J Mol Biol 319:1059–1066
51. Wilkinson KA, Merino EJ, Weeks KM (2006) Selective 2′-hydroxyl acylation analyzed by primer extension (SHAPE): quantitative RNA structure analysis at single nucleotide resolution. Nat Protoc 1:1610–1616
52. Choudhary PK, Gallo S, Sigel RK (2014) Monitoring global structural changes and specific metal-ion-binding sites in RNA by in-line probing and Tb(III) cleavage. Methods Mol Biol 1086:143–158
53. Costa M, Monachello D (2014) Probing RNA folding by hydroxyl radical footprinting. Methods Mol Biol 1086:119–142
54. Biondi E, Burke DH (2014) RNA structural analysis by enzymatic digestion. Methods Mol Biol 1086:41–52
55. Sachsenmaier N, Handl S, Debeljak F et al (2014) Mapping RNA structure in vitro using nucleobase-specific probes. Methods Mol Biol 1086:79–94
56. Seetin MG, Kladwang W, Bida JP et al (2014) Massively parallel RNA chemical mapping with a reduced bias MAP-seq protocol. Methods Mol Biol 1086:95–117
57. Robertson MP, Chi YI, Scott WG (2010) Solving novel RNA structures using only secondary structural fragments. Methods 52:168–172
58. Robertson MP, Scott WG (2008) A general method for phasing novel complex RNA crystal structures without heavy-atom derivatives. Acta Crystallogr D 64:738–744
59. Scott WG (2012) Challenges and surprises that arise with nucleic acids during model building and refinement. Acta Crystallogr D 68:441–445
60. Emsley P, Lohkamp B, Scott WG et al (2010) Features and development of Coot. Acta Crystallogr D 66:486–501
61. Murshudov GN, Vagin AA, Dodson EJ (1997) Refinement of macromolecular structures by the maximum-likelihood method. Acta Crystallogr D 53:240–255
62. Brunger AT, Adams PD, Clore GM et al (1998) Crystallography & NMR system: a new software suite for macromolecular structure determination. Acta Crystallogr D 54:905–921
63. Zwart PH, Afonine PV, Grosse-Kunstleve RW et al (2008) Automated structure solution with the PHENIX suite. Methods Mol Biol 426:419–435
64. Read RJ, McCoy AJ, Storoni LC (2007) Automated structure determination with Phenix. In: Sussman JL, Read RJ (eds) Evolving methods for macromolecular crystallography. Springer, Amsterdam, pp 91–100
65. Navaza J (2001) Implementation of molecular replacement in AMoRe. Acta Crystallogr D 57:1367–1372
66. Long F, Vagin AA, Young P et al (2008) BALBES: a molecular-replacement pipeline. Acta Crystallogr D 64:125–132
67. Vagin A, Teplyakov A (2010) Molecular replacement with MOLREP. Acta Crystallogr D 66:22–25
68. Vagin A, Teplyakov A (1997) MOLREP: an automated program for molecular replacement. J Appl Crystallogr 30:1022–1025
69. Panjikar S, Parthasarathy V, Lamzin VS et al (2005) Auto-rickshaw: an automated crystal structure determination platform as an efficient tool for the validation of an X-ray diffraction experiment. Acta Crystallogr D 61:449–457
70. Reddy V, Swanson SM, Segelke B et al (2003) Effective electron-density map improvement and structure validation on a Linux multi-CPU web cluster: The TB Structural Genomics Consortium Bias Removal Web Service. Acta Crystallogr D 59:2200–2210
71. Claude J-B, Suhre K, Notredame C et al (2004) CaspR: a web server for automated molecular replacement using homology modelling. Nucleic Acids Res 32:W606–W609
72. Strokopytov BV, Fedorov A, Mahoney NM et al (2005) Phased translation function revisited: structure solution of the cofilin-homology domain from yeast actin-binding protein 1 using six-dimensional searches. Acta Crystallogr D 61:285–293
73. Schwarzenbacher R, Godzik A, Jaroszewski L (2008) The JCSG MR pipeline: optimized alignments, multiple models and parallel searches. Acta Crystallogr D 64:133–140
74. Kissinger CR, Gehlhaar DK, Fogel DB (1999) Rapid automated molecular replace-

ment by evolutionary search. Acta Crystallogr D 55:484–491

75. Glykos NM, Kokkinidis M (2000) A stochastic approach to molecular replacement. Acta Crystallogr D 56:169–174
76. Jamrog DC, Zhang Y, Phillips GN Jr (2003) SOMoRe: a multi-dimensional search and optimization approach to molecular replacement. Acta Crystallogr D 59:304–314
77. Jogl G, Tao X, Xu Y et al (2001) COMO: a program for combined molecular replacement. Acta Crystallogr D 57:1127–1134
78. Chothia C, Lesk AM (1986) The relation between the divergence of sequence and structure in proteins. EMBO J 5:823–826
79. Luo D, Ding SC, Vela A et al (2011) Structural insights into RNA recognition by RIG-I. Cell 147:409–422
80. McCoy AJ (2007) Solving structures of protein complexes by molecular replacement with Phaser. Acta Crystallogr D 63:32–41
81. Grosse-Kunstleve RW, Schneider TR (2007) Substructure determination in isomorphous replacement and anomalous diffraction experiments. Methods Mol Biol 364:197–214
82. Smith GD, Lemke CT, Howell PL (2007) Substructure determination in multiwavelength anomalous diffraction, single anomalous diffraction, and single isomorphous replacement with anomalous scattering data using Shake-and-Bake. Methods Mol Biol 364:183–196
83. Vonrhein C, Blanc E, Roversi P et al (2007) Automated structure solution with autoSHARP. Methods Mol Biol 364:215–230
84. Das U, Chen S, Fuxreiter M et al (2001) Checking nucleic acid crystal structures. Acta Crystallogr D 57:813–828
85. Tronrud DE (2007) Introduction to macromolecular refinement. Methods Mol Biol 364:231–254
86. Keating KS, Pyle AM (2012) RCrane: semi-automated RNA model building. Acta Crystallogr D 68:985–995
87. Keating KS, Pyle AM (2010) Semiautomated model building for RNA crystallography using a directed rotameric approach. Proc Natl Acad Sci U S A 107:8177–8182
88. Winn MD, Murshudov GN, Papiz MZ (2003) Macromolecular TLS refinement in REFMAC at moderate resolutions. Methods Enzymol 374:300–321
89. Pyle AM (2002) Metal ions in the structure and function of RNA. J Biol Inorg Chem 7:679–690
90. Draper DE (2004) A guide to ions and RNA structure. RNA 10:335–343
91. Leipply D, Lambert D, Draper DE (2009) Ion-RNA interactions thermodynamic analysis of the effects of mono- and divalent ions on RNA conformational equilibria. Methods Enzymol 469:433–463
92. Draper DE (2013) Folding of RNA tertiary structure: linkages between backbone phosphates, ions, and water. Biopolymers 99: 1105–1113
93. Holbrook SR, Sussman JL, Warrant RW et al (1978) Crystal structure of yeast phenylalanine transfer RNA. II. Structural features and functional implications. J Mol Biol 123:631–660
94. Auffinger P, Grover N, Westhof E (2011) Metal ion binding to RNA. Met Ions Life Sci 9:1–35
95. Cowan JA (1993) Metallobiochemistry of RNA. Co(NH3)6(3+) as a probe for Mg2+(aq) binding sites. J Inorg Biochem 49:171–175
96. Basu S, Strobel SA (1999) Thiophilic metal ion rescue of phosphorothioate interference within the Tetrahymena ribozyme P4-P6 domain. RNA 5:1399–1407
97. Pley HW, Flaherty KM, Mckay DB (1994) 3-Dimensional structure of a hammerhead ribozyme. Nature 372:68–74
98. Edwards AL, Garst AD, Batey RT (2009) Determining structures of RNA aptamers and riboswitches by X-ray crystallography. Methods Mol Biol 535:135–163
99. Ban N, Nissen P, Hansen J et al (1999) Placement of protein and RNA structures into a 5 Å-resolution map of the 50S ribosomal subunit. Nature 400:841–847
100. Brennan S, Cowan PL (1992) A suite of programs for calculating x-ray absorption, reflection, and diffraction performance for a variety of materials at arbitrary wavelengths. Rev Sci Instrum 63:850–853
101. Agarwal RC (1978) A new least-squares refinement technique based on the fast Fourier transform algorithm. Acta Crystallogr A 34:791–809
102. Ten Eyck LF (1973) Crystal physics, diffraction, theoretical and general crystallography. Acta Crystallogr A 29:183–191
103. Mahler J, Persson I (2012) A study of the hydration of the alkali metal ions in aqueous solution. Inorg Chem 51:425–438
104. Harding MM (2001) Geometry of metal-ligand interactions in proteins. Acta Crystallogr D 57:401–411
105. Harding MM (2002) Metal-ligand geometry relevant to proteins and in proteins: sodium and potassium. Acta Crystallogr D 58:872–874

106. Erat MC, Sigel RK (2008) Divalent metal ions tune the self-splicing reaction of the yeast mitochondrial group II intron *Sc*.ai5γ. J Biol Inorg Chem 13:1025–1036
107. Korolev N, Lyubartsev AP, Laaksonen A et al (2002) On the competition between water, sodium ions, and spermine in binding to DNA: a molecular dynamics computer simulation study. Biophys J 82:2860–2875
108. Quigley GJ, Teeter MM, Rich A (1978) Structural analysis of spermine and magnesium ion binding to yeast phenylalanine transfer RNA. Proc Natl Acad Sci U S A 75:64–68
109. Auffinger P, Bielecki L, Westhof E (2004) Anion binding to nucleic acids. Structure 12:379–388
110. Klein DJ, Ferre-D'Amare AR (2006) Structural basis of glmS ribozyme activation by glucosamine-6-phosphate. Science 313:1752–1756
111. Morin A, Eisenbraun B, Key J et al (2013) Collaboration gets the most out of software. Elife 2:e01456
112. Sigel RK, Sashital DG, Abramovitz DL et al (2004) Solution structure of domain 5 of a group II intron ribozyme reveals a new RNA motif. Nat Struct Mol Biol 11:187–192
113. Zhang L, Doudna JA (2002) Structural insights into group II intron catalysis and branch-site selection. Science 295:2084–2088
114. Humphris-Narayanan E, Pyle AM (2012) Discrete RNA libraries from pseudo-torsional space. J Mol Biol 421:6–26

Chapter 16

Helical Symmetry of Nucleic Acids: Obstacle or Help in Structure Solution?

Alexandre Urzhumtsev, Ludmila Urzhumtseva, and Ulrich Baumann

Abstract

Crystallographic molecular replacement method is the key tool to define an atomic structure of nucleic acids. Frequently nucleic acids are packed forming continuous helices in the crystal. This arrangement of individual molecules in "infinite" pseudo helical structures in crystal may be the reason why the molecular replacement fails to find a unique position of the search atomic model as the method requires. The Patterson function, calculated as a Fourier series with diffraction intensities, has auxiliary peaks for such a molecular packing. Those near the origin peak indicate the orientation of the helices. The coordinates of other peaks are related to the molecular position and the rotation angle between two such "infinite" helices. Thus, the peak analysis allows getting molecular position even without a search model. An intelligent selecting and averaging of the phase sets corresponding to multiple probable positions of the search model again result in a unique solution but in the form of a Fourier synthesis and not a model. This synthesis can be used then to build an atomic model as it is the case for usual phasing methods.

Key words Nucleic acids, Atomic structure, Molecular replacement, Pseudo helical symmetry, Patterson analysis, Multisolution approach, Fourier maps

1 Introduction

1.1 Crystallographic Molecular Replacement and Its Basic Features

Starting from the first atomic models of nucleic acids [1], the spatial structure provides the researchers with key information about functioning nucleic acids. X-ray macromolecular crystallography is the principal method to get corresponding atomic models. The electron density $\rho(\mathbf{r})$ of a nucleic acid cannot be observed directly. At the same time, being a periodic function in a crystal, it can be presented as a Fourier series:

$$\rho(\mathbf{r}) = \sum_{\mathbf{s}} F_{\text{obs}}(\mathbf{s}) e^{i\phi(\mathbf{s})} e^{-2\pi i(\mathbf{rs})} \quad (1)$$

Eric Ennifar (ed.), *Nucleic Acid Crystallography: Methods and Protocols*, Methods in Molecular Biology, vol. 1320, DOI 10.1007/978-1-4939-2763-0_16, © Springer Science+Business Media New York 2016

Here s are three-dimensional integer indices and $F_{obs}(s)e^{i\phi(s)}$ are complex Fourier coefficients. An X-ray diffraction experiment results in a set of the amplitudes, $F_{obs}(s)$, of the corresponding Fourier coefficients. The routine crystallographic procedure consists in a search for the associated phase values $\phi(s)$ (i.e. in solving the phase problem) using isomorphous replacement, multiwavelength anomalous dispersion or direct phasing, in calculating then the corresponding Fourier synthesis (Eq. 1) with the available set of data and interpreting it. Instead, a priori knowledge of the structure of nucleic acids allows using the so called molecular replacement method (see [2] and references therein) where no phases are needed. The basic hypothesis of this method is that an atomic model is available for a molecule which is relatively similar to the structure under study. Then, at the first approximation, to solve the structure means to find the position of the search model in the crystal. One may try to position the model in all points of the crystallographic unit cell, and in all orientations, and diffraction data can be simulated for each of such configurations. The configuration for which the calculated diffraction amplitudes $F_{calc}(s)$ reproduce the experimental data $F_{obs}(s)$ the best possible is expected to be the correct one.

1.2 Computational and Structural Problems

In practice, calculation of a set of structure factor amplitudes for each of model positions is time consuming. This difficulty and a non-monotonous behavior of the crystallographic target to be optimized,

$$R\left(\{F_{obs}(s)\},\{F_{calc}(s)\}\right) \rightarrow \min \tag{2}$$

make a straightforward search complicate. By this reason, the routine molecular replacement approach starts from the determination of the model orientation, i.e. the rotation search ignoring the model position. Then, with the orientation found, the rotation search is followed by a translation search. This procedure becomes more complicate when the search model consists from several pieces which shall be positioned consecutively and independently of each other. In this work, we will consider only the simplest case of a search with a single rigid-body type model.

Other problems are specific for structural features of the molecules. For example, a difference between the structure and the search model may be large enough to hide the "signal" making the target value (Eq. 2) for the correct model position only marginally better or even worse than for other positions. Another intrinsic problem is that the molecular replacement fails to identify the correct model position because this position is not defined in a unique way. For example, nucleic acids in crystals are often packed making continuous helices (Fig. 1a). It may be relatively easy to find the direction of these helices but extremely difficult to find the correct

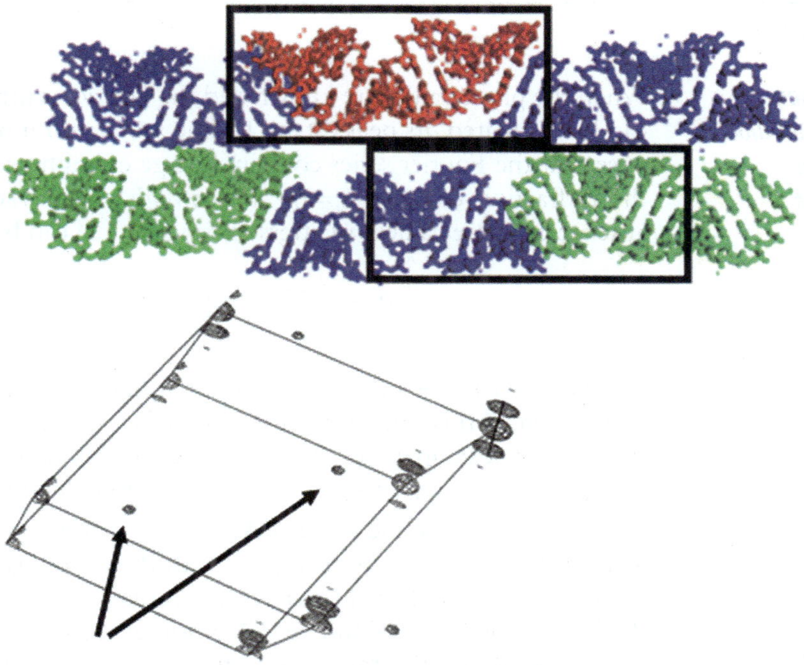

Fig. 1 (**a**) Superposition of RNA models of the 16mer RNA duplex [16]. Two parallel "infinite" pseudo helices are shown as they are in the crystal; neighboring molecules are in different color. The "*red*" molecule can be translationally superposed with the "*blue-green*" molecules of the next pseudo helix. Note a translation shift in the direction of the helix. (**b**) Patterson function calculated for the ribosomal A-site [8]. Note the phantom peaks near the origin indicating the structure in the form of "infinite pseudo helices" and the direction of these helices. *Arrows* indicate the auxiliary Patterson peaks corresponding to a superposition of the RNA molecule with a part of the neighboring helical structure

orientation and position of the search model in this helix. Therefore, the basic assumption of the molecular replacement is not hold. Below we discuss some possible answers to the problems above.

2 Material

Here are some useful programs for the procedures described herein:

1. The PHASER crystallographic software suite.
2. The AMoRe molecular replacement program.
3. The COMPANG program for comparative analysis of rotation angles.
4. The PyMOL molecular visualization system.
5. The Coot macromolecular model building program.

3 Methods

3.1 Patterson Function as a Guide for Molecular Replacement

When the phase values are known and the Fourier synthesis (Eq. 1) can be calculated, its peaks stand for the atomic positions (we suppose that the Fourier series contains a large enough terms, i.e. that the diffraction data set $S = \{\mathbf{s}\}$ has *a high enough resolution*). In absence of phase values, a Patterson function [3] can be calculated

$$P(\mathbf{r}) = \sum_{\mathbf{s}} F_{\text{obs}}^2(\mathbf{s}) e^{-2\pi i(\mathbf{rs})} \quad (3)$$

The peaks of Eq. 3 stand for the interatomic vectors. One classic application of this function is the search for the heavy atom positions. For example, in presence of a twofold screw axis $\mathbf{A} = (x,y,z) \rightarrow \mathbf{B} = \left(-x,-y,z+\frac{1}{2}\right)$, the Patterson function has a peak for the interatomic vector **BA**, in other words at $\left(2x,2y,\frac{1}{2}\right)$. This means that the Patterson peak with the coordinates X and Y in the so-called Harker [4] section $Z = ½$ tells us two coordinates of the heavy atom, $x = ½X$ and $y = ½Y$. Another classical application is an identification of a translation symmetry, when each atom with the coordinates (x, y, z) has its copy at $(x,y,z) \rightarrow (x+t_x, y+t_y, z+t_z)$ giving an interatomic vector (x, y, z). Thus, a corresponding Patterson peak corresponds directly to the translation vector. Franklin and Gosling used the Patterson function for very first studies of nucleic acids [5, 6].

As mentioned above, in the crystals nucleic acids frequently form infinite pseudo helices more or less parallel to each other (Fig. 1a). Each individual molecule, being translated, does not superpose with its copy obtained by a rotational or screw symmetry. However, these pseudo helices can be well superposed translationally as a whole. In this case each molecule does not superpose with itself but with two parts of neighboring molecules (Fig. 1a). Some conformational differences make the superposition less strong than for the structures linked by a pure translation. On the other hand, a usual resolution of diffraction data, that of 2–3 Å or lower, blurs atomic images. Obviously, this masks structural details and improves similarity of superposed imaged. In total, the Patterson peaks are quite strong and by their size are close to or at least comparable with those for a pure translation [7].

Let us consider such a pseudo helix more or less parallel to some rotational or screw axis, as it is often the case. This axis, crystallographic or not, generates another parallel pseudo helix. The shift between symmetry-related copies is calculated through the rotation angle and the period (step) H of the helix along its axis, that approximately of 30 Å. If the rotation angle for a pure rotation is equal to 120°, then a translation of a pseudo helix perpendicular to its axis does not lead to superimposed structures however an extra translation by $H/3$ in the direction of the axis does.

This extra shift is equal to $H/6$ for a rotation by 60°, to $H/2$ for a rotation by 180°, etc. As for the Harker sections, the coordinates of the Patterson peak can identify the position of the helix [7] if the direction of the rotation axis and its order are known. Usually, the height of the Patterson peak slightly decreases with the rotation angle due to imperfect parallelism of the axes. Figure 1b gives an example of such an auxiliary Patterson peak for the *P*1 crystal structure of the *H. sapiens* mitochondrial A site [8] where two molecules were related by a 180° non-crystallographic rotation axis. When the position of the helix is known, the position of the individual molecule in this helix can be searched for.

Figure 1b shows another important feature of the Patterson maps for nucleic acids. Its pseudo periodic structure leads to the phantom peaks near the origin since a shift by about 3.3 Å in the axis of the pseudo helix gives the electron density pretty close to the correct one, especially when they are compared at the resolutions of order of 3 Å or lower. The direction of these phantom peaks corresponds to the direction of the pseudo helix making it known as soon as diffraction data are available.

We may note that this Patterson-guided molecular replacement does not require a search model at all getting molecular position directly from the diffraction amplitudes; obviously the model is needed for further steps. More details of this approach as well as multiple illustrations can be found in [7].

3.2 Multisolution Molecular Replacement

3.2.1 Ambiguity of Molecular Replacement Solutions

Unfortunately, the situation may be less favorable than discussed above. For example, the symmetrically related pseudo helices may be non-parallel to each other. As mentioned in Subheading 1, in presence of a pseudo-helical symmetry molecular replacement cannot distinguish between multiple positions of the search model because for all of them the crystallographic target (Eq. 2) has practically the same value. In other words this means that the search model reproduces equally well the electron density in all these positions.

Usually, a structure solution found by molecular replacement means that a unique position of the atomic search model is determined. Then a Fourier synthesis, usual or a difference one, is calculated with this model allowing its further correction. When we fail to identify the unique position of the search model, we may search for the unique solution of the molecular replacement problem but in a different form, not in direct but in reciprocal space. When several possible solutions found, each of these models fits more or less equally well the electron density of the crystal as the target (Eq. 2) tells. This means that if we calculate Fourier syntheses for these models, "in average" these syntheses will remind the correct one and differs from it in some parts. Therefore, we may take all of them and average. This will reinforce the common, correct part of the data and will remove the noise.

However, a simple averaging would be a wrong procedure since individual syntheses may correspond to different allowed origins of the unit cell. Also, such a collection of possible solutions may contain some occasional outliers, including of which will make the average result worse. Therefore, some kind of clustering procedure [9] with a preliminary alignment of syntheses (or, which is the same, of the corresponding phase sets) can be a solution for both these problems.

3.2.2 Processing of Multiple Solutions

As explained in Introduction, molecular replacement procedure starts from the rotation function analysis. In case of structures with a pseudo helical symmetry, the results of this function may be also confusing, the function has a lot of peaks (each peak corresponds to a possible solution), practically indistinguishable by their size. Persistent peaks or those corresponding to the model rotated around the axis of the helix can be identified by multiple rotation function analysis [10, 11]. Then for each model orientation, the corresponding translation function is calculated and all model positions with a significant value of the crystallographic target (Eq. 2) are selected. For each of them, we calculate a set of structure factors. Corresponding phases and experimental amplitudes are used to calculate the Fourier syntheses which then are aligned and clustered [9]. The principal cluster shall give the map which corresponds to the structure. Then this map shall be interpreted in the same way as this is done for other phasing methods [12].

Switching from the search in direct space to that in reciprocal space opens a way to use more information. Since the phase-and-synthesis analysis does not refer to the search model, the phase sets can be obtained from different search models, and the searches can be done with different condition, in particularly at different resolutions. This eventually can increase the amount of information and therefore improve the results.

While originally the method was illustrated with protein data [12] and contributed to the development of the *archimboldo* procedure [13] for protein structure solution, this approach seems to be even more promising for solving structure of nucleic acids as the example below illustrates.

3.3 Example

1. The crystals of the dodecamer duplex GGCGCCGGCGCC were obtained in space group $P4_12_12$ with unit cell parameters a=43.530, b=43.530, c=81.020 Å. Asymmetric unit contained a single molecule. A highly complete data set at the resolution of 3.0 Å has been collected. The rotation function showed a number of strong peaks quite similar in size, and a straightforward molecular replacement with *PHASER* [14] resulted in a solution recognized as incorrect. Therefore, we tried the approaches described above.

2. The phantom peaks around the origin peak of the Patterson function indicated a packing of the dodecamer molecules in infinite helices and their alignment roughly in the direction of the axis **a** (or **b** in this space group). Unfortunately, auxiliary Patterson peaks did not lead us to a convincing conclusion about the molecular position as Subheading 2 suggests.
3. Then we ran *PHASER* in an automated way and took blindly all the highest-score solutions, 34 in total. For each of them we calculated structure factors, aligned them and applied the clustering procedure [9] to this ensemble of the phase sets.
4. When comparing the phase sets at 4 or at 5 Å, a few clear clusters were observed one of them including more than a half of all sets (Fig. 2a); the analysis at the full resolution of 3 Å showed a poorer clustering.

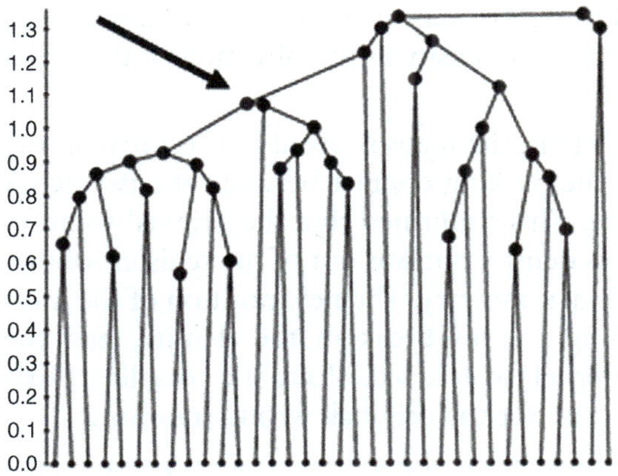

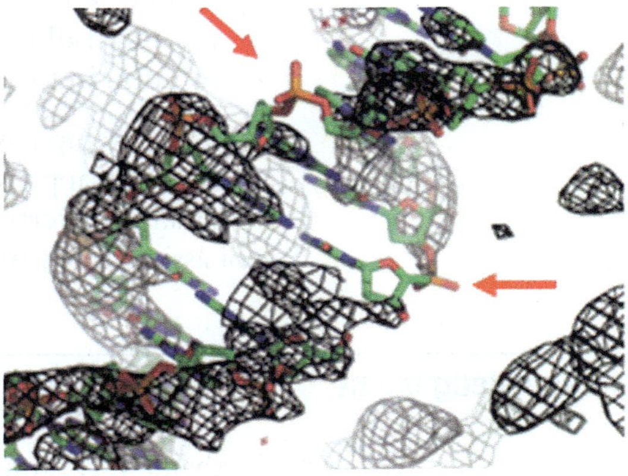

Fig. 2 (**a**) Cluster tree for the selected phase sets for the dodecamer duplex GGCGCCGGCGCC (Subheading 3.3). Data set resolution is equal to 4 Å. The lower the cluster, the mutually closer included phase sets. The *arrow* indicates the summit of the principal cluster. (**b**) The Fourier map calculated with the averaged phase set for the principal cluster. The map is superposed with the original solution with a wrong choice of register. The *arrows* show a gap in the density where the model shall not be present

Table 1
Phase set obtained after alignment and averaging the phase sets from the principal cluster

Resolution shell, Å	∞-6.93	6.93–5.47	5.47–4.79	4.79–4.33	4.33–4.00	∞-4.00 (total)
$N_{reflections}$	155	155	155	156	156	777
Phase error, °	44	54	67	80	79	65
FOM	0.55	0.46	0.29	0.13	0.14	0.31

For each reflection, its figure of merit is obtained from the dispersion of the phase values for this reflection in the selected phase sets; FOM stands for the figure of merit averaged for all reflections

5. Averaging of the phases in the principal cluster gave the phase values and the corresponding weights which reflected the dispersion of the phase values over the phase sets (Table 1). The corresponding weighted Fourier synthesis with the average phase values and experimental structure factor amplitudes showed clearly the helical density in the direction of the axis **a** (or, equivalent, **b**).
6. What is more important, the maps showed a clear break in the density indicating the packing contact between the two RNA molecules (Fig. 2b). This confirmed that the original solution was in overall in this density but was out of the register following the pseudo helical symmetry. The new position of the molecule, corresponding to this break in the density, had the direction of the internal twofold axis along the coordinate axis which was not the case for the original, wrong solution.
7. This choice was confirmed by the self-rotation function which had no peaks except those for the crystallographic axes. Thus, the multisolution molecular replacement resulted in a single solution in the form of a map and not of a molecular model.

This example shows a feasibility of non-classical molecular replacement approaches to solving the structure of nucleic acids essentially converting an obstacle of the pseudo symmetry into a tool assisting the structure solution.

Acknowledgement

We thank V.Y. Lunin, J. Kondo, A. Bühler and O. Sobolev for help in different parts of the relevant projects. All figures except 2a have been done using *PyMOL* [15]. AU thanks the French Infrastructure for Integrated Structural Biology (FRISBI) ANR-10-INSB-05-01 and Instruct, part of the European Strategy Forum on Research Infrastructures (ESFRI) and supported by national member subscription.

References

1. Watson JD, Crick FH (1953) Molecular structure of nucleic acids; a structure for deoxyribose nucleic acid. Nature 171:737–738
2. Arnold E, Himmel DM, Rossmann MG (2012) In: International tables for crystallography. Vol. F: crystallography of biological macromolecules. Wiley, Chichester, pp 333–366
3. Patterson AL (1935) A direct method for the determination of the components of interatomic distances in crystals. Z Kristallogr 90: 517–542
4. Harker D (1936) The application of the three-dimensional Patterson method and the crystal structure of proustite, Ag_3AsS_3, and pyrargyrite, Ag_3SbS_3. J Chem Phys 4:381–390
5. Franklin RE, Gosling RG (1953) The structure of sodium thymonucleate fibres. II. The cylindrically symmetrical Patterson function. Acta Cryst 6:678–685
6. Franklin RE, Gosling RG (1955) The structure of sodium thymonucleate fibres. III. The three-dimensional Patterson function. Acta Cryst 8:151–156
7. Kondo J, Urzhumtseva L, Urzhumtsev A (2008) Patterson-guided ab initio analysis of structures with helical symmetry. Acta Cryst D 64:1078–1091
8. Kondo J, Westhof E (2008) The bacterial and mitochondrial ribosomal A-site molecular switches possess different conformational substates. Nucleic Acids Res 36:2654–2666
9. Sobolev OV, Lunina NL, Lunin VY (2010) The use of cluster analysis methods for the study of a set of feasible solutions of the phase problem in biological crystallography. Comp Res Modelling 2:91–101 (russ)
10. Urzhumtsev A, Urzhumtseva L (2002) Multiple rotation function. Acta Cryst D 58:2066–2075
11. Urzhumtseva L, Urzhumtsev A (2002) COMPANG : program for comparative analysis of rotation angles. J Appl Crystallogr 35: 644–647
12. Bühler A, Urzhumtseva L, Lunin VY, Urzhumtsev A (2009) Cluster analysis for phasing with molecular replacement. A feasibility study. Acta Cryst D 65:644–650
13. Rodríguez DD, Grosse C, Himmel S, González C, de Ilarduya IM, Becker S, Sheldrick GM, Usón I (2009) Crystallographic ab initio protein structure solution below atomic resolution. Nat Methods 6:651–653
14. McCoy AJ, Grosse-Kunstleve RW, Adams PD, Winn MD, Storoni LC, Read RJ (2007) J Appl Crystallogr 40:658–674
15. DeLano WL (2002) The PyMOL molecular graphics system. DeLano Scientific, San Carlos, CA, USA, http://www.pymol.org
16. Pan B, Mitra SN, Sundaralingam M (1998) Structure of a 16-mer RNA duplex r(GCAGACUUAAAUCUGC)2 with wobble C.A+ mismatches. J Mol Biol 283:977–984

Chapter 17

RNA Structure Refinement Using the ERRASER-Phenix Pipeline

Fang-Chieh Chou, Nathaniel Echols, Thomas C. Terwilliger, and Rhiju Das

Abstract

The final step of RNA crystallography involves the fitting of coordinates into electron density maps. The large number of backbone atoms in RNA presents a difficult and tedious challenge, particularly when experimental density is poor. The ERRASER-Phenix pipeline can improve an initial set of RNA coordinates automatically based on a physically realistic model of atomic-level RNA interactions. The pipeline couples diffraction-based refinement in Phenix with the Rosetta-based real-space refinement protocol ERRASER (*E*numerative *R*eal-Space *R*efinement *AS*sisted by *E*lectron density under *R*osetta). The combination of ERRASER and Phenix can improve the geometrical quality of RNA crystallographic models while maintaining or improving the fit to the diffraction data (as measured by R_{free}). Here we present a complete tutorial for running ERRASER-Phenix through the Phenix GUI, from the command-line, and via an application in the Rosetta On-line Server that Includes Everyone (ROSIE).

Key words RNA structure, Structure prediction, X-ray crystallography, Refinement, Force field

1 Introduction

Over the last decade, fruitful progress in RNA X-ray crystallography has revealed three-dimensional all-atom models of numerous riboswitches, ribozymes, and ribonucleoprotein machines [1–3]. Due to the difficulty of manually fitting RNA backbones into experimental density maps, many of these crystallographic models contain myriad unlikely conformations and unusually close contacts as revealed by automated MolProbity tools for geometric evaluation of models [4, 5]. Inspired by recent advances in ab initio RNA structure prediction [6–8] and successful applications of the Rosetta modeling suite in crystallographic and electron microscopy density fitting problems [9, 10], we developed the ERRASER method and integrated it with Phenix diffraction-based refinement [11]. In our previous publication [12], we demonstrated that the ERRASER-Phenix pipeline resolves the majority of steric clashes

Eric Ennifar (ed.), *Nucleic Acid Crystallography: Methods and Protocols*, Methods in Molecular Biology, vol. 1320, DOI 10.1007/978-1-4939-2763-0_17, © Springer Science+Business Media New York 2016

and anomalous backbone and bond geometries assessed by MolProbity in a benchmark of 24 RNA crystal structures. Furthermore, this method led to models with similar or better R_{free}. This chapter describes the details of using ERRASER in three easily accessible ways: by a GUI in the Phenix package, the command-line, and the ROSIE server [13].

2 Materials

The ERRASER-Phenix pipeline relies on two software toolkits: the Rosetta modeling suite [14] and the Phenix package [11]. Both toolkits are currently officially supported on Linux and Mac-OS X platforms. (Phenix is available on Windows; Rosetta might be compiled in Windows using Cygwin but is not officially supported and well-tested.) To run the pipeline locally, the user needs to have the following versions of the above toolkits installed on their computer:

Rosetta (version 3.5) http://www.rosettacommons.org/.

Phenix (version 1.8.3) http://www.phenix-online.org/.

Both Rosetta and Phenix are freely available to academic and non-profit institutions. Details of downloading, licensing, and the installation instructions can be found in the above listed websites. Phenix installation instructions can be found at http://www.phenix-online.org/documentation/install.htm. On Mac OS X systems, the installation simply consists of downloading a .dmg file and double-clicking the icon. On Linux systems it consists of unpacking a tar archive and running an installation script. Instructions for Rosetta installation compatible with Phenix and ERRASER can be found at http://www.phenix-online.org/documentation/erraser.htm.

It is also possible to run the ERRASER part of the pipeline online and privately using the ROSIE server (http://rosie.rosettacommons.org/).

3 Methods

The standard ERRASER-Phenix pipeline consists of three major stages: an initial Phenix refinement, followed by iterative ERRASER refinement, and a final Phenix refinement (Fig. 1). Here the initial Phenix refinement can be skipped if the input structure has already been refined with all hydrogen atoms included in the model. In general, we find that maintaining hydrogen atoms during diffraction-based refinement tends to give models with better geometrical quality, particularly with regards to steric interactions, as assessed by the MolProbity clashscore. Since ERRASER

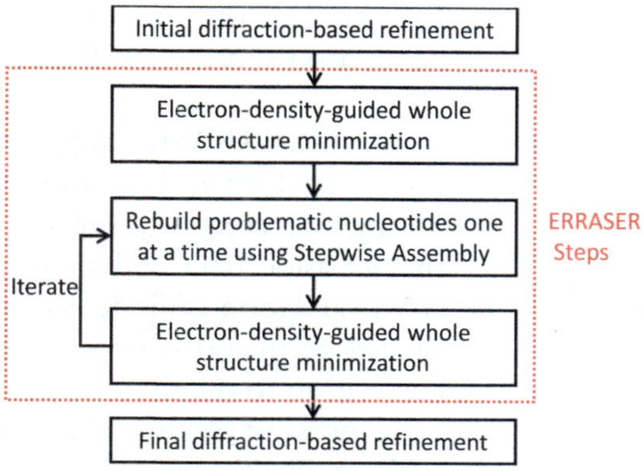

Fig. 1 Flow chart of the ERRASER-Phenix pipeline

performs only real-space refinement, a final diffraction-based refinement is necessary to fit the model directly to the original data and evaluate R_{free} statistics. We have carried out tests using the Phenix refinement tool for these two refinement stages [15], but users should be able to substitute in refinement tools if preferred (e.g. SHELXL [16], Refmac [17], CNS [18], etc.).

In the sections below, we will focus on the details of the ERRASER refinement stage. We will mainly briefly describe how to run ERRASER using the Phenix GUI interface, and briefly describe how to run ERRASER using shell command lines and ROSIE web server. Finally, we discuss some settings and options we found useful in the Phenix refinement of RNA.

3.1 Set Up the Phenix-Rosetta Connection

After both Phenix and Rosetta are properly installed and compiled on the user's local computer, the user should set the path so that Phenix can locate the Rosetta applications. Suppose you have Rosetta installed at "/home/user/rosetta-3.5." If using the bash or sh shells, add the following line into "~/.profile" or "~/.bashrc":

```
export PHENIX_ROSETTA_PATH=/home/user/rosetta-3.5
```

Or if using C-shell, put the following line into "~/.cshrc":

```
setenv PHENIX_ROSETTA_PATH /home/user/rosetta-3.5
```

3.2 Prepare the ERRASER Input Files

The following files need to be prepared before running ERRASER. Note that ERRASER is designed for the final fine-tuning of the RNA models, and has only been tested for such cases. Applying ERRASER to models from earlier refinement stages might lead to unexpected behavior and poor models.

3.2.1 Input PDB File

ERRASER accepts PDB files of RNA with the standard format as input. The users should ensure the input file follows the standard PDB format, with correct residue and atom naming conventions.

In addition, the user should ensure that there are no missing heavy atoms in the model, or ERRASER may not run properly.

3.2.2 Density Map File

Currently, ERRASER supports input $2mF_o$–DF_c density maps in the CCP4 format. It is important to exclude R_{free} diffraction data during map creation to ensure that ERRASER is not influenced by the set-aside data and that final R_{free} values are appropriate for cross-validation. All density data covering the entire unit cell should be included to allow ERRASER to correctly evaluate the correlation to density during the sampling. Here, we demonstrate the process of creating the density map using the "calculate maps" GUI in Phenix [19]. In the GUI window, click the "CCP4 or XPLOR Maps" button, and a window will pop out. In the new window, select the following options (Fig. 2):

Map type: 2mFo-DFc.

File format: CCP4.

Map region: Unit cell.

Enable the following three options:

1. "Kicked."
2. "Fill missing f obs."
3. "Exclude free r reflections."

Other options are kept default. Here the "kicked" option makes use of the kicked map algorithm to improve the map quality and to reduce map bias [19]. The "Fill missing f obs" option will

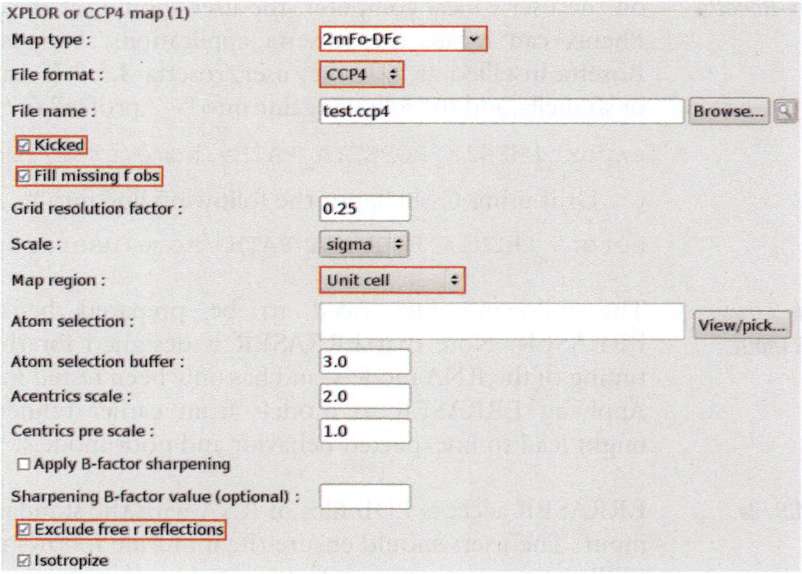

Fig. 2 Snapshot and useful options of the phenix.maps utility

allow Phenix to fill missing experimental diffraction data (F_{obs}) with calculated diffraction (F_{calc}), thereby avoiding Fourier truncation errors. We found that these two options lead to better ERRASER results (e.g., in terms of final R_{free}) in many cases. "Exclude free r reflections" ensures that the test set of reflections is not included. If you fail to check this option then your final R_{free} may be biased and be misleadingly low.

3.3 A Simple ERRASER Job Example

It is straightforward to run ERRASER using the Phenix GUI. Click on "Refinement"->"ERRASER" in the main Phenix GUI to access the ERRASER GUI (Fig. 3). As a quick start, put in a PDB file and a corresponding CCP4 map using the "Input PDB" and "CCP4 map" boxes. Then click on the "Run" button. ERRASER will run in a new tab. Descriptions of ERRASER options that the user may wish to explore are given below in Subheading 3.5.

After refinement is complete, ERRASER runs several validation metrics from MolProbity [4, 5] as implemented in Phenix. Results are summarized along with output files in a new tab (Fig. 4), with buttons to load the results in Coot and PyMOL. The validation display is identical to the all-atom contact and RNA-specific components in the Phenix GUI, which interact directly with Coot [20]. These include:

- Steric clashes [21], defined as atomic overlaps of at least 0.4 Å when explicit hydrogen atoms are present.
- RNA bond length and angle geometry outliers, which has values > 4 s.d. from the Phenix reference values.

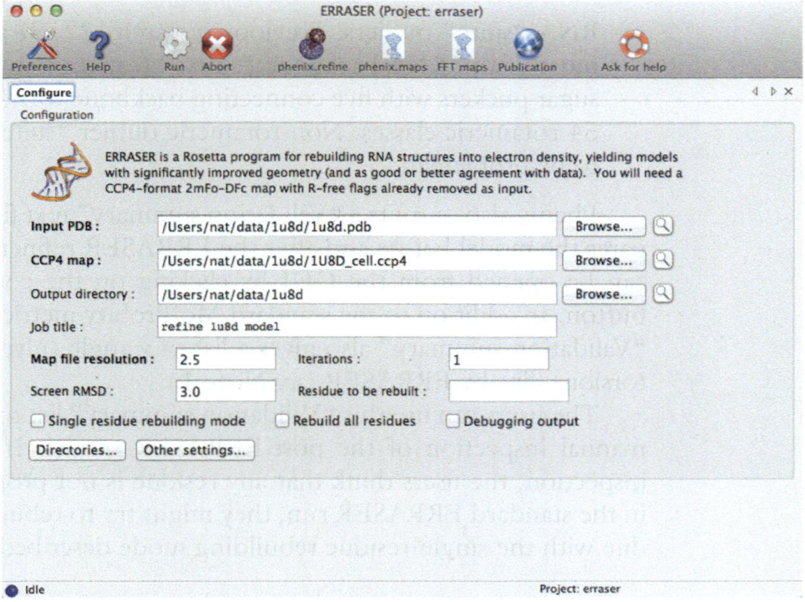

Fig. 3 The main window of the Phenix ERRASER GUI

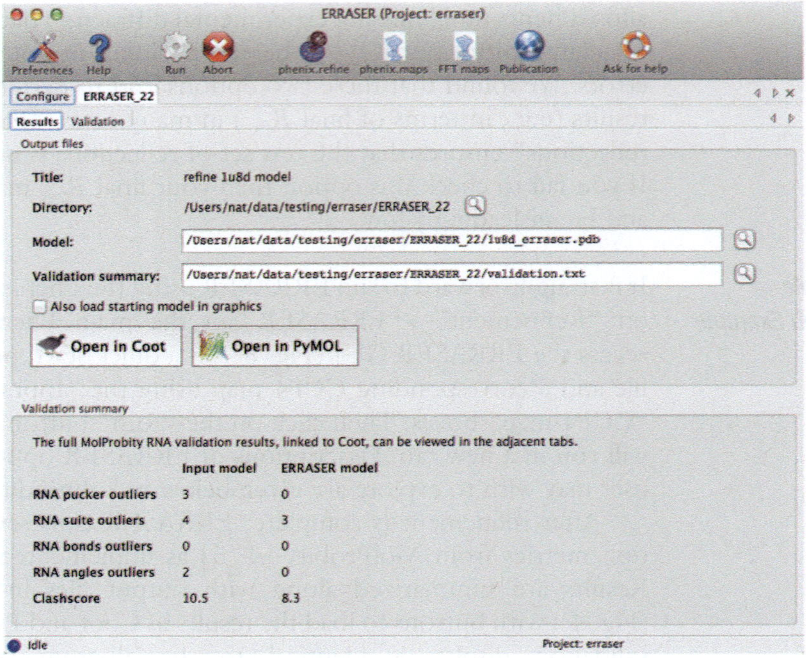

Fig. 4 Output window of the ERRASER GUI

– RNA pucker outliers. RNA sugar rings typically adopt either C2′-endo or C3′-endo conformation, known as the pucker. MolProbity can confidently identify erroneous puckers by measuring the perpendicular distance between the 3′-phosphate and the glycosidic bond vector.

– RNA "suite" outliers. Previous research [22, 23] shows that most of the RNA backbone "suites" (sets of two consecutive sugar puckers with five connecting backbone torsions) fall into 54 rotameric classes. Non-rotameric outlier "suites" are likely to be problematic.

Phenix also outputs a "Validation summary" text file that compares the model before and after the ERRASER refinement, which can be opened from the GUI by clicking on the corresponding button. In addition to the standard MolProbity metrics above, the "Validation summary" also gives a list of χ angle (glycosidic bond torsion) flips by ERRASER (*see* **Note 1**).

The users can use this "Validation summary" list to guide their manual inspection of the post-ERRASER model. If during the inspection, the users think that any residue is not properly rebuilt in the standard ERRASER run, they might try to rebuild such residue with the single-residue rebuilding mode described below.

3.4 Single-Residue Rebuilding Mode

For model regions with low density or for which other information is known (e.g., based on homology to a high-resolution structure), ERRASER's top picks for residue conformations may not be optimal. In addition, ERRASER might not always resolve all errors in the models and might, in a few cases, lead to new errors in the model while fixing other errors. In such cases, the user may want to inspect other possibilities for residue conformations. Therefore, in addition to the standard ERRASER protocol described above, there is a single-residue rebuilding mode available in ERRASER. In this mode, the user selects a particular residue in the complete structure and rebuilds it using the ERRASER algorithm. Up to ten different models, sorted by their relative ERRASER score, are returned as output which the user can then inspect.

The user can run the single-residue mode under the same GUI interface similar to the standard ERRASER application. After inputting the PDB file and CCP4 map file, click on the "single residue rebuilding mode" checkbox, and input the residue to be rebuilt in the "Residue to be rebuilt" box. The format of input residue is chain ID followed by residue number, e.g., "A35" or "F152." Then click "Run" button to execute the program. After completion, the generated models and validation results are displayed in the summary tab. The validation includes the "suite," pucker, and glycosidic torsion assignments, as well as the final ERRASER scores.

3.5 Available Options in ERRASER

The following options can be found in the main window of the ERRASER GUI:

3.5.1 Map File Resolution

It is usually a good idea to provide the resolution of the input density map, so that ERRASER can give a more accurate estimation of the fit between the model and the input map. If not provided, ERRASER will assume a default value of 2.5 Å.

3.5.2 Iterations

The total number of iterations of the ERRASER cycles (Fig. 1). By default it is 1. The user can increase the number to perform multicycle ERRASER refinement. We recommend not to use any value greater than 3. Beyond 3 iterations, the ERRASER model will most likely be converged, and it is a waste of time to continue the iterations.

3.5.3 Screen RMSD

To speed up the ERRASER rebuilding process, by default ERRASER only considers conformations within 3.0 Å RMSD of the input model during the sampling of each residue. The user can change this cutoff RMSD value by editing this option. If the screen RMSD is set to be larger than 10.0 Å, ERRASER will simply skip the RMSD screening step.

3.5.4 Residue to be Rebuilt

This is for the single-residue rebuilding mode only. *See* Subheading 3.4.

3.5.5 Single-Residue Rebuilding Mode

This is for the single-residue rebuilding mode only. *See* Subheading 3.4.

3.5.6 Rebuild all Residues

By default ERRASER will only rebuild residues that are assessed by MolProbity as having outlier pucker, suite, bond length or bond angle, or residues that moved significantly during the Rosetta minimization step. Selecting this option forces ERRASER to rebuild all residues in the model. In general, however, we recommend using the default rebuilding protocol. Rebuilding all residues is likely to be a time-consuming process without obvious model improvement. In some cases such excessive rebuilding might even lead to worse models, due to artifacts in the input density map and the Rosetta scoring function. If the user wishes to rebuild some specific residues, use the "Extra residues to be rebuilt" option below.

3.5.7 Debugging Output

This option is for debugging only. Turning on this option will generate a lot of output and slow down the code. If you find a bug, it might be a good idea to run this mode and send all the output to the developers to aid diagnosis (bugs@phenix-online.org).

The following options are accessible by clicking the "Other settings" button in the GUI.

3.5.8 Fixed Residues

This option allows the user to specify residues that should be kept fixed throughout the ERRASER refinement. The format is "A32", where A is the ID of the chain and 32 is the residue number. The user can also input "A22-30" to fix all residues from A22 to A30. This option is especially useful when the RNA contains ligands, modified nucleobases, or strong crystal contacts, none of which are currently modeled in ERRASER. In these cases, residues near to these unmodeled parts of the molecule might get rebuilt into unreasonable conformations due to absence of critical interactions to the unmodeled parts. It is therefore necessary to fix these residues using the "fixed residues" option. *See* **Note 2** for general suggestions on determining the necessary fixed residues.

3.5.9 Extra Residues to Rebuild

This option allows the user to specify particular residues in the model that ERRASER must rebuild regardless of whether it has apparent errors. The format is the same as "fixed residues" above.

3.5.10 Native Syn Only Pyrimidine

Syn conformers of pyrimidines (U and C) are rare in RNA structures. By default, ERRASER only samples syn conformers of pyrimidine residue if the user supplies an input residue with the syn conformer. The purpose is, first, to speed up the computation, and second, to avoid possibly problematic syn pyrimidine conformers that show up in the final models due to artifacts of the electron density map. Turning off this option will allow ERRASER to sample syn pyrimidine in all cases. In practice we found that this option is advantageous for low-resolution models (>2.5 Å, as a rule of thumb); for high-resolution models where it is possible to distinguish syn/anti conformers from the density map, the user may explore turning off this option. The user should carefully examine all the syn–anti pyrimidines flips in the final model if this option is turned off, to make sure there are no suspicious syn pyrimidines in the models (*see* **Note 1**).

3.5.11 Constrain χ

By default, ERRASER applies a weak constraint on the χ angle (glycosidic torsion) to favor χ conformers that are similar to the input model. The purpose is also to avoid suspicious syn–anti flips during rebuilding; only alternative χ conformers with obvious energy bonuses will be accepted into the final model. Similar to "native syn only pyrimidine" option above, this option is worth exploring for high-resolution data sets, but the user should carefully examine the resulting output (*see* **Note 1**).

The following options are accessible by clicking the "Directories" button in the main GUI.

3.5.12 Rosetta Path

If you have not followed Subheading 3.1 to set up the PHENIX_ROSETTA_PATH environmental variable, you can enter the path to the root Rosetta folder here. Otherwise just leave it blank.

3.5.13 Rosetta Eraser Directory Name

This option specifies the location of ERRASER scripts in the Rosetta root folder. By default Phenix should automatically figure out the correct path for ERRASER, but if the folder organization in future Rosetta releases change, the user can modify it here to make Phenix and ERRASER compatible.

3.6 Run ERRASER from Command Line

It is also easy to run ERRASER from command-line using the phenix.erraser command line tool (Fig. 5). As a simple example, just run

phenix.erraser model.pdb map.ccp4

Here "model.pdb" is the input pdb model, and "map.ccp4" is the density map file. For a complete list of all available options in phenix.erraser, simply run

```
phenix.erraser --help
```

For example, one can use the following command line:

```
phenix.erraser model.pdb map.ccp4 single_res_
   mode=True rebuild_res_pdb=A25 map_reso=2.0
```

This allows you to run ERRASER in the single-residue mode to rebuild residue A25. The input map resolution is also given as 2.0 Å.

```
fcchou@minthome ~ $ phenix.eraser --help


                    eraser

 ERRASER (Enumerative Real-Space Refinement ASsitted
 by Electron density under Rosetta)

# Type phenix.doc for help


usage: phenix.erraser model.pdb mapfile.ccp4 (standard erraser)
phenix.erraser model.pdb mapfile.ccp4 single_res_mode=True rebuild_res_pdb
=A30 (single_res rebuilding mode)

Values of all params:
erraser {
  input_files {
    pdb_in = None
      .help = "PDB file with starting model"
    map_file = None
      .help = "Map file (CCP4 format) 2mFo-DFc map file in CCP4 format. Rf
ree"
              "should be excluded."
    map_reso = 2.5
```

Fig. 5 The phenix.erraser command line application

3.7 Running ERRASER Using the ROSIE Server

An alternative way of running ERRASER is to use to the ROSIE online server (Fig. 6). In this way, the user does not have to install Phenix and Rosetta locally. The ROSIE ERRASER app can be found at http://rosie.rosettacommons.org/erraser/. Simply follow the online documentation to run the job. ROSIE can run ERRASER in both standard mode and single-residue rebuilding mode. Note that the ROSIE server has fewer tunable options than local installation. For example, users cannot currently carry out multiple iterations of an ERRASER run (it is still possible manually; *see* **Note 3**). One additional warning: the memory on each core of the ROSIE backend is limited; therefore the job may crash if the user inputs a very large electron density map file in ROSIE.

3.8 Advice for Phenix RNA Refinement

The refinement tool in Phenix can be accessed by the phenix.refine GUI from the main Phenix window. Comprehensive documentation for the phenix.refine GUI can be found in the Phenix website (http://www.phenix-online.org/documentation/refinement.htm). Here we will discuss some non-default options that we found useful in refining RNA crystallographic models produced by ERRASER. All of the options discussed below can be found under the "refinement setting" tab in the phenix.refine GUI. Since the refinement setting is sensitive to the initial model and diffraction data, there is no generic rule in setting up the refinement that works in all cases. The user can monitor the change of R, R_{free} and other geometric validation results of the refinement outputs to help decide whether the selected refinement strategy is appropriate.

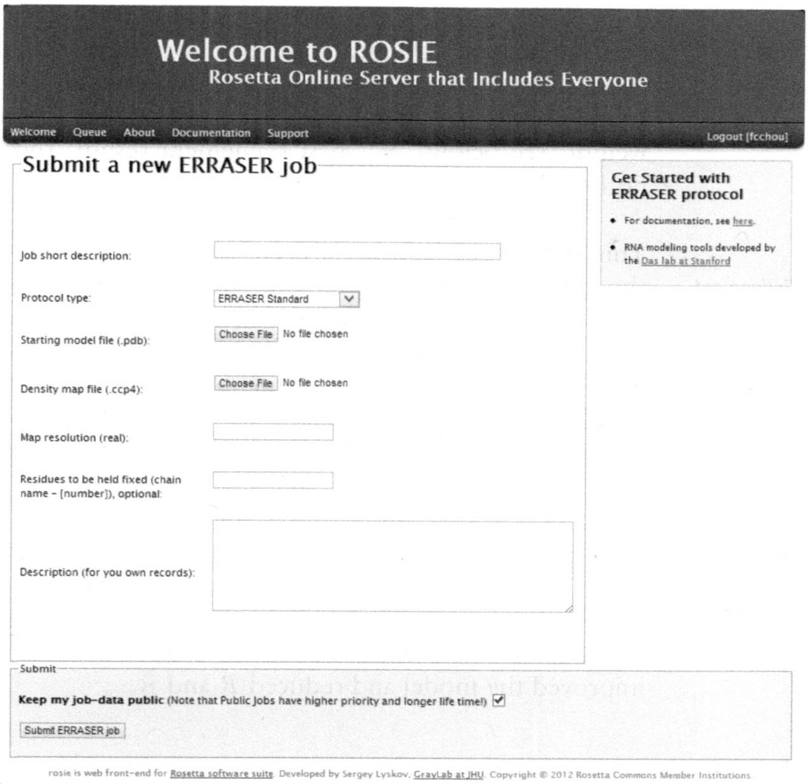

Fig. 6 The ROSIE ERRASER server application

3.8.1 Refinement Strategy: Real-Space

We recommend turning off this option during refinement of ERRASER models. By default Phenix alternates between real-space and diffraction-based refinement during a refinement cycle. As the ERRASER models have already gone through extensive real-space refinement in the ERRASER protocol, further real-space refinement in Phenix usually leads to no model improvement, or even leads to worse models in terms of clashes or other quality measures. For the exact parameters being used in refining the models in the 24 ERRASER benchmark set, please refer to the supplementary information of the original ERRASER paper.

3.8.2 Refinement Strategy: Number of Cycles

While the default of three cycles of refinement is usually good enough to refine the ERRASER models, sometimes additional cycles of refinement leads to better results. The user may try a few different numbers (usually 3–10) for best results.

3.8.3 Targets and Weighting: Optimize X-ray/ADP Weights

This option enables Phenix to perform a grid search to find the best X-ray/ADP (Atomic Displacement Parameter, also known as B-factor) weight ratio during each refinement cycle. In practice we found using this option generally leads to final models with better R and R_{free}.

3.8.4 Targets and Weighting: Optimize X-ray/Stereochemistry Weights

This option enables Phenix to perform a grid search to find the best X-ray/Stereochemistry weight ratio during each refinement cycle. In some cases this option leads to better R and R_{free}, but in other cases it is better to use a constant X-ray/Stereochemistry weight (see below). For best refinement results, the user can try both and select the best model.

3.8.5 Targets and Weighting: Refinement Target Weights: wxc_scale

This option sets constant X-ray/Stereochemistry weight during the refinement. In practice we found that it is better to use a lower wxc_scale for models with lower resolution. For example, we used the following heuristic rule in determining the wxc_scale in refining the ERRASER benchmark: wxc_scale = 0.5 for Resolution < 2.3 Å, wxc_scale = 0.1 for 2.3 Å ≤ Resolution < 3 Å, wxc_scale = 0.05 for 3 Å ≤ Resolution ≤ 3.6 Å, and wxc_scale = 0.03 for Resolution > 3.6 Å.

3.8.6 Other Options: Update Waters

The option enables automatic water picking during Phenix refinement. This procedure removes crystallographic waters lacking evident electron density and adds waters to locations with strong unoccupied electron density. In general we found this algorithm improved the model and reduced R and R_{free}.

4 Notes

1. ERRASER can introduce flips of the χ angles (glycosidic torsion) during the run; we have shown previously that many of these flips are accurate when comparisons can be made of ERRASER-refined models to higher resolution data sets. If "constrain chi" or "native syn only pyrimidine" options are turned off, χ angle flips are made more often, and some of these flips may be incorrect. In low-resolution density maps (>2.5 Å), it can be hard to determine the glycosidic conformer, especially for pyrimidines. While most χ angle flips are reasonable, there are cases where the flips are likely to be incorrect. We recommend the user to inspect visually all the χ angle flips before proceeding with the final model. Here are some general suggestions on examining the χ angle flips: First, syn conformers are rare, especially for syn pyrimidines. Any anti-to-syn flips should be closely examined. Unless there is strong electron density evidence or new hydrogen bond interactions, the flips are likely to be problematic. Syn-to-anti flips are more likely to be correct, but the user should still examine the models to make sure no important hydrogen-bonding interactions are broken during the flips.
2. Currently ERRASER does not handle crystal contacts, ligands and modified residues during the refinement. To address these

issues, the user can use the "fixed residues" option. Residues contacting ligands and residues in crystal contacts can be fixed by adding them to the list of "fixed residues." For modified residues, one should further constrain the residues which the modified residue is directly bonded to. For circular RNA or lariat connections, one can constrain the residues with additional bonds to ensure the bond does not break during ERRASER refinement. An alternative way to handle crystal contacts is to manually add in the crystal packing partners during ERRASER step; this can be achieved using the symexp utility in PyMol. Usually it is unnecessary to add in the entire molecules of the crystal packing partners; one can cut out just the relevant residues nearby. These extra add-in residues should be removed from the final ERRASER models before Phenix refinement.

3. To manually iterate ERRASER refinement in ROSIE, simply use the output pdb file of the first job as the input pdb file for a second job, and use the same input density map and options as the first iteration.

5 Acknowledgements

The authors would like to thank Sergey Lyskov for implementing and maintaining the ROSIE Rosetta server, and members of the Rosetta and the Phenix communities for discussions and code sharing. The authors acknowledge support from Howard Hughes International Student Research Fellowship (F.C.), Stanford BioX graduate student fellowship (F.C.), Burroughs-Wellcome Career Award at Scientific Interface (R.D.), NIH grants R21 GM102716 (R.D.) and P01 GM063210 (P.D. Adams, PI, to N.E. and T.T.).

References

1. Golden BL, Kim H, Chase E (2005) Crystal structure of a phage Twort group I ribozyme-product complex. Nat Struct Mol Biol 12: 82–89
2. Serganov A, Huang L, Patel DJ (2008) Structural insights into amino acid binding and gene control by a lysine riboswitch. Nature 455:1263–1267
3. Dunkle JA, Wang L, Feldman MB, Pulk A, Chen VB, Kapral GJ, Noeske J, Richardson JS, Blanchard SC, Cate JH (2011) Structures of the bacterial ribosome in classical and hybrid states of tRNA binding. Science 332:981–984
4. Davis IW, Leaver-Fay A, Chen VB, Block JN, Kapral GJ, Wang X, Murray LW, Arendall WB 3rd, Snoeyink J, Richardson JS, Richardson DC (2007) MolProbity: all-atom contacts and structure validation for proteins and nucleic acids. Nucleic Acids Res 35(suppl 2): W375–W383
5. Chen VB, Arendall WB 3rd, Headd JJ, Keedy DA, Immormino RM, Kapral GJ, Murray LW, Richardson JS, Richardson DC (2010) MolProbity: all-atom structure validation for macromolecular crystallography. Acta Cryst D 66:12–21
6. Das R, Baker D (2007) Automated de novo prediction of native-like RNA tertiary structures. Proc Natl Acad Sci U S A 104: 14664–14669

7. Das R, Karanicolas J, Baker D (2010) Atomic accuracy in predicting and designing noncanonical RNA structure. Nat Methods 7: 291–294
8. Sripakdeevong P, Kladwang W, Das R (2011) An enumerative stepwise ansatz enables atomic-accuracy RNA loop modeling. Proc Natl Acad Sci U S A 108:20573–20578
9. DiMaio F, Terwilliger TC, Read RJ, Wlodawer A, Oberdorfer G, Wagner U, Valkov E, Alon A, Fass D, Axelrod HL, Das D, Vorobiev SM, Iwaï H, Pokkuluri PR, Baker D (2011) Improved molecular replacement by density- and energy-guided protein structure optimization. Nature 473:540–543
10. DiMaio F, Tyka MD, Baker ML, Chiu W, Baker D (2009) Refinement of protein structures into low-resolution density maps using Rosetta. J Mol Biol 392:181–190
11. Adams PD, Afonine PV, Bunkoczi G, Chen VB, Davis IW, Echols N, Headd JJ, Hung LW, Kapral GJ, Grosse-Kunstleve RW, McCoy AJ, Moriarty NW, Oeffner R, Read RJ, Richardson DC, Richardson JS, Terwilligen TC, Zwart PH (2010) PHENIX: a comprehensive Python-based system for macromolecular structure solution. Acta Cryst D 66: 213–221
12. Chou FC, Sripakdeevong P, Dibrov SM, Hermann T, Das R (2013) Correcting pervasive errors in RNA crystallography through enumerative structure prediction. Nat Methods 10:74–76
13. Lyskov S, Chou FC, Conchuir SO, Der BS, Drew K, Kuroda D, Xu J, Weitzner BD, Renfrew PD, Sripakdeevong P, Borgo B, Havranek JJ, Kuhlman B, Kortemme T, Bonneau R, Gray JJ, Das R (2013) Serverification of molecular modeling applications: the Rosetta online server that includes everyone (ROSIE). PLoS One 8:e63906
14. Leaver-Fay A, Tyka M, Lewis SM, Lange OF, Thompson J, Jacak R, Kaufman K, Renfrew PD, Smith CA, Sheffler W, Davis IW, Cooper S, Treuille A, Mandell DJ, Richter F, Ban YE, Fleishman SJ, Corn J, Kortemme T, Gray JJ, Kuhlman B, Baker D, Bradley P (2011) ROSETTA3: an object-oriented software suite for the simulation and design of macromolecules. Methods Enzymol 487:545–574
15. Afonine PV, Grosse-Kunstleve RW, Echols N, Headd JJ, Moriarty NW, Mustyakimov M, Terwilliger TC, Urzhumtsev A, Zwart PH, Adams PD (2012) Towards automated crystallographic structure refinement with phenix. refine. Acta Cryst D 68:352–367
16. Sheldrick G (2008) A short history of SHELX. Acta Cryst A 64:112–122
17. Vagin AA, Steiner RA, Lebedev AA, Potterton L, McNicholas S, Long F, Murshdov GN (2004) REFMAC5 dictionary: organization of prior chemical knowledge and guidelines for its use. Acta Cryst D 12:2184–2195
18. Brunger AT (2007) Version 1.2 of the crystallography and NMR system. Nat Protocols 2:2728–2733
19. Praznikar J, Afonine PV, Guncar G, Adams PD, Turk D (2009) Averaged kick maps: less noise, more signal…and probably less bias. Acta Cryst D 65:921–931
20. Echols N, Grosse-Kunstleve RW, Afonine PV, Bunkoczi G, Chen VB, Headd JJ, McCoy AJ, Moriarty NW, Read RJ, Richardsson DC, Richardson JS, Terwillerger TC, Adams PD (2012) Graphical tools for macromolecular crystallography in PHENIX. J Appl Crystallogr 45:581–586
21. Word JM, Lovell SC, LaBean TH, Taylor HC, Zalis ME, Presley BK, Richardson JS, Richardson DC (1999) Visualizing and quantifying molecular goodness-of-fit: small-probe contact dots with explicit hydrogen atoms. J Mol Biol 285:1711–1733
22. Murray LJW, Arendall WB, Richardson DC, Richardson JS (2003) RNA backbone is rotameric. Proc Natl Acad Sci U S A 100: 13904–13909
23. Richardson JS, Schneider B, Murray LW, Kapral GJ, Immormino RM, Headd JJ, Richardson DC, Ham D, Hershkovits E, Williams LD, Keating KS, Pyle AM, Micallef D, Westbrook J, Berman HM, RNA Ontology Consortium (2008) RNA backbone: consensus all-angle conformers and modular string nomenclature (an RNA Ontology Consortium contribution). RNA 14:465–481

Chapter 18

Neutron Nucleic Acid Crystallography

Toshiyuki Chatake

Abstract

The hydration shells surrounding nucleic acids and hydrogen-bonding networks involving water molecules and nucleic acids are essential interactions for the structural stability and function of nucleic acids. Water molecules in the hydration shells influence various conformations of DNA and RNA by specific hydrogen-bonding networks, which often contribute to the chemical reactivity and molecular recognition of nucleic acids. However, X-ray crystallography could not provide a complete description of structural information with respect to hydrogen bonds. Indeed, X-ray crystallography is a powerful tool for determining the locations of water molecules, i.e., the location of the oxygen atom of H_2O; however, it is very difficult to determine the orientation of the water molecules, i.e., the orientation of the two hydrogen atoms of H_2O, because X-ray scattering from the hydrogen atom is very small.

Neutron crystallography is a specialized tool for determining the positions of hydrogen atoms. Neutrons are not diffracted by electrons, but are diffracted by atomic nuclei; accordingly, neutron scattering lengths of hydrogen and its isotopes are comparable to those of non-hydrogen atoms. Therefore, neutron crystallography can determine both of the locations and orientations of water molecules. This chapter describes the current status of neutron nucleic acid crystallographic research as well as the basic principles of neutron diffraction experiments performed on nucleic acid crystals: materials, crystallization, diffraction experiments, and structure determination.

Key words Hydration, Hydrogen bond, Neutron crystallography, Crystallization, Deuteration

1 Introduction

Polynucleotides, such as DNA and RNA, are flexible biopolymers composed of negatively charged nucleotides; therefore, their interactions with water molecules via hydrogen bonds (H-bonds) and metal cations contribute to stabilizing their double helical structures. Moreover, their higher-order structures arise from H-bonds between the helical structures, which are essential for their biological functions. Water molecules mediate not only folding but also function of nucleic acids (e.g., molecular recognition, enzymatic reactions) through the formation of H-bonds.

Eric Ennifar (ed.), *Nucleic Acid Crystallography: Methods and Protocols*, Methods in Molecular Biology, vol. 1320,
DOI 10.1007/978-1-4939-2763-0_18, © Springer Science+Business Media New York 2016

X-ray crystallography is the most popular method for determining the three-dimensional structures of nucleic acids and their hydration shells. Numerous crystal structures of nucleic acids have been determined since the early period of DNA and RNA structural studies; however, X-ray crystallography cannot provide the complete information on the nature of macromolecular H-bonds because of the low X-ray scattering power of hydrogen (H) atoms. In particular, extremely high-resolution X-ray diffraction data are necessary for determining the orientations of water molecules, which correspond to the positions of the two H atoms of H_2O. On the other hand, neutron crystallography has an advantage with respect to the positional/structural determination of H atoms and their networks. Neutrons are not diffracted by electrons, but are diffracted by atomic nuclei, and neutron scattering of H and its isotope (D: deuterium) are comparable to those from non-hydrogen atoms (*see* Table 1) [1]. To date, three DNA duplexes have been studied by neutron crystallographic analyses (*see* **Note 1**) [2–4]. In these analyses, H atoms of water molecules could be localized even at medium resolution; hence, H-bonding networks could be studied in detail. Currently, neutron crystallography is expected to become an important analytical tool for elucidating various H-bonding networks, which are closely related to nucleic acid structure and function. In this chapter, the basic principles of neutron diffraction experiments of nucleic acid crystals have been described (*see* **Note 2**).

A flow chart illustrating the steps involved in neutron crystallographic analysis of nucleic acids is shown in Fig. 1. It consists of three experimental steps: crystallization, diffraction experiment, and structure determination in a manner similar to X-ray crystallography. However, some other techniques specific to

Table 1
Neutron scattering length and incoherent cross section of elements in nucleic acids

Element symbol	Atomic number	b (fm)	σ_i (barn)
H (^1H)	1	-3.74	80.26
D (^2H)	1	6.67	2.05
C	6	6.48	0.00
N	7	9.36	0.50
O	8	5.81	0.00
P	15	5.13	0.01

b scattering length, σ_i incoherent cross section

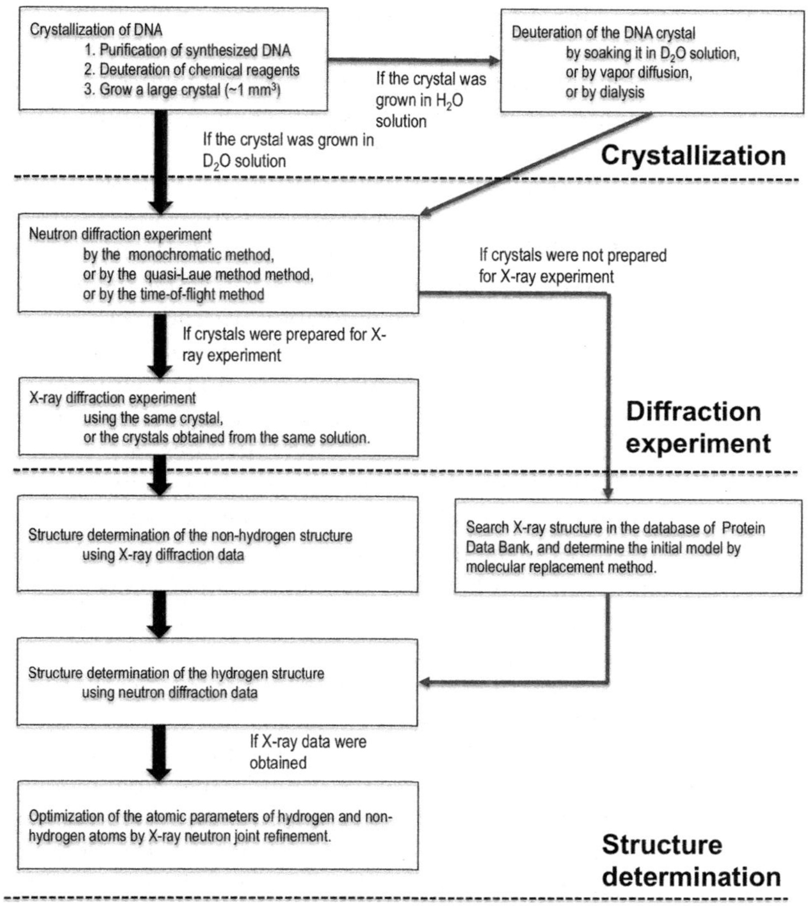

Fig. 1 A flowchart of neutron nucleic acid crystallography. *Black bold arrows* indicate the recommended protocol, which requires that a single crystal of nucleic acid is obtained from D_2O solution and that both of X-ray and neutron diffraction experiments are carried out using the same crystal

neutron crystallography are required since neutron diffraction has two distinctive features, setting it apart from X-ray diffraction. The most striking feature is the isotope effect of neutron scattering of hydrogen. Coherent neutron scattering lengths (b) of H (^{1}H) and D (^{2}H) are −3.74 fm and 6.67 fm, respectively (*see* **Note 3**). When an H atom in a crystal is replaced with a D atom (H/D replacement), the corresponding neutron density in the Fourier map increases by a factor of 1.58. Therefore, the H/D replacement provides the improvement in the identification of D atoms and optimization of their atomic parameters during structure determination. Moreover, the incoherent cross section (σ_i) of the H atom (80.2 barn) is much larger than that of the D atom (2.05 barn). The strong incoherent neutron scattering generated by H atoms gives rise to an undesirably high background of neutron diffraction images, resulting in decrease in the signal-to-noise ratio of neutron reflections. For the above reasons, H atoms in nucleic acid crystals should be replaced with D atoms as much as possible before neutron diffraction experiments (*see* **Note 4**). This replacement is called "deuteration" in neutron crystallography.

Another feature characteristic of neutron crystallography is that the flux of neutron beam is relatively lower than those of X-ray beams. While the resolution limit of X-ray crystallography usually depends on only the nucleic acid crystal quality, the resolution limit of neutron crystallography depends on both of the quality and the volume of the crystal. Crystal volume is an especially important factor in neutron diffraction experiments; small crystals used for X-ray crystallography cannot provide effective diffraction data in neutron crystallography. In cases of protein crystals, their volume should be at least larger than 1 mm^3 (*see* **Note 5**) [5]. Although unit cells of nucleic acid crystals are usually smaller than those of protein crystals, a crystal larger than 0.1 mm^3 is recommended for obtaining sufficient neutron reflections for determining the precise positions of H and D atoms (*see* **Note 6**) [4, 6, 7]. Due to the above reasons, a "deuterated" and "large" nucleic acid crystal is required for neutron nucleic acid crystallography.

In addition, the X-ray diffraction experiment of the crystal is necessary before determining positions of H and D atoms using neutron diffraction data. Currently, the initial model for structural refinement cannot be determined only by neutron crystallography. Therefore, the non-hydrogen structure of nucleic acids and water must first be elucidated by X-ray crystallography. The best method to obtain the non-hydrogen structure is to collect X-ray diffraction data using the crystals that are used in neutron diffraction experiments. In the following sections, the principles and protocols of crystallization, diffraction experiments, and structure determination have been described. These protocols are based on previous neutron studies of DNA, but would be applicable for other nucleic acids such as RNA.

2 Material

1. D_2O, $\geq$ 99.9 %.
2. Chemically synthesized oligonucleotides.
3. Sephadex XK 16/40 G10 column (GE Healthcare, Pittsburgh, PA, USA).
4. Chromatography system (GE Healthcare AKTA or equivalent).
5. Vacuum concentrator.
6. Heating block.

3 Methods

3.1 Purification and Deuteration of Nucleic Acids

As mentioned in the previous section, crystals should be deuterated prior to neutron crystallographic experiments. Materials for crystallization are deuterated by treatment with heavy water (D_2O) or are obtained commercially. Treatment of crystals with D_2O replaces those exchangeable H atoms in DNA, namely, the N–H and O–H groups, with D atoms; H atoms covalently bonded to carbon atoms are not spontaneously replaced with D atoms (*see* **Note 7**). Since most H atoms in nucleic acid crystals are in the solvent, more than half of the H atoms in a crystallization solution can be replaced with D atoms without any D_2O treatment of materials. Therefore, use of a D_2O solution during crystallization is a primary method for facile deuteration.

3.1.1 DNA

1. Synthesize oligonucleotides in a DNA synthesizer or purchase synthesized DNA (HPLC grade) from a commercial source (*see* **Note 8**). Dissolve the oligonucleotide in 1 mL H_2O.
2. In order to purify the synthesized oligonucleotide by gel-filtration chromatography on a fast chromatography system (*see* **Note 9**), equilibrate Sephadex G10 column with 100 mL H_2O.
3. Apply 1 mL DNA solution.
4. Elute DNA, maintaining a flow rate of 1–2 mL/min. Collect DNA fraction by monitoring UV absorption ($\lambda = 260$ or 280 nm) and conductivity.
5. Lyophilize DNA using a freeze-drying equipment.
6. Dissolve the lyophilized DNA in pure D_2O to make twice the concentration in DNA crystallization (2 × DNA solution).
7. Anneal 2 × DNA solution for a few minutes at a temperature higher than the T_m of DNA (*see* **Note 10**) [8], and then cool it to room temperature.

3.1.2 Preparation of Crystallization Solution (see Note 11)

1. Dissolve salt in a D_2O solution to prepare a saturated salt solution.
2. Remove water from the salt by an evaporator.
3. Repeat **steps 1–2** three times.
4. If an alcohol is required as a precipitant during crystallization, e.g., 2-propanol and 2-methyl-2,4-pentanediol (MPD), the deuterated alcohol should be purchased from a commercial source.
5. Dissolve salts, alcohol, and buffering agents in ~40 % of the total D_2O volume required for the crystallization solution.
6. Adjust pH using 1.0 M HCl or 1.0 M NaOH (*see* **Note 12**) [9].

7. Add D_2O up to twice the concentration of chemical reagents (2 × crystallization solution).
8. Mix equal volumes of 2 × DNA solution (3.1.1.4) and 2 × crystallization solution to prepare an aliquot of DNA crystallization solution.

3.2 Crystallization of DNA

Various crystallization techniques have been applied to obtain large single crystals of DNA oligomers. If researchers have already established the crystallization protocol for their target nucleic acid, it is best to try the same crystallization method in D_2O solution instead of H_2O. It is possible that an isotope effect derived from deuteration shifts the optimum crystallization conditions of macromolecules. The isotope effect is usually negligibly small for neutron crystallography [10–14], but is sometimes large [15]. Crystallization conditions ought to be optimized in a D_2O crystallization solution in order to proceed smoothly to large-scale crystallization. A list of DNA oligomer crystallization methods known thus for neutron crystallography have been given below.

3.2.1 A-DNA [6]

A-DNA d(AGGGGCCCCT)$_2$ was crystallized with the sitting-drop vapor diffusion method. Crystallization parameters were optimized from the previously published protocol for crystallization in H_2O solution [16]. In order to obtain improvements in crystallization, vapor diffusion was controlled by application of a layer of Al's oil (1:1 paraffin/silicone oil) [17], wherein the oil on the top of the reservoir solution suppresses vapor diffusion so that crystallization can be carried out under milder conditions. A single crystal with a size of 0.06 mm^3 grew in a 30 μL crystallization droplet (*see* **Note 13**).

3.2.2 B-DNA [7]

First, the best conditions for crystal growth of B-DNA d(CCATTAATGG)$_2$ were acquired using a crystallization phase-diagram technique. The phase diagrams for $MgCl_2$–DNA systems were determined by the microbatch method. Second, the quality of crystals obtained under various conditions was investigated by X-ray diffraction experiments. The best conditions were found near the DNA solubility minimum point on the phase diagram. Based on data from these results, the largest single crystal, which was 2.77 mm^3 in size, was obtained by growth in a large-scale batch (2 mL) (*see* **Note 14**).

3.2.3 Z-DNA [3, 18, 19]

A thermal reversible process was employed for the crystallization of Z-DNA d(CGCGCG)$_2$ [3]. A double-stranded DNA molecule is converted into two single-stranded DNA molecules at high temperature to increase its water solubility. The crystallization solution was heated above the T_m of DNA and was then cooled slowly. Crystallization conditions could be optimized in the same manner as that of the microbatch method with the exception of differences in temperature [18]. Droplets of the DNA solution (4 μL) were applied on the individual wells of a 96-well microbatch plate, and each well was covered with 20 μL of paraffin oil to avoid evaporation. The microbatch plate was incubated at 343 K for 12 h and was then cooled down from 343 K to 293 K at a rate of 5 K/day for 10 day. Large-scale crystallization was carried out using 1 mL in a micro tube crystallization solution to obtain the largest crystal (1.6 mm^3) (*see* **Note 15**).

Slow evaporation method was also applied for obtaining the large single crystals of the Z-DNA for neutron crystallographic study [19]. A silica tube containing 200 μL crystallization solution was plugged at with at both ends with beeswax. The beeswax plugs had a small hole (ϕ~200 μm), cause slow evaporation. The large Z-DNA crystal, which was 0.7 mm^3 in size, was obtained over several weeks (*see* **Note 16**).

3.2.4 Deuteration After Crystallization

If a large nucleic acid crystal is obtained from a H_2O solution, the undeuterated crystal can be converted to the deuterated crystal by treatment with D_2O solution. The simplest way is soaking; a crystallization solution is prepared using a D_2O solution, and the crystal is soaked in the deuterated crystallization solution. The first neutron crystallographic analysis of a macromolecule was carried out using myoglobin crystals soaked in D_2O solution [20]. However, the soaking technique increases the risk of obtaining cracked or degraded crystals. Dialysis using a dialysis button enables gentle deuteration and alleviates crystal degradation. These treatments were usually carried out over a span of several months to years. Another method is the application of vapor diffusion [21]. This method was used for deuteration of crystalline ribonuclease A with a size of 2.5 mm^3. The crystal was placed on the bottom of a 40 mm-length capillary. The capillary was turned upside down, and the end of the capillary was soaked in D_2O solution containing protein and crystallization reagents. After several months, it was confirmed that H_2O molecules and exchangeable H atoms of the protein were replaced with D atoms. To date, a Z-DNA crystal, which was obtained by slow evaporation method, was deuterated by vapor diffusion method [19].

3.3 Diffraction Experiments

3.3.1 Neutron Diffraction Experiment

Neutron diffraction experiments of bimolecular crystals are carried out at neutron research facilities operating research reactors or neutron spallation sources. Neutron diffractometers dedicated to neutron crystallography are summarized in Table 2 [22–29] (*see* **Note 17**). The list of neutron diffractometers is limited, as neutron crystallography of macromolecules requires a high-flux neutron source. Each neutron diffractometer adopts one of three diffraction techniques: monochromatic, quasi-Laue, or time-of-flight (TOF). Each of these techniques has both its advantages and

Table 2
Neutron diffractometers for macromolecular crystallography

Diffractometer	Neutron source	Diffraction techniques	Status
BIX-3	JRR-3, Japan	Monochromatic	Working[a] [22]
BIX-4	JRR-3, Japan	Monochromatic	Working[a] [23]
BioDiff	FRM-II, Germany	Monochromatic	Working[a]
D-19	ILL, France	Monochromatic	Working[a] [24]
BioC	HANARO, Korea	Monochromatic	Under construction
LADI-III	ILL, France	Quasi-Laue	Working[a] [25]
IMAGINE	HFIR, USA	Quasi-Laue	Working[a] [26]
PCS	LANSCE, USA	Time-of-flight	Working[a] [27]
iBIX	J-PARC, Japan	Time-of-flight	Working[a] [28]
ManDi	SNS, USA	Time-of-flight	Under construction [29]
NMX	ESS, Sweden	Time-of-flight	Under construction

[a]Detailed diffractometer specifications and contact address are available at the corresponding Web sites

drawbacks. The principles of these techniques have been described in the following subsections.

When a researcher succeeds in obtaining a large nucleic acid single crystal, he submits the application proposal for his neutron experiment to the neutron facility operating the desired neutron diffractometer. After the acceptance of the proposal, he carries out his experimental plan according to the proposal. If the researcher has no experience performing any neutron experiments, it is recommended that he contacts the person responsible for the neutron diffractometer and takes counsel regarding the plan and proposal of his neutron experiment. Since it usually takes several months from proposal submission to performing the experiment, the crystal must be stored at a stable state until the neutron experiment.

3.3.2 Neutron Diffractometers

1. Neutron diffraction experiments using a monochromatic diffractometer are carried out in a manner similar to general X-ray crystallography. A monochromator is used to isolate the desired neutron wavelength from a white neutron beam originating from a nuclear reactor; the monochromatic neutron beam is then focused through a super-mirror path. Neutron diffractions from a nucleic acid crystal are recorded on a two-dimensional neutron detector (neutron imaging plate or position-sensitive detector). The signal-to-noise ratios of intensities of the diffractions are usually much better than those of the quasi-Laue method, as described in the next

subsection. On the other hand, recording the numerous diffraction images is necessary to collect all neutron reflections, resulting in longer experimental times.

2. Quasi-Laue neutron diffractometer: this type of diffractometer is also installed at nuclear reactors. A continuous neutron beam with a restricted wavelength band is extracted from a white neutron beam through Ni/Ti multilayer band-pass filters [30]. This continuous beam reduces background scattering and reflection overlap compared with use of the full white beam. Quasi-Laue neutron diffractometry allows for shorter data-measurement times, and greater neutron diffractions than those of monochromatic neutron diffractometers, however a disadvantage of this technique is a higher background and reflection overlap.
3. Time-of-flight neutron diffractometer: Time-of-flight neutron experiments for macromolecule crystals are carried out at powerful proton accelerator facilities. Accelerated particles hit a heavy-metal target at the intervals of 1/60–1/20 s (*see* **Note 18**), and pulsed neutron beams are generated by the incidents. The flight time of the neutrons from the heavy-metal target to neutron detectors via the sample is inversely proportional to neutron velocity, which corresponds to the neutron wavelength. Since time-of-flight neutron diffractometers are equipped with time-resolved detectors, the wavelength of each neutron reflections can be distinguished according to the time of arrival of the neutron at the detectors. This type of neutron detector is expected to collect a large number of neutron reflections simultaneously, while maintaining the accuracy of measurements.

3.3.3 Diffraction Experiment Protocols

1. Seal a large nucleic acid single crystal in a capillary (*see* **Note 19**): prepare a capillary suitable for the neutron experiment (*see* **Note 20**). Seal a crystal with sufficient crystallization solution in the capillary. Store the capillary, while keeping the temperature at crystallization temperature.
2. Transport the capillary to a neutron facility (*see* **Note 21**).
3. Conduct a preliminary neutron experiment to the check crystal quality and estimate the total measurement time for the entire data collection.
4. Collect complete neutron diffraction data (*see* **Note 22**).
5. Collect X-ray diffraction data after the neutron experiment (*see* **Note 23**). In a synchrotron facility, the X-ray experiment can be performed using the intact capillary from the neutron experiment, since the X-ray beam is bright enough to permeate the capillary. When a rotating anode is used, crystals must be transferred to a capillary for X-ray experiments. If the size of

the crystal is larger than the inner diameter of the capillary the crystal should be cut into smaller pieces, one of which is sealed with the mother liquor in an X-ray capillary. If an X-ray experiment is not possible, the atomic coordinates may be obtained from the Protein Data Bank online database (www.rcsb.org) (*see* **Note 24**).

3.4 Structure Determination

The initial model for neutron structure determination is derived from the X-ray structure. First, the H and D atoms present in the nucleic acid are placed into the model (with the exception of the D atoms of the –OD functional groups), since the positions of these H and D atoms can be calculated stereochemically. Second, their atomic parameters are optimized using the refinement program modified for neutron crystallography. Neutron refinement can be carried out using the program for X-ray refinement with the exchange of X-ray scattering factors for elements with neutron scattering lengths. In previous neutron crystallographic studies, the refinements were performed in this manner; presently, the module for neutron refinement is implemented in the Phenix program [31] (*see* **Note 25**). Since Phenix has modules for both X-ray and neutron refinements, all refinement process in neutron crystallography can be carried out using this program. After optimization of the atomic parameters, the hydrogen model is built on molecular graphics applications such as Coot [32] or MIFit [33] (*see* **Note 26**). D atoms of –OD functional groups, D_2O molecules, and ligand molecules (such as polyamines and drugs) were included in the model by referring neutron ($2|F_o|-|F_c|$) and ($|F_o|-|F_c|$) Fourier density maps. Next, the atomic parameters are optimized by X-ray-neutron joint refinement in which both X-ray and neutron data are included in the refinement [34]. The final structure, as shown in Fig. 2, is obtained after several repetitions of the model building and X-ray-neutron joint refinements.

3.4.1 Protocols of Structure Determination

1. Determine positions of non-hydrogen atoms using X-ray diffraction data (*see* **Note 27**).
2. First neutron refinement using Phenix:
   (a) Start Phenix, and launch “phenix.refine”.
   (b) Load the coordinate file and the neutron reflection file in the window for input (*see* **Note 28**).
   (c) Launch “ReadySet” by clicking on the bottom in “phenix.refine”.
   (d) Select checkboxes (*see* **Note 29**).
   (e) Run “ReadySet” (*see* **Note 30**).
   (f) Close “ReadySet”.

(g) Select the "Refinement setting" tab of "phenix.refine", and change the scattering table from "n_gaussian" to "neutron".

(h) Run "phenix.refine".

3. Model building using Coot on a graphic workstation:

   (a) After refinement using "phenix.refine", the model and Fourier maps are loaded in "Coot" (*see* **Note 31**).

   (b) Check and modify the nucleic acid model (*see* **Note 32**).

   (c) Assign positive peaks as D atoms by referring ($|F_o|-|F_c|$) and ($2|F_o|-|F_c|$) neutron Fourier maps.

4. X-ray and neutron joint refinement (skip this step, if there are no X-ray data):

   (a) Load the modified coordinate file, X-ray reflection file, and neutron reflection file, and assign the data type of the neutron reflections file as neutron data.

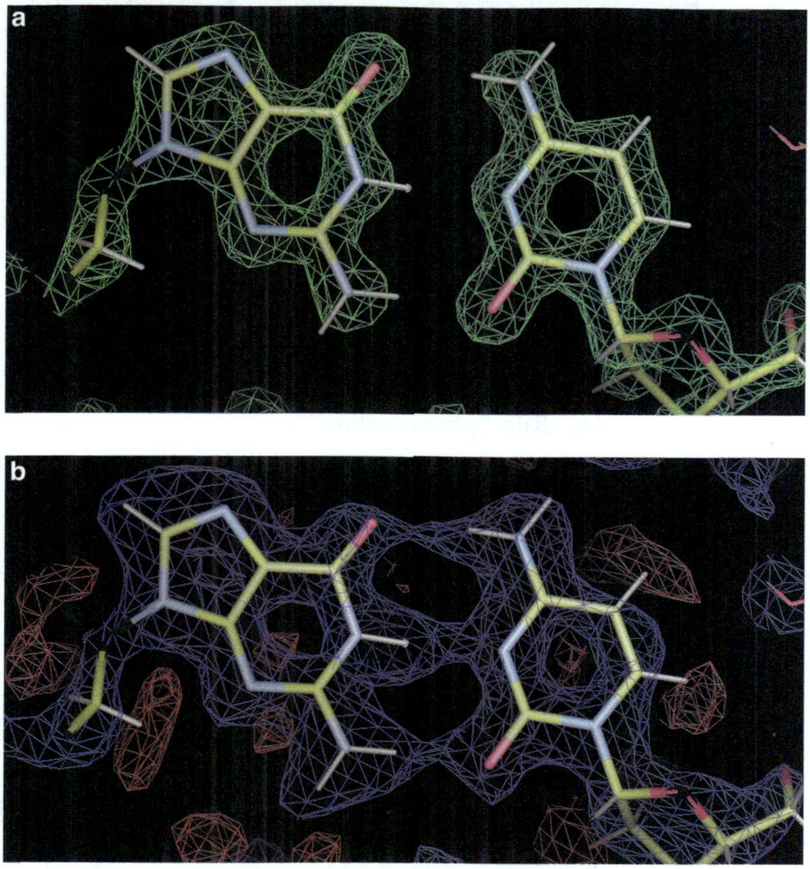

Fig. 2 $2|F_o|-|F_c|$ Fourier maps of DNA d(CGCGCG)$_2$. ((**a**) X-ray map, (**b**) neutron map, (**c**) X-ray and neutron composite map). *Blue*, *red*, and *green* contours indicate positive neutron, negative neutron, and positive X-ray densities. Deuterium and hydrogen atoms have positive and negative densities on neutron maps, respectively. These maps were calculated using 1.6 Å neutron (PDB entry: 1v9g) and 1.4 Å X-ray (PDB entry: 1woe) data of Z-DNA d(CGCGCG)$_2$

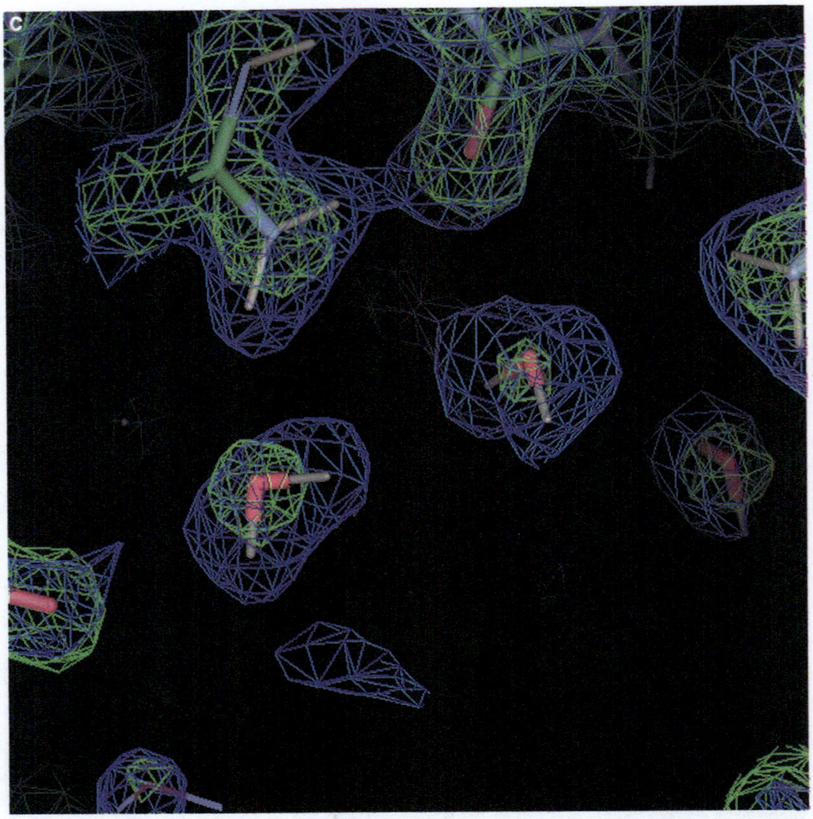

Fig. 2 (continued)

(b) Run "ReadySet" (*see* **Note 33**).

(c) Run "phenix.refine".

5. Repeat **steps 3–4** to reach the final model (in case of no X-ray data are available, **steps 2–3**).

4 Notes

1. Neutron diffraction experiments for A-DNA d(AGGGGC CCCT)$_2$ [2], B-DNA d(CCATTAATGG)$_2$ [3], and Z-DNA d(CGCGCG)$_2$ [4] were carried out at 2.4, 3.0, and 1.8 Å resolutions, respectively.
2. Further details of neutron macromolecular crystallography are described in "Neutron Protein Crystallography: Hydrogen, Protons, and Hydration in Bio-macromolecules," written by Nobuo Niimura and Alberto Podjarny (IUCr Monographs on Crystallography–25, Oxford University Press, 2011, ISBN 978-0-19-957886-3).

3. Neutron scattering length corresponds to the scattering factor in X-ray crystallography, Negative densities are observed at hydrogen atom positions on neutron Fourier maps due to negative scattering length (*see* Fig. 2).
4. Recently built high-performance neutron diffractometers enable the collection of neutron diffraction data without the replacement of H atoms with D atoms: 1.85 Å neutron diffraction images of the Z-DNA d(CGCGCG)$_2$ crystal obtained from H_2O solution were obtained using the iBIX diffractometer installed at J-PARC (unpublished data).
5. Biomacromolecular crystals for which neutron data had been collected were summarized in 2008 (Table 2 in ref. 5).
6. The volumes of DNA crystals, which were used for neutron crystallography, were 0.06 mm^3 for A-DNA [6], 2.77 mm^3 for B-DNA [7], and 1.6 mm^3 for Z-DNA [4], respectively.
7. In the case of Z-DNA d(CGCGCG)$_2$, 26 % of the H atoms of the Z-DNA are deuterated. In order to achieve the complete DNA deuteration, DNA should be prepared from fully deuterated nucleotides by chemical synthesis or enzymatic reactions. Therefore, the fully deuterated DNA molecule is too expensive for general structural studies of nucleic acids.
8. 1–10 mg synthesized DNA is required to obtain a single large crystal in order to perform neutron diffraction experiments. The minimum requirement varies depending on the crystallization technique used.
9. A large amount of DNA is lost due to nonspecific absorption in the Sephadex G10 medium during gel filtration, since pure H_2O is used for elution instead of buffer. Another purification method for avoiding this loss involves the combination of gel filtration performed using a low-salt solution and ionic-exchange chromatography.
10. The T_m value of DNA is predicted by the equation proposed by Sugimoto et al. [8].
11. The water contents of additive salts (e.g., $MgCl_2 \cdot 6H_2O$) and buffering agents (ex sodium cacodylate trihydrate $(CH_3)_2AsO_2Na \cdot 3H_2O$), are negligibly small in comparison with total water content. Hence, it is not essential to deuterate these chemicals for crystallization. However, when ammonium salts such as $(NH_4)_2SO_4$ and NH_4Cl are used for a precipitant, these salts must be deuterated because of their large amount of hydrogen atoms.
12. The pK_a value measured in D_2O solution is 0.4 higher than that in H_2O solution due to the deuterium isotope effect; pD is the pH meter reading plus 0.4 (pD = pH + 0.4) [8].

13. A crystallization solution containing 2 mM DNA, 20 mM sodium cacodylate (pD 6.0), 5 mM hexammine(III) cobalt chloride, 60 mM $MgCl_2$, 40 mM LiCl, and 5 % MPD was equilibrated to 35 % MPD at room temperature.
14. A crystallization solution (2 mL) containing 2 mM DNA, 100 mM sodium cacodylate (pD 6.6), 100 mM $MgCl_2$, and 30 % MPD was stored at 279 K.
15. A crystallization solution (1.0 mL) containing 1.4 mM DNA, 100 mM sodium cacodylate (pD 6.6), 30 mM spermine tetrahydrochloride, 30 mM $MgCl_2$, and 7.5 % MPD was prepared in a micro tube. The tube was then placed into a 353 K heat block for 12 h; the temperature of the heat block was then decreased to 298 K at a rate of 2 K/day.
16. A crystallization solution (200 μL) containing 25 mg/ml DNA, 5 mM $Mg(Ac)_2(3H_2O)$, 25 mM MES pH(5.6), and 1.25 M Ammonium sulfate.
17. BIX-3&4: http://jrr3uo.jaea.go.jp/about/institution/bix3.htm (in Japanese)

    BioDiff: http://www.mlz-garching.de/biodiff

    D-19: http://www.ill.eu/instruments-support/instruments-groups/instruments/d19/

    description/instrument-layout/

    LADI-III:http://www.ill.eu/instruments-support/instruments-groups/instruments/ladi-iii/

    description/instrument-layout/

    IMAGINE: http://neutrons.ornl.gov/imagine/

    iBIX: http://j-parc.jp/researcher/MatLife/en/instrumentation/ns_spec.html#bl03

18. The pulse frequencies are 20 Hz, 25 Hz, 60 Hz, and 60 Hz for LANSCE, J-PARC, SNS, and ESS, respectively.
19. Neutron diffractometers are usually equipped attachments for cryogenic experiments. Neutron diffraction experiments at liquid nitrogen temperatures (~100 K) sometimes improve the effective resolution of neutron data. On the other hand, the thermal dynamic behavior of water molecules in biological environments cannot be observed. All previous neutron structure determinations of nucleic acids were carried out without cryogenic conditions.
20. Since neutron absorption is much lower than X-ray absorption, thicker transparent tubes are better to avoid accidental capillary breakage. Quartz capillaries are considered ideal; borosilicate

glass contains boron that has exceptionally high neutron absorption. NMR sample tube ($\phi = 3–4$ mm) is also used for neutron bimolecular crystallography.

21. Crystallization solution of nucleic acid generally contains cacodylate, which is highly toxic chemical reagent. That is a point to notice in oversea transportation.
22. Indexing and integration of neutron diffractions are carried out using a processing program that is specialized for each neutron diffractometer. Every neutron facility provides adequate support for beginners to neutron diffraction experiments.
23. Atomic coordinates of non-hydrogen atoms of a nucleic acid crystal must be determined before its neutron structure determination, as mentioned in the Introduction. The neutron beam does not damage the crystal; therefore, it is ideal to use the same crystal for both of neutron and X-ray diffraction experiments. If the same crystal cannot be used for X-ray experiments, a crystal grown in the same crystallization solution should be used. The X-ray experiment must be carried out at the same temperature as the neutron experiment, to maintain isomorphism of the crystal.
24. The PDB should be the last choice for use. X-ray and neutron joint refinement (an advanced refinement method for neutron crystallography, described in the next section) cannot be performed without X-ray diffraction data.
25. Phenix is available on the Web (http://www.phenix-online.org/)
26. Coot and MIFit are available at the corresponding Web sites: on the Web sites:

    (http://www2.mrc-lmb.cam.ac.uk/personal/pemsley/coot/; https://github.com/mifit)
27. X-ray structure can be determined by conventional methods. When an X-ray diffraction experiment is not carried out with the nucleic acid crystal prepared in D_2O solution, the isomorphous crystal structure, which has the same space group and cell parameters close to those of the crystals in neutron experiment, must be obtained on the database. In this case, rigid body refinement should be applied before neutron structure determination.
28. The reflection file is assigned as "X-ray data, X-ray R-free", respectively. Do not change the data type of the reflection file.
29. In the first run of "ReadySet", check the following five boxes; "Add hydrogens to model if absent", "Convert exchangeable sites to deuterium", "Generate ligand restraints", "Metal ion coordination restraints", and "Output edits determined by LINK records".

30. After termination of "ReadySet", the coordinate file including hydrogen atoms of the nucleic acid is automatically loaded in "phenix.refine".
31. If the automatic loading is not working, start "Coot", and open the model and maps by "Open coordinate" and "Open MTZ, mmCIF, fcf or phs", respectively.
32. In neutron crystallography, it is possible to find appreciable protonation and H/D exchange of bases. In previous studies, the N^7 position of guanine base of A-DNA was protonated [2], and hydrogen atoms of C_8–H_8 bonds of guanine bases of Z-DNA were partially replaced with deuterium atoms [3]. When H/D exchange is observed, the coordinate file should be modified as follows. The occupancies of the H and D atoms are optimized in "phenix.refine".

    ATOM 57 H8 A DG A 2 21.678 10.733 29.831 0.50 9.11

    ATOM 58 D8 B DG A 2 21.677 10.739 29.829 0.50 9.11
33. When the solvent molecules are treated, the setting is a combination of "Add hydrogens to model if absent", "Add deuteriums to solvent molecules", "Convert all possible sites to deuterium", "Generate ligand restraints, Metal ion coordination restraints", "Output edits determined by LINK records."

Acknowledgement

This work was supported in part by JSPS KAKENHI Grant numbers 18790030 and 23770176.

References

1. Rauch A, Waschkowski W (2003) Neutron scattering length. In: Dianoux A-J, Lander G (eds) ILL neutron data booklet, 2nd edn. Old City publishing, Philadelphia, PA, pp 1.1-8–1.1-9
2. Leal RM, Callow S, Callow P, Blakeley MP, Cardin CJ, Denny WA, Teixeira SC, Mitchell EP, Forsyth VT (2010) Combined neutron and X-ray diffraction studies of DNA in crystals and solutions. Acta Crystallogr D Biol Crystallogr 66:1244–1248
3. Arai S, Chatake T, Ohhara T, Kurihara K, Tanaka I, Suzuki N, Fujimoto Z, Mizuno H, Niimura N (2005) Complicated water orientations in the minor groove of the B-DNA decamer d(CCATTAATGG)$_2$ observed by neutron diffraction measurements. Nucleic Acids Res 33:3017–3024
4. Chatake T, Tanaka I, Umino H, Arai S, Niimura N (2005) The hydration structure of a Z-DNA hexameric duplex determined by a neutron diffraction technique. Acta Crystallogr D Biol Crystallogr 61:1088–1098
5. Myles DDA (2006) Neutron protein crystallography: current status and a brighter future. Curr Opin Struct Biol 16:630–637
6. Leal RM, Teixeira SC, Blakeley MP, Mitchell EP, Forsyth VT (2009) A preliminary neutron crystallographic study of an A-DNA crystal. Acta Crystallogr Sect F Struct Biol Cryst Commun 65:232–235
7. Arai S, Chatake T, Suzuki N, Mizuno H, Niimura N (2004) More rapid evaluation of biomacromolecular crystals for diffraction experiments. Acta Crystallogr D Biol Crystallogr 60:1032–1039

8. Sugimoto N, Nakano S, Yoneyama M, Honda K (1996) Improved thermodynamic parameters and helix initiation factor to predict stability of DNA duplexes. Nucleic Acids Res 24: 4501–4505
9. Glasoe PF, Long FA (1960) Use of glass electrodes to measure acidities in deuterium oxide. J Phys Chem 64:188–193
10. Meilleur F, Dauvergne MT, Schlichting I, Myles DA (2005) Production and X-ray crystallographic analysis of fully deuterated cytochrome P450cam. Acta Crystallogr D Biol Crystallogr 61:539–544
11. Artero JB, Hartlein M, McSweeney S, Timmins P (2005) A comparison of refined X-ray structures of hydrogenated and perdeuterated rat gammaE-crystallin in H_2O and D_2O. Acta Crystallogr D Biol Crystallogr 61:1541–1549
12. Di Costanzo L, Moulin M, Haertlein M, Meilleur F, Christianson DW (2007) Expression, purification, assay, and crystal structure of perdeuterated human arginase I. Arch Biochem Biophys 465:82–89
13. Blum MM, Tomanicek SJ, John H, Hanson BL, Schoenborn BP, Lagan P, Chen JC (2010) X-ray structure of perdeuterated diisopropyl fluorophosphatase (DFPase): perdeuteration of proteins for neutron diffraction. Acta Crystallogr Sect F Struct Biol Cryst Commun 66:379–385
14. Chatake T, Ishikawa T, Yanagisawa Y, Yamada T, Tanaka I, Fujiwara S, Morimoro Y (2011) High-resolution X-ray study of the effects of deuteration on crystal growth and the crystal structure of proteinase K. Acta Crystallogr Sect F Struct Biol Cryst Commun 67: 1334–1338
15. Liu X, Hanson BL, Langan P, Viola RE (2007) The effect of deuteration on protein structure: a high-resolution comparison of hydrogenous and perdeuterated haloalkane dehalogenase. Acta Crystallogr D Biol Crystallogr 63: 1000–1008
16. Gao YG, Robinson H, Wang AH (1999) High-resolution A-DNA crystal structures of d(AGGGGCCCCT). An A-DNA model of poly(dG) x poly(dC). Eur J Biochem 261: 413–420
17. Chayen NE (1997) A novel technique to control the rate of vapour diffusion, giving larger protein crystals. J Appl Crystallogr 30: 198–202
18. Chatake T, Sazaki G, Kikkou T, Fujiwara S, Ishikawa T, Matsumoto O, Morimoto Y (2010) An approach to DNA crystallization using the thermal reversible process of DNA duplexes. Cryst Growth Des 10:1090–1095
19. Langan P, Li X, Hanson BL, Coates L, Mustyakimov M (2006) Synthesis, capillary crystallization and preliminary joint X-ray and neutron crystallographic study of Z-DNA without polyamine at low pH. Acta Crystallogr Sect F Struct Biol Cryst Commun 62: 453–456
20. Schoenborn BP (2010) A history of neutrons in biology: the development of neutron protein crystallography at BNL and LANL. Acta Crystallogr D Biol Crystallogr 66:1262–1268
21. Chatake T, Fujiwara S, Morimoto Y (2010) Crystallization of ribonuclease A and H/D exchange for neutron diffraction experiment. KURRI Progress Rep 2009:164
22. Tanaka I, Kurihara K, Chatake T, Niimura N (2002) A high-performance neutron diffractometer for biological crystallography (BIX-3). J Appl Crystallogr 35:34–40
23. Kurihara K, Tanaka I, Niimura N, Muslih MR, Ostermann A (2004) A new neutron single crystal diffractometer dedicated for biological macromolecules (BIX-4). J Synchrotron Radiat 11:68–71
24. Jogl G, Wang X, Mason SA, Kovalenvsky A, Mustyakimov M, Fisher Z, Hoffman C, Kratky C, Langan P (2011) High-resolution neutron crystallographic studies of the hydration of the coenzyme cob(II)alamin. Acta Crystallogr D Biol Crystallogr 67:584–591
25. Blakeley MP, Teixeira SCM, Petit-Haertlein I, Hazemann I, Mitschler A, Haertlein M, Howard E, Podjarny AD (2010) Neutron macromolecular crystallography with LADI-III. Acta Crystallogr D Biol Crystallogr 66:1198–1205
26. Meilleur F, Munshi P, Robertson L, Stoica AD, Crow L, Kovalevsky A, Koritsanszky T, Chakoumakos BC, Blessing R, Myles DA (2013) The IMAGINE instrument: first neutron protein structure and new capabilities for neutron macromolecular crystallography. Acta Crystallogr D Biol Crystallogr 69:2157–2160
27. Langan P, Greene G (2004) Protein crystallography with spallation neutrons: collecting and processing wavelength-resolved Laue protein data. J Appl Crystallogr 37:253–257
28. Tanaka T, Kusaka K, Hosoya T, Niimura N, Ohhara T, Kurihara K, Yamada T, Ohnishi Y, Tomoyori K, Yokoyama T (2010) Neutron structure analysis using the IBARAKI biological crystal diffractometer (iBIX) at J-PARC. Acta Crystallogr D Biol Crystallogr 66:1194–1197
29. Coates L, Stoica AD, Hoffmann C, Richard J, Cooper R (2010) The macromolecular neutron diffractometer (MaNDi) at the Spallation Neutron Source, Oak Ridge: enhanced optics

design, high-resolution neutron detectors and simulated diffraction. J Appl Crystallogr 43:570–577
30. Schoenborn BP (1992) Multilayer monochromators and supermirrors for neutron protein crystallography using a quasi-Laue technique. Proc SPIE 23:192–198
31. Adams PD, Afonine PV, Bunkóczi G, Chen VB, Davis IW, Echols N, Headd JJ, Hung LW, Kapral GJ, Grosse-Kunstleve RW, McCoy AJ, Moriarty NW, Oeffner R, Read RJ, Richardson DC, Richardson JS, Terwilliger TC, Zwart PH (2010) PHENIX: a comprehensive Python-based system for macromolecular structure solution. Acta Crystallogr D Biol Crystallogr 66:213–221
32. Emsley P, Lohkamp B, Scott WG, Cowtan K (2010) Features and development of coot. Acta Crystallogr D Biol Crystallogr 66: 486–501
33. McRee DE (2004) Differential evolution for protein crystallographic optimizations. Acta Crystallogr D Biol Crystallogr 60:2276–2279
34. Afonine PV, Mustyakimov M, Grosse-Kunstleve RW, Moriarty NW, Langan P, Adams PD (2010) Joint X-ray and neutron refinement with phenix.refine. Acta Crystallogr D Biol Crystallogr 66:1153–1163

Part IV

Functional Examples

Chapter 19

Reconstitution of Functionally Active *Thermus thermophilus* 30S Ribosomal Subunit from Ribosomal 16S RNA and Ribosomal Proteins

Sultan Agalarov, Marat Yusupov, and Gulnara Yusupova

Abstract

In vitro reconstitution systems of ribosomal subunits from free ribosomal RNA and ribosomal proteins are helpful tool for studies on the structure, function and assembly of ribosome. Using this system mutant or modified ribosomal proteins or ribosomal RNA can be incorporated into ribosomal subunits for studying ribosome structure and function. Developing the protocol for reconstitution of 30S subunits from an extreme thermophilic bacterium *Thermus thermophilus* can be beneficial especially for structural studies, as proteins and nucleic acids from this organism are very stable and crystallize easier than those from mesophilic organisms.

Key words Ribosome, 30S subunit, In vitro reconstitution, *Thermus thermophilus*, Ribosomal proteins, RNA, Polyamines

1 Introduction

The ribosome is the giant ribonucleoprotein assembly that translates the genetic code into protein in all living cells. Ribosomes from bacteria and archaea consist of a large (50S) and a small (30S) subunit, which together compose the 2.5 megadalton 70S ribosome. The small subunit contains 21 ribosomal proteins (designated S1–S21) and a 16S ribosomal RNA (rRNA), which is 1,542 nucleotides (nt) in length, whereas the large subunit is made up of 33 proteins (designated L1–L36) and two rRNAs, the 23S, which is 2,904 nt in length, and 5S, which is 120 nt in length.

Over the last decade, remarkable advances have been made in the areas of ribosome crystallography, such that now it is possible to obtain at medium or high resolution not only the structure of the ribosome subunit, full 70S ribosome, but also structures of the ribosome with key components bound such as messenger RNA (mRNA), transfer RNAs (tRNA), and various protein translation

Eric Ennifar (ed.), *Nucleic Acid Crystallography: Methods and Protocols*, Methods in Molecular Biology, vol. 1320, DOI 10.1007/978-1-4939-2763-0_19, © Springer Science+Business Media New York 2016

factors. This progress in elucidation of X-ray crystal structures of ribosome shed light on some important aspects of the mechanism of translation. But it is still far from understanding how the atomic structure of the ribosome ultimately governs its function. Many questions stay unanswered, and limited information is available, for example, on how the ribosome assembles into a stable multi-component complex.

In vitro reconstitution systems of ribosomal subunits from free ribosomal RNA and ribosomal proteins can be very helpful tool for structural studies. For example using this system can enable mutant or modified proteins or ribosomal RNA to be incorporated into subunits for studying ribosome structure and function. In vitro reconstitution systems also have proven invaluable for the study of ribosomal subunit assembly.

In vitro reconstitution of functional ribosomal subunits from *Escherichia coli* was reported almost 40 years ago. This milestone occurred, when Traub and Nomura [1] demonstrated that an active 30S subunit from *E. coli* could be assembled in vitro from free ribosomal RNA and ribosomal proteins without any additional components. These in vitro reconstitution experiments demonstrated that all of the information necessary for assembly is encoded within the rRNA and proteins themselves. Active 30S particles from *E. coli* can also be reconstituted from 16S rRNA in combination with individually purified proteins [2, 3], recombinant ribosomal proteins [4, 5], and unmodified 16S rRNA transcribed in vitro [6]. The conditions for reconstitution of the small and large ribosomal subunits from their natural components were found for several other organisms [7–10].

In this chapter we propose the optimized protocol for reconstitution of 30S subunits from free *Thermus thermophilus* 16S rRNA and total proteins from 30S subunit (TP30). Three decades ago an extreme thermophilic bacterium *T. thermophilus* has been chosen as a source for the isolation of components of the protein-synthesizing system to investigate their structures by X-ray crystallographic methods. *T. thermophilus* is a gram-negative extreme thermophile bacterium with an optimal growth temperature of +75 °C, and was first isolated by Dr. Oshima (Japan) from a hot spring [11]. Proteins and nucleic acids from this organism are very stable and crystallize easier than those from mesophilic organisms. We can say that *T. thermophilus* is a leader in the number of crystallized components of the protein synthesizing systems [12–14]. A huge number of *T. thermophilus* full 70S ribosome or 30S subunit complexes in complex with functional ligands are published. That is why we believe that optimized system of reconstitution of *T. thermophilus* 30S subunits, which yield in particles with the same functional efficiency as native, can be a useful tool for studying ribosome structure and function.

Table 1
Activity of *T. thermophilus* 30S in poly(U)-dependent cell-free translation system

Small ribosomal subunit component (conditions of reconstitution)	Amount of [^{14}C] Phe polymerized into poly (Phe), pmol (% activity)
Native 30S	50.6 (100 %)
30S reconstituted in Buffer G, incubation 50 min at 37 °C (without polyamine)	10.1 (19.9 %)
30S reconstituted in Buffer G, incubation 50 min at 60 °C (without polyamine)	19.2 (37.9 %)
30S reconstituted in Buffer G with 10 mM thermine in 1 step: incubation 50 min at 60 °C	26.8 (52.9 %)
30S reconstituted in Buffer G with 10 mM spermine in 2 steps: incubation 50 min at 60 °C, then 50 min at 70 °C	27.1 (53.5 %)
30S reconstituted in Buffer G with 10 mM thermine in 2 steps: incubation 50 min at 60 °C, then 50 min at 70 °C	52.8 (104.4 %)

The poly(U)-dependent cell-free translation reaction mixture contained 5 pmol of 30S ribosomal subunit in addition to 5 pmol of native 50S subunit and other components of cell-free system (*see* **Note 9**)

We found that the conditions for physical reconstitution of *T. thermophilus* 30S characterized by sedimentation property and proteins composition were very closed as in the experiments described by Traub and Nomura for *E. coli* particles. Another important criterion for correct assembly of ribosomal subunit is its functional activity in protein synthesis. We examined several factors, which might affect efficiency of reconstitution of functionally active *T. thermophilus* 30S particle, using cell-free system from *T. thermophilus* as a main tool (*see* Table 1). One of the obvious parameters influencing the correct assembly is the temperature to which the reconstitution mixture is exposed. As it was already mentioned, full physical reconstitution of *T. thermophilus* 30S subunit can be obtained at ionic conditions closed to Traub and Nomura, but carried at high temperatures, which are optimal for the extreme thermophile *T. thermophilus*. However, obtained reconstituted 30S particles still did not exhibit functional activity in the *T. thermophilus* cell-free system in comparison with native 30S subunits. It was known that in vivo assembly of *E. coli* 30S subunit was stimulated by polyamines, as in *E. coli* polyamine-deficient cells the 30S assembly was disturbed. Though in vitro influence on the reconstitution functionally active *E. coli* 30S subunit was not demonstrated [15–17], the polyamines were used in in vitro reconstitution study of 50S subunits from other organisms [10, 18]. Our study showed that such polyamines as spermidine or spermine did not fully recovered functional activity of reconstituted *T. thermophilus* 30S subunit. But two-step incubation at high temperatures in the presence of thermine [19], specific for the extreme thermophile bacterium polyamine, stimulated reconstitution of functionally active *T. thermophilus* 30S subunit.

Here we describe the protocol of in vitro reconstitution system for *T. thermophilus* 30S subunit from ribosomal proteins and natural 16S rRNA under optimized conditions. This procedure includes two-step incubations of *T. thermophilus* rRNA with ribosomal proteins in the presence of specific polyamine thermine at two different high temperatures, which result in efficient physical reconstitution of small subunit and in fully restoration of functional activity compared to the native one. As a control for 30S subunit physical reconstitution the method of analytical centrifugation in the sucrose gradient is suggested. In addition, protocol for cell-free system from *T. thermophilus* is also described as a tool for analysis of *T. thermophilus* 30S ribosome functional reconstitution.

2 Materials

2.1 Cell Culture and Lysis

1. French Press.
2. Fermentor.
3. Centrifugal evaporator.
4. Low speed centrifuge.
5. Spectrophotometer set at a wavelength 600 nm.
6. *T. thermophilus* HB8 cells.
7. Thermophilus vitamin-mineral stock 1,000× (TYE Media, Castenholz media).
8. Yeast extract.
9. Polypeptone Peptone.
10. Glycerol 100 %.

2.2 Purification of *T. thermophilus* 70S Ribosome, 30S and 50S Subunits

1. Gradient maker (SG15, GE Healthcare Life Sciences, Pittsburgh, PA, USA).
2. Peristaltic pump.
3. Capillary tubes.
4. Fraction collector.
5. Ultracentrifuge and rotors SW28, SW41, Ti-50.2 rotors (Beckman Coulter, Fullerton, CA, USA).
6. Tube Thinwall Ultra-Clear, 38.5 ml (Beckman Coulter, Fullerton, CA, USA).
7. Resin Toyopearl Butyl 650S (Tosoh Biosciences GMBH, Stuttgart, GERMANY).

8. Centrifugal Filter Units Amicon Ultracel-10 or -100 K membrane (Merck Millipore, Billerica, MA, USA).
9. Buffer A: 150 mM $MgCl_2$, 500 mM NH4Cl, 40 mM Tris–HCl $pH_{20°C}$ 7.5, 1.5 mM EDTA-Na_2, 1 mM DTT.
10. Buffer B: 50 mM $MgCl_2$, 150 mM NH_4Cl, 20 mM Tris–HCl $pH_{20°C}$ 7.5, 0.5 mM EDTA-Na_2, 1 mM DTT.
11. Cushion 1: 1.5 M sucrose, 0.68 M CsCl, 150 mM $MgCl_2$, 20 mM Tris–HCl $pH_{20°C}$ 7.5, 1.5 mM EDTA-Na_2, 1 mM DTT.
12. Cushion 2: 1.8 M sucrose, 0.8 M CsCl, 150 mM $MgCl_2$, 20 mM Tris–HCl $pH_{20°C}$ 7.5, 1.5 mM EDTA-Na_2.
13. Buffer C: 10 mM $MgCl_2$, 400 mM NaCl, 20 mM Tris–HCl $pH_{20°C}$ 7.5, 0.5 mM EDTA-Na_2, 1 mM DTT.
14. Buffer D: 20 mM Tris–HCl, $pH_{20°C}$ 7.5, 1 mM $MgCl_2$, 100 mM NH_4Cl, 0.5 mM EDTA-Na_2, 2 mM DTT.
15. Buffer D1: 5 % sucrose in Buffer D.
16. Buffer D2: 20 % sucrose in Buffer D.
17. Buffer E: 20 mM Tris–HCl, $pH_{20°C}$7.5, 50 mM $MgCl_2$, 150 mM NH_4Cl, 2 mM DTT.
18. DNAse 1 U/ml.
19. Phenylmethanesulfonylfluoride (PMSF) protease inhibitor.

2.3 Preparation of 16S rRNA and Total Ribosomal Proteins of 30S

1. Sephadex G-25 Fine gel filtration (GE Healthcare Life Sciences, Pittsburgh, PA, USA).
2. 12 M LiCl.
3. Buffer F: 6 M LiCl, 10 mM Tris–HCl, $pH_{20°C}$ 7.5, 25 mM $MgCl_2$, 75 mM NH_4Cl, 1 mM DTT.
4. Buffer G: 20 mM Tris–HCl, $pH_{20°C}$7.5, 20 mM $MgCl_2$, 35 mM KCl, 1 mM DTT.
5. 1 M spermine adjusted by 5 M HCl to pH 7.0.
6. 1 M thermine (N, N′-Bis (3-aminopropyl-1,3-propanediamine)), adjusted by 5 M HCl to pH 7.0.

2.4 30S Subunit Reconstitution

1. Buffer H : 20 mM HEPES pH 7.5, 20 mM $MgCl_2$, 100 mM NH_4Cl, 1 mM DTT.

2.5 Isolation of the S100 Enzyme Fraction from *T. thermophilus*

1. 40 ml diethylaminoethyl-cellulose (DEAE-cellulose) column.
2. Buffer I: 20 mM Tris–HCl, $pH_{20°C}$7.5, 15 mM $MgCl_2$, 50 mM NH_4Cl, 1 mM DTT.
3. Buffer J: 20 mM Tris–HCl, $pH_{20°C}$7.5, 15 mM $MgCl_2$, 250 mM NH_4Cl, 1 mM DTT.

2.6 Isolation of the Total tRNA from *T. thermophilus*

1. Ethanol 100 %.
2. Ethanol 70 %.
3. Phenol.
4. Chloroform.
5. Buffer K: 20 mM Tris–HCl, $pH_{20°C}$ 7.5, 15 mM $MgCl_2$, 800 mM KCl, 1 mM DTT.
6. Buffer L: 100 mM Tris–HCl, $pH_{37°C}$ 8.8.

2.7 Poly(U)-Dependent Cell-Free Translation

1. 8 mM GTP solution (pH adjusted to pH 7.0).
2. 60 mM ATP (pH adjusted to pH 7.0).
3. 20 μg Poly (U).
4. 0.5 nmol [^{14}C] phenylalanine (sp. act. 513 mCi/mmol).
5. 5 % trichloroacetic acid (TCA).
6. 2 mg/ml solution of bovine serum albumin.
7. Glass fiber filters.
8. Buffer H: 20 mM HEPES pH 7.5, 20 mM $MgCl_2$, 100 mM NH_4Cl, 1 mM DTT (*see* **Note 1**).

3 Methods

3.1 Reconstitution of 30S Subunit from *T. thermophilus*

3.1.1 Preparation of 70S Ribosome from *T. thermophilus*

1. A 1 l preculture of *T. thermophilus* HB8 cells starting from 1 ml of saturated culture in 25 % glycerol (kept at − 80 °C) was made growing the bacteria overnight at 75 °C in a medium containing 0.5 % Polypeptone, 0.2 % Yeast extract, 0.2 % NaCl, Vitamin-mineral solution 1 ml.l^{-1} media. The pH was adjusted with KOH to 7.2–7.4. Then the cells were grown in 100 l fermentor at 75 °C with aeration in the same medium. Cells were collected at 1 O.D. at 600 nm, and were frozen at − 80 °C.
2. Cells (100 g) were washed with 1 l of buffer A (*see* **Note 2**), suspended in 100 ml of the same buffer. DNase was then added to 1 U/ml together with PMSF 1 μg/ml (final).
3. Cells were disrupted in French Press and debris were removed by 30 min centrifugation at 30,000 × *g* (4 °C).
4. Supernatant (S30) was layered on the Cushion 1 (29 ml S30 on 7 ml of cushion 1) and centrifuged on SW28 rotor at 27,000 rpm (105,000 × *g*) during 20 h at 4 °C.
5. A fraction of 5 ml from the bottom of the cushion was collected and diluted three times with buffer B. Ribosomes were layered on Cushion 2 (29 ml S30 on 7 ml of cushion 2) and centrifuged on SW28 rotor at 27,000 rpm (40 h at 4 °C). A fraction of 4 ml from the bottom of cushion were collected and dialyzed against buffer CT.

6. Ribosomes (about 900 mg) were loaded on 200 ml column of Toyopearl Butyl 650S equilibrated in buffer C containing 1 M ammonium sulfate. The column was washed with two volumes of buffer C with 0.8 M of ammonium sulfate. Ribosomes were eluted by 900 ml of a reverse gradient of ammonium sulfate (from 80 to 40 %), keeping the other components of buffer C constant. The flow rate was 6 ml/min, and the fraction volume was 12 ml. The 70S peak was collected and ribosomes were dialyzed against buffer B. Solutions of the 70S ribosome was concentrated on CentriPrep-100 until the concentration of 20 mg/ml (1 absorption unit at 260 nm, A_{260}, corresponds to 66 μg/ml of ribosomes) and stored in aliquots at − 80 °C (*see* **Note 3**).

3.1.2 Preparation of 30S and 50S Ribosome Subunits from T. thermophilus

1. The 70S ribosomes (8 mg per gradient) were extensively dialyzed in Buffer D at 4 °C.
2. Sucrose gradient was prepared directly in Thinwall tubes for the SW28 rotor. 18 ml of Buffer D1 and 18 ml of Buffer D2 were mixed with a gradient maker in order to have a 36 ml of 20 % to 5 % (top) sucrose gradient. Six gradients are usually done. The gradients were kept on ice for 2 h prior to ultracentrifugation.
3. The dialyzed 70S ribosomes were carefully loaded on the top of the tube (no more than 8 mg per gradient) and ultracentrifugation was performed in SW28 rotor at 15 400 rpm (70,000 × *g*) for 17 h at 4 °C.
4. The sucrose gradients were fractionated (1 ml per fraction) and fractions containing 30S and 50S were collected separately.
5. The fractions containing pure 30S and 50S subunits were dialyzed in Buffer E at 4 °C overnight. The 30S and 50S were then concentrated using Centricon 100 K to a final concentration of 10 mg/ml.
6. Obtained 30S and 50S were used directly, or were frozen and stored at − 80 °C in aliquots.

3.1.3 Preparation of 16S rRNA and Total Proteins from 30S Subunit (TP30) from T. thermophilus

1. 1 ml of 30S subunits (10 mg/ml) was gently mixed with 1 ml of 12 M LiCl at 4 °C for 10–14 h.
2. The incubation mixture was centrifuged at 20,000 × *g* during 30 min. The supernatant contained TP30 and pellet contained 16S rRNA.
3. The pellet (16S rRNA) was washed out by suspending in 2 ml of buffer F, and after it was pelleted at 20,000 × *g* during 30 min.
4. The pellet was suspended in 1 ml of the buffer G for reconstitution, and dialyzed against the buffer G overnight. Obtained sample of 16S rRNA was used immediately for the reconstruction

essays. The concentration of the 16S rRNA was estimated. Optimal concentration for future reconstitution experiments was 4 mg/ml.

5. The supernatant (1 ml) (from Subheading 3.1.3, **step 2**), containing TP30, was transferred in buffer G by gel-filtration through 5 ml of Sephadex G-25 equilibrated in buffer G. Optimal final concentration was 2–3 mg/ml (taking in account an average coefficient of extinction for the ribosomal proteins the concentration of the TP30 was estimated as follows: 1 unit of A_{280} corresponds to 2 mg of TP30).
6. Solution of the TP30 (usually up to 1 ml) was centrifuged using a low speed centrifuge during 30 min (*see* **Note 4**). Obtained sample of TP30 was used immediately for the reconstruction essays.

3.1.4 Reconstitution of 30S Subunit from T. thermophilus

1. 4 mg of the sample of 16S rRNA (4 mg/ml) after dialysis (Subheading 3.1.2, **step 4**) was incubated at 40 °C during 30 min, and then centrifuged using a low speed centrifuge during 30 min to remove the aggregates. The concentration of 16S rRNA in supernatant was checked by UV analysis.
2. The reconstitution of 30S subunit was carried in Buffer G. The reaction mixture with final volume 4 ml contained 4 mg of 16S rRNA, 4 mg of TP30 (initial concentration 2 mg/ml) and 10 mM thermine (or no polyamine either 10 mM spermine for the control experiments).
3. Incubation mixture was treated at 65 °C for 50 min (or at 60 °C for 20 min as a control) with following heating at 70 °C during 30 min.
4. The solution of reconstituted 30S subunits was cooled down at room temperature for 20 min.
5. Sample was loaded on the top of 36 ml of a 5–20 % sucrose gradient in buffer (0.5–1 ml, not more than 5 mg of 30S per gradient). Sucrose gradient was prepared directly in Thinwall tubes for the SW28 rotor as it was described in Subheading 3.1.2, **step 2**.
6. Centrifugation was during 23 h at 15 400 rpm at 4 °C in SW28 rotor. The gradient was fractionated (1.5 ml per faction), and A_{260} was measured.
7. The fraction of 30S subunits fractions, which contained high value of absorption (A_{260}) (*see* Fig. 1) were collected and dialyzed in buffer H (*see* **Note 5**).
8. Dialyzed 30S subunits were concentrated using Centricon 100 K to a final concentration of 5–10 mg/ml, and were used for functional or structural tests directly or stored at − 80 °C in aliquots.

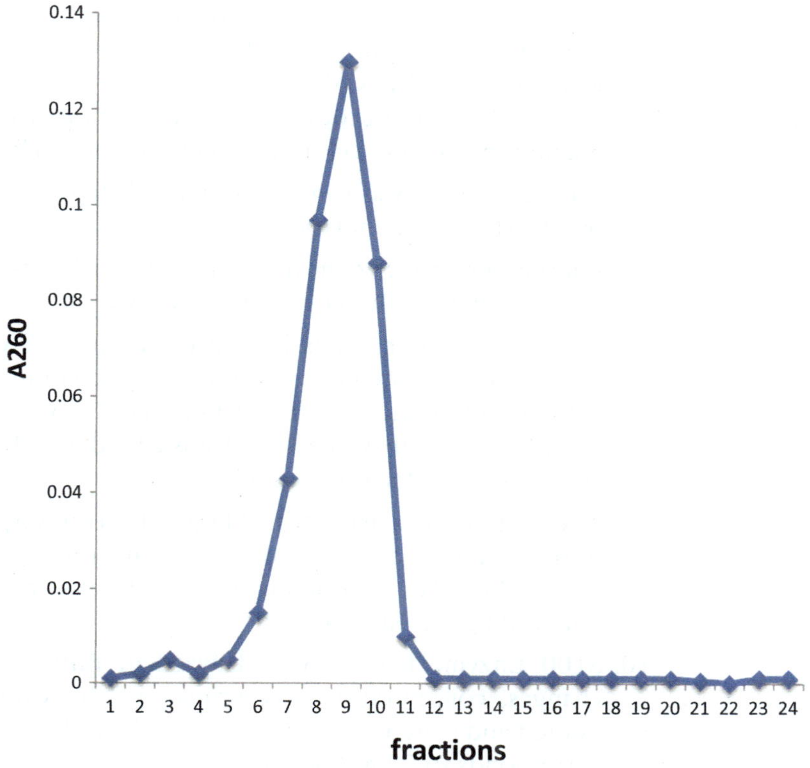

Fig. 1 Purification of 30S ribosomal subunit from *Thermus thermophilus* reconstituted in vitro by centrifugation in sucrose gradient. Fractions 8–10 were combined, and treated as indicated in Subheading 3.1.4, **step 6**

3.1.5 Analytical Centrifugation of Reconstituted 30S Subunits in Sucrose Gradient

1. Sucrose 5–20 % linear sucrose gradient in Buffer H was prepared directly in Thinwall tubes for the SW41 rotor (11 ml final volume).
2. Approximately 1 A_{260} (66 μg/ml) of reconstructed or native 30S subunits were loaded onto a sucrose gradient, and were centrifuged at 4 °C for 2 h at 40 000 rpm. Gradients were fractionated (0.5 ml per fraction) and A_{260} was measured.

3.2 Isolation of the S100 Enzyme Fraction from T. thermophilus

1. 10 g of *T. thermophilus* cell mass (Subheading 3.1.1, and *see* also **Note 1**) were washed with 0.1 l of Buffer H, and suspended in 10 ml of the same buffer. DNase was then added to 1 U/ml together with PMSF 1 μg/ml (final).
2. Cells were disrupted in French Press and debris were removed by 30 min centrifugation at 30,000 × *g* at 4 °C for 40 min.
3. Crude S100 extract was prepared by centrifugation in a Beckman Ti-50.2 rotor at 49 000 rpm for 3 h at 4 °C.
4. The supernatant (crude S100 extract) was loaded onto a 40 ml DEAE-cellulose column, equilibrated in the buffer I and washed with 400 ml of the same buffer until the A_{280} will be closed to 0, and then the enzyme fraction was eluted with buffer J. The volume of fraction was 3 ml (*see* **Note 6**).

5. The yellowish fractions, which contained high value of absorption (A_{280}), were combined. Final concentration (in mg/ml) was estimated by UV measurements as followed (A_{228nm}–A_{234nm})/3. The most convenient concentration of S100 enzyme fraction for using in cell-free system was about 10 mg/ml.
6. If concentration was too low, additional step of concentration on DEAE-cellulose was added.
7. For this obtained enzyme fraction (in Buffer J), first was diluted in three times by Buffer J without NH_4Cl.
8. The diluted sample was loaded on 10 ml DEAE-cellulose column, equilibrated in the Buffer I, and column was washed with 100 ml of the same buffer until the A_{280} will be closed to 0, and then the enzyme fraction was eluted with Buffer J. The volume of each fraction was 3 ml.
9. The fractions, which contained high value of absorption (A_{280}), were combined, and final concentration (in mg/ml) was estimated by UV measurements (as it was mentioned in Subheading 3.2, **step 5**).
10. S100 Enzyme fraction was dialyzed in Buffer H with two changes, concentration was estimated. The sample was aliquoted and stored at − 80 °C. Usually the yield from 10 g cells corresponds to 30 mg of S100.

3.3 Isolation of the Total tRNA from *T. thermophilus*

The sample of the total tRNA from *T. thermophilus* was obtained by continuation of fractionation of the crude S100 extract on DEAE cellulose column (*see* **Note 6**).

1. The total tRNA was eluted with Buffer K from DEAE-cellulose column after elution of the enzyme fraction (*see* Subheading 3.2, **step 4**). The volume of the fraction was 3 ml.
2. The fractions, which contained high value of absorption (A_{260}), were combined (usually 1,800–2,000 A_{260} or 88 mg in 40–50 ml were collected).
3. tRNA was precipitated with three volumes of cold ethanol.
4. Obtained pellet of total tRNA was dissolved in 30 ml of H_20, and then was treated three times by phenol–chloroform, and finally was precipitated by ethanol.
5. The pelleted tRNA was washed with 70 % ethanol and vacuum dried (*see* **Notes 7, 8**).

3.4 Test of Functional Activity of the Reconstituted 30S Subunits: Poly(U)-Dependent Cell-Free Translation

1. All the reaction mixtures were prepared in Buffer H.
2. The incubation mixture consisted of 5 pmol of reconstituted or native 30S subunit, 5 pmol of 50S *T. thermophilus*, 100–150 μg of total tRNA from *T. thermophilus*, 20 μg of poly(U), 50 μg of fraction S100 from *T. thermophilus*, 0.3 pmol ATP, 40 nmol GTP, 0.5 nmol [^{14}C] phenylalanine. The final volume was from 50 to 75 μl. Incubation was 20 min at 65 °C (plateau).

3. Incorporation was estimated by measuring the radioactivity in hot TCA-insoluble materials. For this to the reaction mixture 3 ml of 5 % TCA and 50 μl of bovine serum albumin (as a carrier) were added.
4. The sample was hydrolyzed for 20 min at 90 °C, and then cooled. The precipitated were placed onto glass fiber filter and washed with 5 % TCA. The filters were dried and counted.

4 Notes

1. It was shown that such an unusual high concentration of Mg^{2+} ions is optimal for the cell-free system from *T. thermophilus* [20].
2. Washed cells of *T. thermophilus* can be fractionated (for example by 10 g), frozen and kept at − 80 °C. Frozen cells were used for reconstitution experiments, obtaining S100 enzyme fraction and total tRNA.
3. The ribosomes were flash freeze into liquid nitrogen and stored at − 80 °C.
4. This procedure was introduced to remove the aggregates from the TP30 sample.
5. Collecting only the pick of the A_{260} distribution permits to select "healthy" 30S particles. Asymmetrical distribution of A_{260} in the region closer to the bottom shows the presence of possible dimers of 30S subunit or not well reconstituted particles.
6. After this stage of fractionation of crude S100 extract, i.e., after elution of the Enzyme S100 fraction, all ribonucleic acids, mostly total tRNA, stayed attached to the DEAE-resin.
7. Optional: To obtain fully deacylated tRNA, the sample of total tRNA was incubated in the Buffer L at 37 °C during 1 h. Then tRNA was precipitated with three volumes of cold ethanol and the pellet was vacuum dried.
8. The sample of total tRNA can be dissolved in water (optimal concentration was 8–10 mg/ml) and be kept frozen at − 80 °C.
9. Background radioactivity of the control sample containing only 50S subunits (about 6.6 % of the radioactivity) was subtracted from all activity values.

Acknowledgements

We are grateful to A.S. Spirin for his support of early stages on the study of *Thermus thermophilus* ribosomes; V.A. Shirokov for stimulating discussions. This work was supported by Russian Academy of Science.

References

1. Traub P, Nomura M (1968) Structure and function of *E. coli* ribosomes. V. Reconstitution of functionally active 30S ribosomal particles from RNA and proteins. Proc Natl Acad Sci U S A 59:777–784
2. Mizushima S, Nomura M (1970) Assembly mapping of 30S ribosomal proteins from *E. coli*. Nature 226:1214
3. Held WA, Mizushima S, Nomura M (1973) Reconstitution of *Escherichia coli* 30S ribosomal subunits from purified molecular components. J Biol Chem 248:5720–5730
4. Culver GM, Noller HF (1999) Efficient reconstitution of functional *Escherichia coli* 30S ribosomal subunits from a complete set of recombinant small subunit ribosomal proteins. RNA 5:832–843
5. Culver GM, Noller HF (2000) In vitro reconstitution of 30S ribosomal subunits using complete set of recombinant proteins. Methods Enzymol 318:446–460
6. Krzyzosiak W, Denman R, Nurse K, Hellmann W, Boublik M, Gehrke CW, Agris PF, Ofengand J (1987) In vitro synthesis of 16S ribosomal RNA containing single base changes and assembly into a functional 30S ribosome. Biochemistry 26:2353–2364
7. Cohlberg JA, Nomura M (1976) Reconstitution of Bacillus stearothermophilus 50S ribosomal subunits from purified molecular components. J Biol Chem 251:209–221
8. Fahnestock SR (1979) Reconstitution of active 50S ribosomal subunits from *Bacillus licheniformis* and *Bacillus subtilis*. Methods Enzymol 59:437–443
9. Sanchez ME, Urena D, Amils R, Londei P (1990) In vitro reassembly of active large ribosomal subunits of the halophilic archaebacterium *Haloferax mediterranei*. Biochemistry 29:9256–9261
10. Londei P, Teixido J, Acca M, Cammarano P, Amils R (1986) Total reconstitution of active large ribosomal subunits of the thermoacidophilic archaebacterium *Sulfolobus solfataricus*. Nucleic Acids Res 14:2269–2285
11. Oshima T (1974) Comparative studies on biochemical properties of an extreme thermophile, *Thermus thermophilus* HB 8 (author's transl). Seikagaku 46:887–907
12. Trakhanov SD, Yusupov MM, Agalarov SC, Garber M, Ryazancev SN, Tischenko SV, Shirokov VA (1987) Crystallization of 70S ribosomes and 30S ribosomal subunits from *Thermus thermophilus*. FEBS Lett 220:319–322
13. Garber M, Agalarov C, Eliseikina I, Tischenko S, Shirokov V, Yusupov M, Reshetnikova L, Trakhanov S, Tukalo M, Yaremchuk A (1991) Purification and crystallization of components of the protein-synthesizing system from *Thermus thermophilus*. J Cryst Growth 110:228–236
14. Yusupova G, Yusupov M, Spirin A, Ebel JP, Moras D, Ehresmann C, Ehresmann B (1991) Formation and crystallization of *Thermus thermophilus* 70S ribosome/tRNA complexes. FEBS Lett 290:69–72
15. Igarashi K, Kashiwagi K, Kishida K, Watanabe Y, Kogo A, Hirose S (1979) Defect in the split proteins of 30-S ribosomal subunits and under-methylation of 16-S ribosomal RNA in a polyamine-requiring mutant of *Escherichia coli* grown in the absence of polyamines. Eur J Biochem 93:345–353
16. Kakegawa T, Hirose S, Kashiwagi K, Igarashi K (1986) Effect of polyamines on in vitro reconstitution of ribosomal subunits. Eur J Biochem 158:265–269
17. Igarashi K, Kashiwagi K, Kishida K, Kakegawa T, Hirose S (1981) Decrease in the S1 protein of 30-S ribosomal subunits in polyamine-requiring mutants of *Escherichia coli* grown in the absence of polyamines. Eur J Biochem 114:127–131
18. Khaitovich P, Tenson T, Kloss P, Mankin AS (1999) Reconstitution of functionally active *Thermus aquaticus* large ribosomal subunits with in vitro-transcribed rRNA. Biochemistry 38:1780–1788
19. Oshima T (1975) Thermine: a new polyamine from an extreme thermophile. Biochem Biophys Res Commun 63:1093–1098
20. Gogia ZV, Yusupov MM, Spirina TN (1986) Structure of *Thermus thermophilus* ribosomes method of isolation and purification of the ribosomes. Molekul Biol (USSR) 20:519–526

Chapter 20

Crystallographic Studies of the Ribosomal A-Site Molecular Switches by Using Model RNA Oligomers

Jiro Kondo

Abstract

An RNA molecular switch in the aminoacyl-tRNA decoding site (A site) of the ribosome plays a key role in the decoding process of the protein biosynthesis. The switch discriminates a single cognate-tRNA from near-cognate tRNAs by changing its conformation from "off" to "on" states and recognizing the first two base pairs of codon–anticodon mini-helix to check whether these base pairs are of the canonical Watson–Crick type or not. Aminoglycoside antibiotics specifically target the "on" state of the bacterial A-site molecular switch and disturb the fidelity of the decoding process, resulting to cell death. If it occurs in human who was given aminoglycosides, it can lead to undesirable side effects. In order to understand the molecular bases of the decoding and the antibacterial and toxic side effects of aminoglycosides, it is necessary to determine the three-dimensional structures of the A-site molecular switches both in the presence and absence of aminoglycosides. This chapter focuses on methods in crystallographic studies of the A-site switches by using model RNA oligomers. The methods can be utilized in crystallographic studies of any DNA/RNA oligomers.

Key words RNA, Antibiotics, Ribosome, Aminoglycosides

1 Introduction

The aminoacyl-tRNA decoding site (A site) is one of the active sites of the ribosome, in which a single cognate tRNA is discriminated from near-cognate tRNAs. In this decoding process, an RNA molecular switch composed of 15 nucleotide residues located at the top of helix 44 of the 16S rRNA plays a key role (*see* Fig. 1) [1–4]. In the absence of aminoacyl-tRNA, the switch is conformationally dynamic, and two consecutive adenines A1492 and A1493 in an asymmetrical internal loop adopt various conformations called "off" states [5]. When an aminoacyl-tRNA is delivered to the A site, the switch changes its conformation to a unique "on" state, in which both A1492 and A1493 bulge out from the A-site internal loop toward the mRNA–tRNA complex. These adenines interact with the first two base pairs of the codon–anticodon mini-helix

Eric Ennifar (ed.), *Nucleic Acid Crystallography: Methods and Protocols*, Methods in Molecular Biology, vol. 1320, DOI 10.1007/978-1-4939-2763-0_20, © Springer Science+Business Media New York 2016

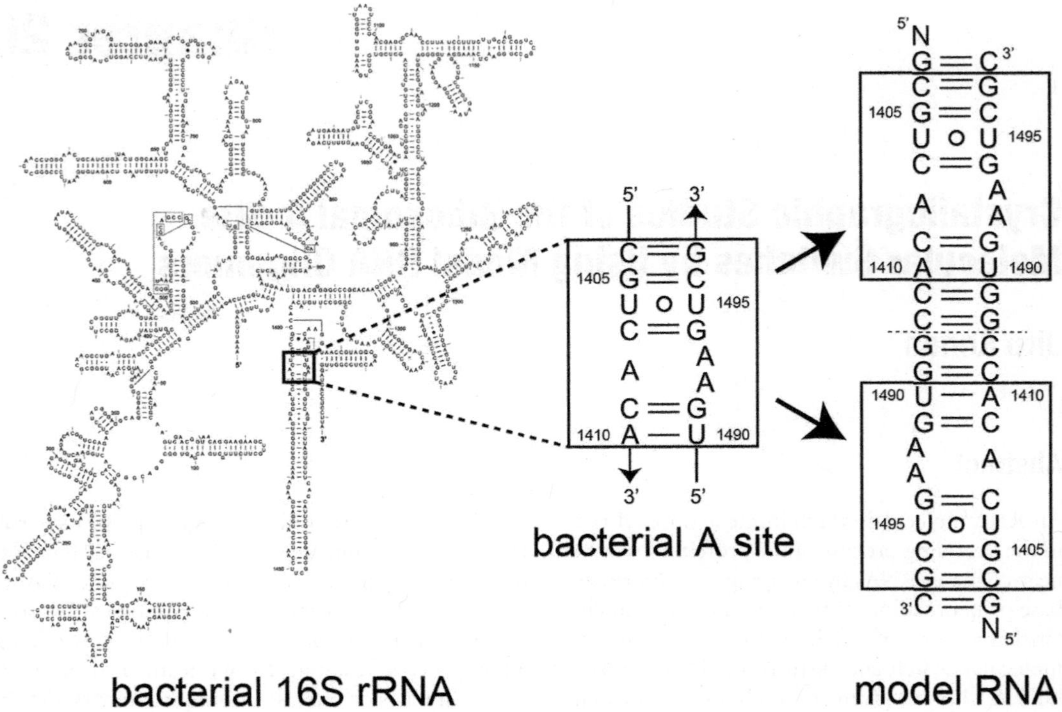

Fig. 1 Secondary structures of the bacterial 16S rRNA (*left*) and A site molecular switch (*center*), and the model RNA oligomers used in our crystallographic studies (*right*). The symbol *N* represents any overhanging nucleotide

through A-minor motifs and geometrically check whether these base pairs are of the canonical Watson–Crick type or not [1–4].

Aminoglycoside antibiotics with broad-spectrum antibacterial activity against a variety of gram-negative and certain gram-positive bacteria specifically bind to the bacterial A-site molecular switch and stabilize it in the "on" state. Therefore, the A-site RNA loses its role as the molecular switch, leading to disturbing the fidelity of protein synthesis [6]. It should also occur in eukaryotes if aminoglycosides bind to their A-site molecular switch. For example, some aminoglycosides display antiprotozoal activity [7–10] and also undesirable side effects to human [10–12].

The molecular mechanisms of the decoding process and the antibacterial activity of aminoglycosides were initially revealed by X-ray analyses of the 30S ribosome [1, 6]. Around the same time, Vicens and Westhof determined the interaction mode of aminoglycosides to the A-site molecular switch by using an RNA model oligomer [13]. There are several advantages of using the model RNA system than the ribosome itself; (1) ease of sample preparation by using the solid-phase synthesis, (2) much chance of obtaining better resolution X-ray data due to the fact that molecules with lower molecular weight tend to present higher

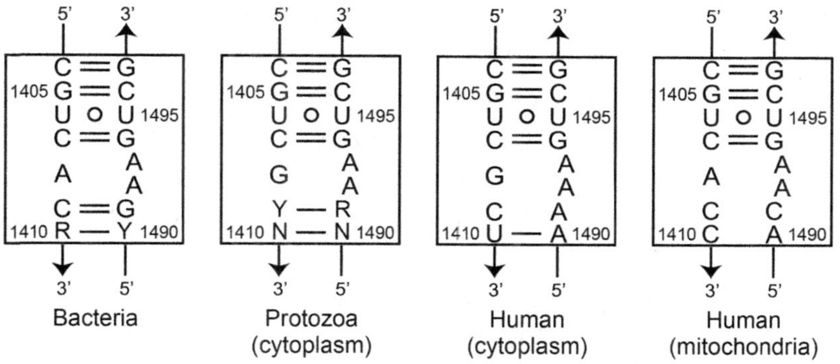

Fig. 2 Secondary structures of the A-site molecular switches. The rRNA residues are numbered according to the numbering used in *E. coli* 16S rRNA. The symbols *R*, *Y*, and *N* represent purine (A or G), pyrimidine (C or U), and any nucleotide, respectively

structural homogeneity in crystal, and (3) excellent applicability to structural studies of the ribosomal A site of any organism other than bacteria, such as protozoa and human, both cytoplasmic and mitochondrial ones (*see* Fig. 2). Therefore, the model system has been extensively applied to high-resolution structural studies of not only the bacterial [14–23] but also eukaryotic [24–29] A-site molecular switches both in the presence and absence of aminoglycosides. As a result, the molecular mechanisms of antiprotozoal activities and a part of the toxic side effects of aminoglycosides have recently been revealed by our X-ray analyses.

In this chapter, methods of sample design, sample preparation, crystallization and structure determination used in our crystallographic studies of the A-site molecular switches in the presence and absence of aminoglycosides are described. The methods can be utilized in crystallographic studies of any DNA/RNA oligomers.

2 Materials

2.1 Design of RNA Constructs

1. The *m*-fold Web Server (http://mfold.rna.albany.edu/?q=mfold)

2.2 RNA Preparation

2.2.1 Deprotection of 2′-ACE-Protected RNA

1. Chemically synthesized RNA oligonucleotides (GE Healthcare Dharmacon, Lafayette, CO, USA). Store at −20 °C in a dark place.
2. 2′-Deprotection buffer: 100 mM acetic acid, adjusted to pH 3.8 with *N, N, N′,N′*-tetramethylethylenediamine (TEMED).
3. Hot dry bath.
4. Loading buffer: 0.025 % bromophenol blue (BPB), 0.025 % xylene cyanol (XC), 10 mM ethylenediaminetetraacetic acid (EDTA), 7.4 M Urea.

2.2.2 Denaturing Polyacrylamide Gel Electrophoresis (PAGE)

1. Polyacrylamide gel solution: 40 % acrylamide–bis-acrylamide solution 19:1 (w/w). Store at 4 °C in a dark place.
2. 8 M urea solution. Store in a dark place.
3. 10× Tris/borate/EDTA (TBE) buffer: 890 mM tris base, 890 mM boric acid, 20 mM EDTA.
4. 1× TBE buffer: 89 mM tris base, 89 mM boric acid, 2 mM EDTA.
5. 10 % ammonium persulfate (APS). Store at 4 °C in a dark place up to 4 weeks.
6. TEMED. Store at 4 °C in a dark place.
7. Electrophoresis equipment with 30 × 40 cm^2 glass plates, spacers, and a comb with at least 1 mm thickness and an aluminum back plate.
8. Scotch tape.
9. Thin layer chromatography (TLC) plates.
10. Polyethylene sheet.
11. Handy UV lamp.
12. Razor blade.
13. 50 ml centrifuge tube.
14. Disposable messpipette or sterilized glass rod.
15. Shaker.
16. Rapid-Flow 50 ml conical tube filter with 0.2 μm aPES membrane (Thermo Scientific Nalgene, Rochester, NY, USA).
17. Diaphragm vacuum pump.

2.2.3 Desalting (Reversed-Phase Chromatography)

1. 500 mM sodium cacodylate buffer (pH 7.0). Store at 4 °C.
2. Sep-Pak classic C18 reversed-phase chromatography column (Waters, Milford, MA, USA).
3. 10 ml syringe.
4. 15 ml centrifuge tube.
5. 1.5 ml micro tube.
6. Ethanol, ≥99.8 %.
7. Acetonitrile 80 % (v/v).
8. Millex-LG 0.20 μm syringe filter unit (Millipore, Billerica, MA, USA).
9. SpeedVac system.
10. Micro UV-visible fluorescence spectrophotometer.

2.2.4 Desalting (Ultrafiltration)

1. Centrifugal concentrator, 3,000 Da molecular weight cut off.
2. Micro UV-visible fluorescence spectrophotometer.
3. Centrifuge with swing bucket rotors or fixed angle rotors.

2.3 Complex Formation

1. Annealing buffer: 200 mM Sodium cacodylate (pH 7.0).
2. 4 mM aminoglycoside solution.
3. Hot dry bath.

2.4 Crystallization

1. Homemade and/or commercial screening kits such as Nucleic Acid Mini Screen and Natrix (Hampton Research, Aliso Viejo, CA, USA), or JBScreen Nuc-Pro (Jena Bioscience, Jena, Germany).
2. EasyXtal 15-Well Tools crystallization plates (QIAGEN, Venlo, The Netherlands).
3. 96× 2-wells crystallization plates for sitting-drop vapor diffusion method with sealing film.
4. Incubator for crystallization plates.
5. Stereoscopic microscope with polarizing filter.
6. Digital camera with C-mount for microscope.

3 Methods

3.1 Design of RNA Constructs

3.1.1 Sequence Design

1. Chemical probing experiments on the 16S rRNA [30] and aminoglycoside-binding experiments using A-site model RNAs [31] suggested that the A-site molecular switch could be isolated from the rRNA and inserted into a small oligonucleotides without losing its structure, function, and aminoglycoside-binding properties. Indeed, structural identity of the A-site switch in the whole ribosome and model RNAs have been confirmed by crystallographic studies [6, 13]. In the crystal structure of our RNA model (*see* Fig. 1), the crystal packing contact perfectly mimics the A-minor recognition between two bulged-out adenines from the A site and the codon–anticodon stem of the tRNA–mRNA complex occurring in the ribosome (*see* Fig. 3). Binding mode of aminoglycosides to the A site are also well conserved in the ribosome and the model RNA (*see* Fig. 3).
2. Two asymmetrical internal loops of the A site were inserted between Watson–Crick base pairs in sequence designed to fold as double helices (*see* Fig. 1). These RNA sequences containing different lengths and ends can be checked for alternative two-dimensional foldings using the m-fold Web Server. The advantage of this RNA model system is that we can perform structural studies of the A-site molecular switches of any organism, such as bacteria, protozoa and human, both cytoplasmic and mitochondrial ones (*see* Fig. 2).
3. In order to facilitate crystal packing, some overhanging nucleotides, such as UU and C, can be introduced. Indeed, these overhanging nucleotides are often observed at crystal packing interface.

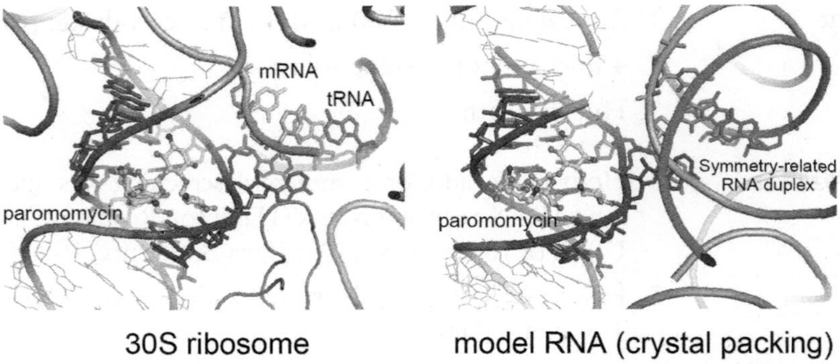

Fig. 3 The A-minor recognition between two bulged-out adenines from the A site (A1492 and A1493) and the codon–anticodon stem of the tRNA–mRNA complex observed in the 30S ribosomal particles (*left*). In the crystal structure of model RNA oligomer, crystal packing interaction perfectly mimics the A-minor recognition (*right*)

3.1.2 Preparation of Heavy-Atom Derivatives for Phase Determination

1. Since many of RNA molecules have well conserved right-handed antiparallel helical structures called the A-form, one may think that the phase problem in RNA crystallography can be easily resolved by the molecular replacement method using model RNA duplexes. However, in practice, the simple helical structure complicates the identification of the correct solution among a number of candidates. Imagine how not simple it is to determine the correct orientation around the helical axis and position of the molecule within the crystal where RNA duplexes are packed side-by-side, stacked head-to-tail and make a monotonous infinite helix in a certain direction. Therefore, as in the case of the protein phasing, various heavy-atom methods such as MAD, SAD and MIR are often used for the RNA phasing. For that, nucleotides covalently modified with bromine, iodine or selenium atom can be introduced into RNA sequences [32–35].
2. Soaking is an alternative technique for obtaining heavy-atom derivatives. Hexammine of cobalt, osmium and iridium are useful cations for the RNA phasing, because they preferentially bind to the deep/major groove of guanine in RNA helix [36, 37].
3. Besides these heavy atom methods, phasing techniques based on structural features of RNA helix have been recently developed [38–40].

3.2 RNA Preparation

3.2.1 Deprotection of 2′-ACE-Protected RNA

1. RNA oligonucleotide purchased from Dharmacon has the 2′-ACE protecting group at the 2′ position of each nucleotide. Usually, requisite amount of 2′-Deprotection buffer is provided with 2′-ACE-protected RNA, but it can be easily prepared in your lab.

2. Centrifuge tubes containing RNA pellet briefly, add 400 μl of 2′-Deprotection buffer to each tube, then completely dissolve RNA pellet by pipetting and vortex.
3. Incubate tubes at 60 °C for 30 min on a hot dry bath.
4. Add 400 μl of loading buffer to each tube and mix by a vortex mixer.

3.2.2 Denaturing Polyacrylamide Gel Electrophoresis (PAGE)

1. RNA oligonucleotides of up to approximately 30 nucleotides can be purified by 20 % denaturing PAGE. A sequencing electrophoresis apparatus with 30×40 cm^2 glass plates together with an aluminum back plate for homogenizing glass plates' temperature is recommended. The running temperature is set at 50–60 °C. The gel with at least 1 mm thickness allows the separation of RNA oligonucleotides with different lengths.
2. Clean glass plates, position two spacers (1 mm thick) along the sides of a plate, place another glass plate on top, clamp the plates together, and seal the bottom of plates with scotch tape.
3. Prepare 250 ml of 20 % polyacrylamide gel by mixing 125 ml of 40 % acrylamide–bis-acrylamide solution 19:1 (w/w), 25 ml of 10× TBE and 100 ml of 8 M Urea. Add 2.5 ml of 10 % APS and 250 μl of TEMED to the gel, mix well but not aerated, pour the gel into the space of the glass plates, insert a comb into the gel, and allow the gel to polymerize for 1 h.
4. Clamp an aluminum plate against one surface of the gel plate, place the gel plate on an electrophoresis apparatus, pour approximately 700–1,000 ml of 1× TBE into electrophoresis chambers (top and bottom of the gel plate), and connect the apparatus to an electric power supply.
5. Preheat the gel at 500–700 V for 1 h.
6. Load the RNA sample (prepared at **step 4** in Subheading 3.2.1) onto the gel and run the gel at 500–700 V for 3–6 h. Progress of the migration can be followed by the course of the BPB and XC dyes contained in the loading buffer.
7. After finishing the PAGE purification, place the gel plate on the lab bench, remove the top glass plate, cover the gel with a polyethylene sheet, flip the gel onto a fluorescent TLC plate, remove another glass plate, and cover the gel with another polyethylene sheet.
8. In a dark room, use a UV lamp to observe RNA bands as dark shadows and mark the major band by an indelible marker. Clean a razor blade using ethanol and Kimwipes, cut the marked major band out of the gel, and transfer it to a 50 ml centrifuge tube.

9. Crush the gel by a sterilized disposable messpipette or a glass rod, add approximately 30 ml of sterilized water, and shake at 37 °C overnight to elute the RNA oligonucleotide from the crushed gels.
10. Remove the crushed gels from the sample solution using a 50 ml conical tube filter with 0.2 μm aPES membrane connected to a diaphragm vacuum pump.

3.2.3 Desalting (Reversed-Phase Chromatography)

1. Reversed-phase chromatography could be a first choice for desalting RNA oligonucleotides. Prior to the process, add approximately 1 ml of 500 mM sodium cacodylate buffer (pH 7.0) to the sample solution (obtained at **step 10** of Subheading 3.2.2) and preheat at 37 °C for a better yield.
2. Attach a Sep-Pak classic C18 cartridge to a 10 ml syringe with the plunger removed, add 10 ml of ethanol to the syringe and use the plunger to wash the ethanol through the cartridge slowly.
3. Remove the syringe from the cartridge, remove the plunger, reattach the cartridge to the tip of the syringe, add 10 ml of sterilized water, and use the plunger to wash the water through the cartridge slowly.
4. Remove the cartridge and plunger from the syringe, reattach the cartridge to the syringe, add the sample solution to the syringe, and collect the drops from the cartridge in a 15 ml centrifuge tube. Check the UV absorption of the collected solution to ensure that the RNA oligonucleotide binds to and not be leaked from the cartridge.
5. Remove the cartridge from the syringe, attach the cartridge to a new 10 ml syringe with plunger removed, and add 10 ml of sterilized water to wash the C18 resin in the cartridge. Collect the drops in another 15 ml centrifuge tube to check the UV absorption to confirm that no RNA is leaked.
6. Once again remove the cartridge and plunger, reattach the cartridge, add 5 ml of 80 % acetonitrile to the syringe, and use the plunger to elute the RNA oligonucleotide from the C18 resin in the cartridge to a 15 ml centrifuge tube.
7. Attach a filter unit to the syringe, add 5 ml of the collected RNA solution, use the plunger to filter a piece of the C18 resin leaked from the cartridge, and collect 1 ml fractions in five 1.5 ml microtubes.
8. Concentrate the RNA sample using a vacuum concentrator. Resuspend the obtained RNA pellet in sterilized water so that the concentration of the RNA is 4 mM. The concentration of RNA can be monitored by a UV spectrophotometer. Store the RNA stock solution at −70 °C.

3.2.4 Desalting (Ultrafiltration)

1. For RNA oligonucleotides larger than 12 nucleotides (molecular weight is approximately 4,000 Da), ultrafiltration can be an alternative choice for desalting.
2. Fill concentrators (MWCO 3,000) with up to the maximum volumes (20 ml and 14 ml for centrifuges with swing bucket rotors and fixed angle rotors, respectively) of the sample solution (obtained at **step 10** of Subheading 3.2.2), and insert assembled concentrators into a centrifuge in correct orientation.
3. Centrifuge at recommended speeds (3,000 × *g* and 6,000 × *g* for centrifuges with swing bucket rotors and fixed angle rotors, respectively) at 20 °C for approximately 0.5 – 1 h.
4. Remove assembly, empty filtrate container and refill the concentrators with sterilized water.
5. Concentrate the sample again and repeat the process until the concentration of contaminating microsolute is sufficiently reduced.
6. Recover the sample from the bottom of the concentrate pocket with a pipette in a 1.5 ml microtube.
7. Concentrate the RNA sample using a vacuum concentrator. Resuspend the obtained RNA pellet in sterilized water so that the concentration of the RNA is 4 mM. Store the RNA stock solution at −70 °C.

3.3 Complex Formation

1. Before setting up crystallization experiments, the RNA oligonucleotide has to be properly folded. Mix 10 μl of 4 mM RNA stock solution and 10 μl of annealing buffer in a 1.5 ml microtube. To anneal the RNA, heat the mixed solution at 70–90 °C (depending on the melting temperature of RNA) for 2 min followed by slow cooling to 37 °C in a programmable hot dry bath.
2. Add 20 μl of 4 mM aminoglycoside solution or water (for crystallizations in the presence and absence of aminoglycosides, respectively) prewarmed at 37 °C to the RNA solution and then cooled the mixture slowly to room temperature in the switched-off hot dry bath.
3. The sample solution contains 1 mM RNA oligonucleotide (in our case, *see* **Note 1**), 2 mM aminoglycoside and 50 mM sodium cacodylate.

3.4 Crystallization

3.4.1 General Protocol

1. To determine preliminary crystallization conditions, we usually employ two homemade screening kits containing 2-methyl-2,4-pentanediol (MPD) and polyethylene glycol 3350 (PEG3350) as a precipitant, respectively. Both kits commonly contain 50 mM sodium cacodylate (pH 7.0), 1–10 mM spermine tetrahydrochloride or hexammine cobalt(III) chloride, 10–250 mM

monovalent or divalent cations and 1–20 % (v/v) MPD or PEG3350. Several commercial kits dedicated to nucleic acid crystallization are available to the initiate screening of crystallization condition.

2. The initial screening is often carried out using the hanging- or sitting-drop vapor diffusion methods. Hereafter, we describe the former method using EasyXtal 15-Well Tools and our homemade screening kits.
3. Add 250 μl of reservoir solution (40 % MPD or PEG3350) to each well in the crystallization plate.
4. Pipette 1 μl of crystallization buffer into a small cavity (for 2 μl droplet) of a crystallization support, take 1 μl of the sample solution (prepared at **step 2** in Subheading 3.3), add to the droplet of crystallization buffer in the small cavity and mix the droplet gently by pipetting.
5. Repeat **step 4** for two other conditions (one crystallization support has three small cavities).
6. Screw and fasten the crystallization support to the plate's well.
7. Repeat **steps 4–6** for the remaining reservoirs.
8. Store the crystallization plate in an incubator at constant temperature (usually at 20 °C, *see* **Note 2**). The plate can be stored for some weeks to months.
9. Observe all crystallization droplets carefully using a stereoscopic microscope with polarizing filters and take pictures by a digital camera hooked up to the microscope. Usually, it may take some days to weeks to obtain results. The results can range from large single crystals to microcrystals, precipitates, phase separations, or clear drops.
10. To obtain crystals with better size and quality, optimize several parameters, such as concentration of salts and precipitants, temperature, and pH (*see* **Note 3**).

4 Notes

1. The RNA concentration for the initial crystallization screening is determined according to the following equation proposed by Alexander Rich and coworkers [41]: Concentration (mM) = 24 (or 20) / Number of nucleotides. In the case of our RNA oligonucleotide (23 nt), the concentration is approximately 1 mM.
2. Temperature is one of the most important parameters that strongly affect crystallization. Therefore, it is recommended to run the same screening experiment in parallel for example at 4, 20, and 37 °C conditions.

3. Crystal polymorphism is the phenomenon of a molecule to be crystallized into several different crystal forms and can be observed when a molecule has different conformers. Therefore, RNA molecules with flexible bulges or loops have a great tendency to exhibit the phenomenon. The bacterial A-site RNA duplex with two asymmetric internal loops is the most prominent example. Due to high flexibility of the bacterial A-site internal loop, the RNA duplex has been crystallized into 11 different forms so far, $P1$, $P2_1$ (2 forms), $C2$, $P2_12_12$ (2 forms), $P2_12_12_1$ (3 forms), $P4_1$ and $I4$ space groups ([13–23], *and unpublished data*). As a result, we found a total of 12 different conformations of the bacterial A site, from which the motion of the A-site molecular switch during the decoding process can be studied. These crystals were obtained in different crystallization conditions, in which concentrations and types of cations, aminoglycosides and precipitants are altered. Surprisingly, the number of crystal forms obtained for the A-site RNA duplex is larger than that obtained for a hen egg-white lysozyme, the most studied biological macromolecule in crystallography ([42, 43]: 5 crystal forms, $P1$, $P2_1$, $P2_12_12_1$, $P4_32_12$, and $P6_122$, have been reported so far). In order to understand the biological function of RNA molecules with flexible bulges or loops, it is therefore important to perform crystallization under wide range of conditions.

References

1. Ogle JM, Brodersen DE, Clemons WM Jr, Tarry MJ, Carter AP, Ramakrishnan V (2001) Recognition of cognate transfer RNA by the 30S ribosomal subunit. Science 292:897–902
2. Ogle JM, Carter AP, Ramakrishnan V (2003) Insights into the decoding mechanism from recent ribosome structures. Trends Biochem Sci 28:259–266
3. Ogle JM, Murphy FV, Tarry MJ, Ramakrishnan V (2002) Selection of tRNA by the ribosome requires a transition from an open to a closed form. Cell 111:721–732
4. Ogle JM, Ramakrishnan V (2005) Structural insights into translational fidelity. Annu Rev Biochem 74:129–177
5. Kondo J, Westhof E (2007) Structural comparisons between prokaryotic and eukaryotic ribosomal decoding A sites free and complexed with aminoglycosides. In: Arya DP (ed) Aminoglycoside antibiotics from chemical biology to drug discovery. Wiley-Interscience, Hoboken, NJ, pp 209–223
6. Carter AP, Clemons WM, Brodersen DE, Morgan-Warren RJ, Wimberly BT, Ramakrishnan V (2000) Functional insights from the structure of the 30S ribosomal subunit and its interactions with antibiotics. Nature 407:340–348
7. Waitz JA, Sabatelli F, Menzel F, Moss EL Jr (1974) Biological activity of antibiotic G-418, a new micromonospora-produced aminoglycoside with activity against protozoa and helminths. Antimicrob Agents Chemother 6:579–581
8. Loebenberg D, Counelis M, Waitz JA (1975) Antibiotic G-418, a new micromonospora-produced aminoglycoside with activity against protozoa and helminths: antiparasitic activity. Antimicrob Agents Chemother 7:811–815
9. El-On J, Bazarsky E, Sneir R (2007) Leishmania major: in vitro and in vivo anti-leishmanial activity of paromomycin ointment (Leshcutan) combined with the immunomodulator Imiquimod. Exp Parasitol 116:156–162
10. Hobbie SN, Kalapala SK, Akshay S, Bruell C, Schmidt S, Dabow S, Vasella A, Sander P, Böttger EC (2007) Engineering the rRNA decoding site of eukaryotic cytosolic ribosomes in bacteria. Nucleic Acids Res 35: 6086–6093

11. Hobbie SN, Akshay S, Kalapala SK, Bruell CM, Shcherbakov D, Böttger EC (2008) Genetic analysis of interactions with eukaryotic rRNA identify the mitoribosome as target in aminoglycoside ototoxicity. Proc Natl Acad Sci U S A 105:20888–20893
12. Guthrie OW (2008) Aminoglycoside induced ototoxicity. Toxicology 249:91–96
13. Vicens Q, Westhof E (2001) Crystal structure of paromomycin docked into the eubacterial ribosomal decoding a site. Structure 9:647–658
14. Vicens Q, Westhof E (2002) Crystal structure of a complex between the aminoglycoside tobramycin and an oligonucleotide containing the ribosomal decoding a site. Chem Biol 9:747–755
15. Vicens Q, Westhof E (2003) Crystal structure of geneticin bound to a bacterial 16S ribosomal RNA A site oligonucleotide. J Mol Biol 326:1175–1188
16. François B, Szychowski J, Adhikari SS, Pachamuthu K, Swayze EE, Griffey RH, Migawa MT, Westhof E, Hanessian S (2004) Antibacterial aminoglycosides with a modified mode of binding to the ribosomal-RNA decoding site. Angew Chem Int Ed Engl 43:6735–6738
17. François B, Russell RJM, Murray JB, Aboul-ela F, Masquida B, Vicens Q, Westhof E (2005) Crystal structures of complexes between aminoglycosides and decoding A site oligonucleotides: role of the number of rings and positive charges in the specific binding leading to miscoding. Nucleic Acids Res 33:5677–5690
18. Kondo J, François B, Russell RJ, Murray JB, Westhof E (2006) Crystal structure of the bacterial ribosomal decoding site complexed with amikacin containing the γ-amino-α-hydroxybutyryl (haba) group. Biochimie 88:1027–1031
19. Hanessian S, Szychowski J, Adhikari SS, Vasquez G, Kandasamy P, Swayze EE, Migawa MT, Ranken R, François B, Wirmer-Bartoschek J, Kondo J, Westhof E (2007) Structure-based design, synthesis, and A-site rRNA cocrystal complexes of functionally novel aminoglycoside antibiotics: C2″ ether analogues of paromomycin. J Med Chem 50:2352–2369
20. Kondo J, Pachamuthu K, François B, Szychowski J, Hanessian S, Westhof E (2007) Crystal structure of the bacterial ribosomal decoding site complexed with a synthetic doubly functionalized paromomycin derivative: a new specific binding mode to an a-minor motif enhances in vitro antibacterial activity. Chem Med Chem 2:1631–1638
21. Hanessian S, Pachamuthu K, Szychowski J, Giguère A, Swayze EE, Migawa MT, François B, Kondo J, Westhof E (2010) Structure-based design, synthesis and A-site rRNA co-crystal complexes of novel amphiphilic aminoglycoside antibiotics with new binding modes: a synergistic hydrophobic effect against resistant bacteria. Bioorg Med Chem Lett 20:7097–7101
22. Szychowski J, Kondo J, Zahr O, Auclair K, Westhof E, Hanessian S, Keillor JW (2011) Inhibition of aminoglycoside-deactivating enzymes APH(3′)-IIIa and AAC(6′)-Ii by amphiphilic paromomycin O2″-ether analogues. Chem Med Chem 6:1961–1966
23. Kondo J, Koganei M, Kasahara T (2012) Crystal structure and specific binding mode of sisomicin to the bacterial ribosomal decoding site. ACS Med Chem Lett 3:741–744
24. Kondo J, Urzhumtsev A, Westhof E (2006) Two conformational states in the crystal structure of the Homo sapiens cytoplasmic ribosomal decoding a site. Nucleic Acids Res 34: 676–685
25. Kondo J, François B, Urzhumtsev A, Westhof E (2006) Crystal structure of the Homo sapiens cytoplasmic ribosomal decoding site complexed with apramycin. Angew Chem Int Ed Engl 45:3310–3314
26. Kondo J, Hainrichson M, Nudelman I, Shallom-Shezifi D, Barbieri CM, Pilch DS, Westhof E, Baasov T (2007) Differential selectivity of natural and synthetic aminoglycosides towards the eukaryotic and prokaryotic decoding A sites. Chembiochem 8:1700–1709
27. Kondo J (2012) A structural basis for the antibiotic resistance conferred by an A1408G mutation in 16S rRNA and for the antiprotozoal activity of aminoglycosides. Angew Chem Int Ed Engl 51:465–468
28. Kondo J, Koganei M, Maianti JP, Ly VL, Hanessian S (2013) Crystal structures of a bioactive 6′-hydroxy variant of sisomicin bound to the bacterial and protozoal ribosomal decoding sites. ChemMedChem 8:733–739
29. Shalev M, Kondo J, Kopelyanskiy D, Jaffe CL, Noam A, Baasov T (2013) Identification of the molecular attributes required for aminoglycoside activity against *Leishmania*. Proc Natl Acad Sci USA 110:13333–13338
30. Moazed D, Noller HF (1987) Interaction of antibiotics with functional sites in 16S ribosomal RNA. Nature 327:389–394
31. Purohit P, Stern S (1994) Interactions of a small RNA with antibiotic and RNA ligands of the 30S subunit. Nature 370:659–662
32. Anderson AC, O'Neil RH, Filman DJ, Frederick CA (1999) Crystal structure of a brominated RNA helix with four mismatched

base pairs: an investigation into RNA conformational variability. Biochemistry 38: 12577–12585

33. Ennifar E, Carpentier P, Ferrer JL, Walter P, Dumas P (2002) X-ray-induced debromination of nucleic acids at the Br K absorption edge and implications for MAD phasing. Acta Crystallogr D Biol Crystallogr 58:1262–1268
34. Du Q, Carrasco N, Teplova M, Wilds CJ, Egli M, Huang Z (2002) Internal derivatization of oligonucleotides with selenium for X-ray crystallography using MAD. J Am Chem Soc 124:24–25
35. Olieric V, Rieder U, Lang K, Serganov A, Schulze-Briese C, Micura R, Dumas P, Ennifar E (2009) A fast selenium derivatization strategy for crystallization and phasing of RNA structures. RNA 15:707–715
36. Cate JH, Doudna JA (1996) Metal-binding sites in the major groove of a large ribozyme domain. Structure 4:1221–1229
37. Keel AY, Rambo RP, Batey RT, Kieft JS (2007) A general strategy to solve the phase problem in RNA crystallography. Structure 15: 761–772
38. Robertson MP, Scott WG (2008) A general method for phasing novel complex RNA crystal structures without heavy-atom derivatives. Acta Crystallogr D Biol Crystallogr 64:738–744
39. Kondo J, Urzhumtseva L, Urzhumtsev A (2008) Patterson-guided ab initio analysis of structures with helical symmetry. Acta Crystallogr D Biol Crystallogr 64:1078–1091
40. Robertson MP, Chi YI, Scott WG (2010) Solving novel RNA structures using only secondary structural fragments. Methods 52: 168–172
41. Berger I, Kang CH, Sinha N, Wolters M, Rich A (1996) A highly effic ient 24-condition matrix for the crystallization of nucleic acid fragments. Acta Crystallogr D Biol Crystallogr 52:465–468
42. Vaney MC, Broutin I, Retailleau P, Douangamath A, Lafont S, Hamiaux C, Prangé T, Ducruix A, Riès-Kautt M (2001) Structural effects of monovalent anions on polymorphic lysozyme crystals. Acta Crystallogr D Biol Crystallogr 54:767–780
43. Brinkmann C, Weiss MS, Weckert E (2006) The structure of the hexagonal crystal form of hen egg-white lysozyme. Acta Crystallogr D Biol Crystallogr 62:349–355

Chapter 21

Structure of the HCV Internal Ribosome Entry Site Subdomain IIa RNA in Complex with a Viral Translation Inhibitor

Sergey M. Dibrov and Thomas Hermann

Abstract

The internal ribosome entry site (IRES) in the 5′ untranslated region (UTR) of the hepatitis C virus (HCV) RNA genome is responsible for initiation of viral protein synthesis. The IRES RNA contains autonomously folding domains that are potential targets for antiviral translation inhibitors. Here, we describe the experimental crystal structure determination of the IRES subdomain IIa in complex with a previously discovered benzimidazole translation inhibitor. The structure of an inhibitor complex of the highly conserved IRES subdomain IIa holds promise for structure-based design of new anti-HCV drugs.

Key words Antiviral drugs, Hepatitis C virus, Internal ribosome entry site, Translation inhibition, X-ray crystallography

1 Introduction

Among potential targets for HCV drugs, the IRES element, a structured RNA in the 5′ UTR of the viral RNA genome, is attractive because of its high conservation in clinical isolates [1]. The IRES recruits host cell 40S ribosomal subunits as well as eukaryotic initiation factor 3 (eIF3) and initiates translation of viral proteins trough a 5′ cap-independent mechanism [2]. The subdomain IIa of the HCV IRES, which adopts a unique bent RNA fold [3, 4], is the target site for benzimidazole translation inhibitors that suppress viral replicon at low micromolar concentration [5, 6]. We previously demonstrated that the subdomain IIa acts as a conformational switch whose state is affected by the binding of benzimidazole inhibitors [6]. Capture of an extended conformation of the subdomain IIa by binding of a benzimidazole inhibitor may facilitate undocking of other IRES domains from the ribosome and thereby inhibit viral translation in HCV-infected cells.

Eric Ennifar (ed.), *Nucleic Acid Crystallography: Methods and Protocols*, Methods in Molecular Biology, vol. 1320,
DOI 10.1007/978-1-4939-2763-0_21, © Springer Science+Business Media New York 2016

Here, we describe the use of X-ray crystallography to determine the high resolution structure of the subdomain IIa RNA target in complex with a benzimidazole HCV translation inhibitor. Therefore, a model oligonucleotide corresponding to the subdomain IIa was co-crystallized with a racemic mixture of the inhibitor and the structure was determined by X-ray diffraction (*see* Fig. 1) [7]. Structure analysis revealed that, in the complex, the RNA adopts an overall linear architecture with coaxially stacked helices flanking an internal loop that forms a tightly fitting cavity

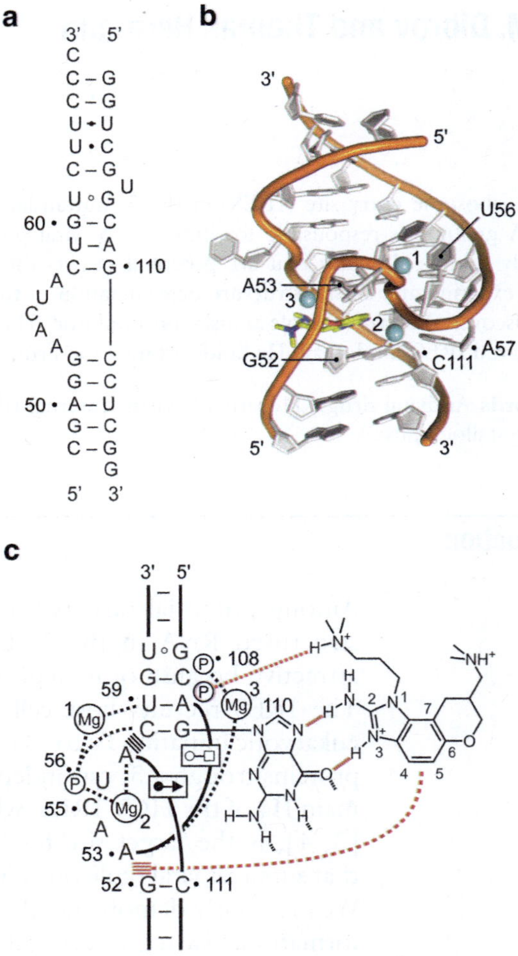

Fig. 1 The HCV IRES subdomain IIa target in complex with a benzimidazole translation inhibitor. (**a**) Oligonucleotide model of the subdomain IIa used for co-crystallization. Residue numbering refers to the HCV genome sequence. (**b**) Overall view of the binding site. The ligand is shown in *yellow* sticks. Mg^{2+} ions are indicated by *light blue* spheres. (**c**) Interactions in the ligand binding site. Hydrogen bonds are shown as *dashed lines. Stacked lines* (≡) indicate stacking of bases and intercalation of the ligand. Formation of non-Watson-Crick base pairs is indicated with solid lines and symbols according to Leontis and Westhof

encapsulating the benzimidazole ligand. The inhibitor docks at the subdomain IIa RNA through hydrogen bonding to a guanine Hoogsteen edge in the C58-G110 base pair and stacks between A53 and the G52-C111 pair. Additional intermolecular hydrogen bond formation is observed between the protonated dimethyl-amino-propyl side chain of the ligand and the phosphate group of A109. The architecture of the inhibitor binding site provides a valuable starting point for the structure-based design of novel HCV inhibitors.

2 Materials

Prepare all solutions using ultrapure water, which is prepared by purifying deionized water to attain a conductivity of 18 MΩ cm at 25 °C, and analytical grade reagents. Prepare and store all reagents at room temperature unless indicated otherwise.

1. Dissolving buffer: 10 mM Sodium cacodylate, pH 6.5, 5 mM $MgCl_2$. Mix 10 μL of 1 M sodium cacodylate buffer with 5 μL of 1 M $MgCl_2$ in a microcentrifuge tube. Add 985 μL of water.
2. Chemically synthesized RNA with the following sequences: 5′ CGA GGA ACU ACU GUC UUC CC 3′ and 5′ GGU CGU GCA GCC UCG G 3′.
3. Ligand solution: 10 mM solution of the benzimidazole inhibitor in water (*see* **Note 1**). The ligand was synthesized as previously described [8].
4. Crystallization solution: 10 mM magnesium sulfate, 50 mM sodium cacodylate pH 6.5, 2.0 M ammonium sulfate.
5. 24-Well plates for crystallization.
6. 22 mm diameter glass slides.
7. Cryoloops for crystal mounting.
8. X-ray data processing program (HKL2000 or equivalent).
9. The PHENIX software suite.
10. The CCP4 software suite for macromolecular crystallography.
11. The program Coot.

3 Methods

3.1 RNA Preparation

1. Purify RNA sequences on gel.
2. To each single strand of the RNA sequence add appropriate amount of dissolving buffer to make 0.8 mM solution (*see* **Note 2**).
3. Vortex for 5 min.

4. Centrifuge for 30 s.
5. Check the concentration by UV spectrometry according to each strand extinction coefficient.
6. Mix equivalent amounts of both strands to obtain 0.4 mM of the double-stranded RNA solution.
7. Anneal obtained RNA solution at 75 °C for 5 min.
8. Cool the solution on the bench top for 5 min to reach the room temperature. This RNA stock solution should be kept frozen at −20 °C.
9. Mix RNA and ligand solutions in 20:1 ratio by volume to obtain a crystallization mix with a final RNA concentration of 0.33 mM and a final ligand concentration of 2 mM.

3.2 Setting Up Crystallization

1. In the well of a 24-well plate put 700 μL of a crystallization solution.
2. Clean a 22 mm diameter glass slide with clean compressed air and pipet 1 μL of the well solution on it.
3. Add 1 μL of the 1:6 RNA-ligand mix to this drop
4. Carefully seal the obtained 2 μL drop over the reservoir solution in a hanging drop experiment.
5. Keep the plate with this well at 16 °C.
6. Check in 1 week intervals for crystal formation (*see* **Note 3**).

3.3 Crystal Mounting

1. Extract and mount a crystal from the crystallization drop on an appropriately sized cryoloop. Flash-freeze the crystal in liquid nitrogen.
2. Transfer the crystal to the goniometer head of the diffractometer within a cold nitrogen stream and collect diffraction data.

3.4 Data Collection and Processing

1. 256 Frames have been collected on a RIGAKU rotating anode X-ray generator ($\lambda = 1.54$ Å) equipped with a mar345 imaging plate detector system.
2. Data collection has been carried with crystal-to-detector distance of 200 mm and with 0.5° oscillations.
3. Dataset has been processed, integrated, and scaled with the HKL2000 package [9] (*see* **Note 4**).
4. Run the Xtriage program (part of the Phenix suite [10]) to see the important statistics on the dataset.

3.5 Structure Solution

1. The program Phaser [11] (part of the CCP4 suite of programs [12]) has been used to solve the three-dimensional structure by molecular replacement.

2. The initial search model has been obtained by modifying the previously solved structure of the ligand-free subdomain IIa RNA molecule (code 2NOK in the PDB database).
3. Modification for use in molecular replacement included introduction of the correct sequence of the crystallization construct, removal of flipped out nucleotide U106 along with the L-shaped corner portion of the structure (A53-A57) and terminal residues to fit the construct used here.
4. The obtained 2 separate double helixes (residues C48-G52, C111-G115 for the first helix and C58-C66, G101-G110 for the second helix) were positioned co-linearly under each other and their coordinates were used in a single file.
5. Molecular Replacement search produced a clear solution for this construct (*see* **Note 5**).

3.6 Structure Refinement

1. The Molecular Replacement solution was subject to rigid body refinement by Refmac [13] (part of the CCP4 package). The R_{work}/R_{free} values obtained (0.38/0.40) indicated a clear solution as they were in sharp contrast with values obtained for other models (0.53/0.54).
2. The resulting model and the electron density map were examined in the program Coot. The model was manually rebuilt by adding missing nucleotides and removing moieties not covered by the electron density map.
3. Refine the improved model with program Refmac using the restrained refinement method (*see* **Note 6**). Trace the backbone of the RNA molecule and assign residues correctly.
4. Based on the electron density map, insert the ligand in the working model (*see* **Note 7**).
5. Continue the iterative rounds of refinement (Refmac) and manual rebuilding (Coot) until a satisfactory model is obtained and no significant changes in R_{work}/R_{free} are observed (*see* **Note 8**).

4 Notes

1. It is advisable to keep the ligand stock solution in aliquots frozen at −20 °C since it degrades over time indicated by the colorless solution turning pink.
2. It is useful to obtain solutions of higher concentrations initially and then dilute them with the dissolving buffer to get the desired concentration for crystallization. Therefore, various RNA concentrations can be tested to avoid excessive precipitation.

3. Crystals should appear within 2–3 months.
4. Based on the indexing results, the space group has been assigned to $P2_12_12_1$ with the unit cell parameters of $a=32.05$, $b=33.58$, $c=81.05$. Upon completion of the data processing step the data obtained had a 2.2 Å resolution limit with 99.15 % completeness (75.6 % for the highest resolution shell), 6.7 (4.5) redundancy, and 32.5 (6.0) signal-to-noise ratio.
5. Other similar constructs such as 14 base pair ideal RNA helix or different variations of variable-length helixes imported from two separate files did not produce clear and unique solution in molecular replacement trials in either CCP4 or the Phenix suite of programs. The obtained molecular replacement solution had TFZ = 21.1 and LLG = 660 with PAK = 0.
6. After the first round of Refmac refinement a continuous electron density was obtained in the region where the complete RNA would fit. The space between the two adjacent molecules remained free of electron density. If the program Xtriage indicates significant twinning, use the determined twin law in the refinement rounds.
7. Initially, coordinate models of the *R* and *S* enantiomers have been built for the benzimidazole ligand in the program Chem3D. A PDB file for the compound along with the corresponding CIF file have been obtained by the electronic Ligand Builder and Optimisation Workbench (eLBOW) program, which is part of a PHENIX suite.
8. During these rounds of refinement, sulfate and metal ions have been placed based on the geometrical considerations according to the electron density map. Both *R* and *S* enantiomers of 1 have been tested and the results based on the refinement statistics were very slightly in favor of the *R* form. The final refinement rounds were performed in PHENIX using individual isotropic atomic displacement parameters and automatic water picking. The final $R_{work}/R_{free} = 0.18/0.24$.

Acknowledgement

This work was supported by the National Institutes of Health, grant AI072012.

References

1. Davis DR, Seth PP (2011) Therapeutic targeting of HCV internal ribosomal entry site RNA. Antivir Chem Chemother 21:117–128
2. Kieft JS, Grech A, Adams P, Doudna JA (2001) Mechanisms of internal ribosome entry in translation initiation. Cold Spring Harb Symp Quant Biol 66:277–283
3. Lukavsky PJ, Kim I, Otto GA, Puglisi JD (2003) Structure of HCV IRES domain II determined by NMR. Nat Struct Biol 10:1033–1038

4. Dibrov SM, Johnston-Cox H, Weng YH, Hermann T (2007) Functional architecture of HCV IRES domain II stabilized by divalent metal ions in the crystal and in solution. Angew Chem Int Ed Engl 46:226–229
5. Seth PP, Miyaji A, Jefferson EA, Sannes-Lowery KA, Osgood SA, Propp SS, Ranken R, Massire C, Sampath R, Ecker DJ, Swayze EE, Griffey RH (2005) SAR by MS: discovery of a new class of RNA-binding small molecules for the hepatitis C virus: internal ribosome entry site IIA subdomain. J Med Chem 48:7099–7102
6. Parsons J, Castaldi MP, Dutta S, Dibrov SM, Wyles DL, Hermann T (2009) Conformational inhibition of the hepatitis C virus internal ribosome entry site RNA. Nat Chem Biol 5:823–825
7. Dibrov SM, Ding K, Brunn ND, Parker MA, Bergdahl BM, Wyles DL, Hermann T (2012) Structure of a hepatitis C virus RNA domain in complex with a translation inhibitor reveals a binding mode reminiscent of riboswitches. Proc Natl Acad Sci U S A 109:5223–5228
8. Parker MA, Satkiewicz E, Hermann T, Bergdahl BM (2011) An efficient new route to dihydropyranobenzimidazole inhibitors of HCV replication. Molecules 16:281–290
9. Otwinowski Z, Minor W (1997) Processing of X-ray diffraction data collected in oscillation mode. Methods Enzymol 276:307–326
10. Adams PD, Afonine PV, Bunkoczi G, Chen VB, Davis IW, Echols N, Headd JJ, Hung LW, Kapral GJ, Grosse-Kunstleve RW, McCoy AJ, Moriarty NW, Oeffner R, Read RJ, Richardson DC, Richardson JS, Terwilliger TC, Zwart PH (2010) PHENIX: a comprehensive Python-based system for macromolecular structure solution. Acta Cryst D 66: 213–221
11. Adams PD, Afonine PV, Bunkoczi G, Chen VB, Echols N, Headd JJ, Hung LW, Jain S, Kapral GJ, Grosse Kunstleve RW, McCoy AJ, Moriarty NW, Oeffner RD, Read RJ, Richardson DC, Richardson JS, Terwilliger TC, Zwart PH (2011) The Phenix software for automated determination of macromolecular structures. Methods 55:94–106
12. Collaborative Computational Project, Number 4 (1994) The CCP4 suite: programs for protein crystallography. Acta Cryst D 50:760–763
13. Murshudov GN, Vagin AA, Dodson EJ (1997) Refinement of macromolecular structures by the maximum-likelihood method. Acta Cryst D 53:240–255

Chapter 22

Anions in Nucleic Acid Crystallography

Luigi D'Ascenzo and Pascal Auffinger

Abstract

Nucleic acid crystallization buffers contain a large variety of chemicals fitting specific needs. Among them, anions are often solely considered for pH-regulating purposes and as cationic co-salts while their ability to directly bind to nucleic acid structures is rarely taken into account. Here we review current knowledge related to the use of anions in crystallization buffers along with data on their biological prevalence. Chloride ions are frequently identified in crystal structures but display low cytosolic concentrations. Hence, they are thought to be distant from nucleic acid structures in the cell. Sulfate ions are also frequently identified in crystal structures but their localization in the cell remains elusive. Nevertheless, the characterization of the binding properties of these ions is essential for better interpreting the solvent structure in crystals and consequently, avoiding mislabeling of electron densities. Furthermore, understanding the binding properties of these anions should help to get clues related to their potential effects in crowded cellular environments.

Key words X-ray crystallography, Anion, Chloride, Sulfate, Phosphate, Data mining, CSD, Cambridge Structural Database, PDB, Protein Data Bank

1 Introduction

Much has been written on optimization of nucleic acid crystallization conditions [1–10], with a frequent emphasis on the cationic surrounding of these negatively charged biopolymers [11]. However, knowledge related to anionic conditions in crystallization buffers and their effects on nucleic acid crystallization is limited to a few empirical facts. A 2004 survey of 18 PDB crystal structures demonstrated that anions can, in certain conditions, intrude the nucleic acid first hydration shell [12]. Present work summarizes our current knowledge on the biology and biophysics of anions related to nucleic acid crystallization. A remarkable historical account on biocrystallogenesis that relates the importance of ionic compounds as crystallization agents has been published recently [13] along with a review [14] covering the multiple and often unanticipated implications of the Hofmeister series in biology [15, 16]. Binding of negatively charged amino acids (Asp, Glu) to nucleic acid systems [12, 17] and base–phosphate interactions [12, 18] will not be discussed.

Eric Ennifar (ed.), *Nucleic Acid Crystallography: Methods and Protocols*, Methods in Molecular Biology, vol. 1320,
DOI 10.1007/978-1-4939-2763-0_22, © Springer Science+Business Media New York 2016

2 General and Specific Considerations

2.1 Which Anions Are Biologically Relevant?

A limited number of anions seem to be biologically relevant and only a few of them are found in the cytosol at high concentration. Among them, the chloride ion is probably the first that comes to mind. Yet, its cytosolic concentration is much smaller (≈4-20 mM) than its extracellular concentration (≈100 mM) [19]. The sulfate (SO_4^{2-}) anion that is often used in crystallization buffers has low cytosolic concentrations. It was estimated from x-ray absorption spectroscopy studies that sulfate represents 0.8 % of the total intracellular concentration of sulfur in horse erythrocytes [20, 21]. A normalized buffer mimicking *Saccharomyces cerevisiae* "in vivo" conditions was designed to contain 2.5–10 mM sulfate di-anions [22]. The halide (Br^- and I^-) as well as the thyocianate (SCN^-) and nitrate (NO_3^-) ions display specialized functions and are toxic at high concentrations in most organisms implying the existence of detoxification pathways similar to those characterized for F^- [23]. Citrate is an important metabolite but produced in small quantities. Formate is found in the venom of insects and acetate as a degradation product of ethanol. Therefore, it is unlikely that these anions contact directly nucleic acid structures under in vivo conditions.

Phosphate and carboxylate-containing compounds are the prevalent cytosolic anions. The former group encompasses nucleotides and other phosphorylated metabolites with $HPO_4^{2-}/H_2PO_4^-$ concentrations close to 1.5 mM [19, 24]. The latter includes proteins, free amino acids with probably a high concentration of glutamate [22, 25], and various metabolites. Among them, a significant concentration of glyoxalate, pyruvate, fumarate, succinate, malate, oxoglutarate, and citrate was reported in *Saccharomyces cerevisiae* cells [25]. The concentration of these metabolites is to a certain extent, organism- and cell type-dependent and might differ in various cell compartments or organelles.

2.2 Which Anions Are Found in Crystallization Buffers?

For designing crystallization buffers, emphasis is primarily put on factors including: "*the identity and concentration of divalent metal ions, polyamines such as spermine, spermidine, putrescine, metal hexammines, and pH*" [7]. In such buffers, anions are essentially used as cationic co-salts, for pH regulating purposes as well as to structure the solvent in a network that will force the macromolecules to aggregate [1]. A non-exhaustive list of anions used in nucleic acid crystallization assays and identified in PDB structures is given in Table 1.

Some non-physiological anions are specifically used for regulating the pH in the crystallization liquors. Cacodylate is a popular anion that contains arsenic and that inhibits the growth of microorganisms in crystallization drops [8]. Given its low pH range (6.0–6.5) cacodylate reduces hydroxide-catalyzed hydrolysis [5].

Table 1
Number of contacts identified in a subset of nucleic acid crystallographic structures from the PDB (December 2014 release; resolution ≤3.0 Å) between nucleotides and common ions along with water coordination distances, pK_a values in water, anion occurrences in all structures and in a subset of structures containing nucleic acid fragments

Ionic species	PDB ligand ID	Coord. dist (Å)[a]	pK_a[b]	PDB[c]	NA[d]	NA contacts[e]
F^-	F	2.7	3.2	34	5	/[f]
Cl^-	CL	3.2	-	7,602	233	931
Br^-	BR	3.4	-	281	9	12
I^-	IOD	3.6	-	505	11	13
SCN^-	SCN	2.8	1.0	231	1	1
NO_3^-	NO3	2.8	-	421	3	2
SO_4^{2-}	SO4	2.8/3.8	-/2.0	12,592	406	311
SeO_4^{2-}	SE4	2.8/3.8	-/2.0	5	1	4
PO_4^{3-}	PO4	2.8/3.8	2.2/7.2/12.4	3656	71	52
HPO_4^{2-}	PI	2.8/3.8	2.2/7.2/12.4	13	/	/
$H_2PO_4^-$	2HP	2.8/3.8	2.2/7.2/12.4	22	9	5
Pyrophosphate^{2-}	POP	2.8/3.8	0.9/2.1/6.7/9.3	204	12	16
Pyrophosphate	PPV	2.8/3.8	0.9/2.1/6.7/9.3	46	29	13
Formate$^-$	FMT	2.8	3.8	723	19	32
Acetate$^-$	ACT/ACY	2.8	4.8	3,669	101	88
Cacodylate$^-$	CAC	2.8	6.2	202	9	6
Bis-TRIS$^-$	BTB	2.8	6.5	122	2	6
MES$^-$	MES	2.8	6.2	730	25	17
HEPES$^-$	EPE	2.8	7.5	570	22	29
Malonate^{2-}	MLI	2.8	2.8/5.7	195	2	5
Succinate^{2-}	SIN	2.8	4.2/5.6	117	11	4
Citrate^{3-}	FLC/CIT	2.8	3.1/4.8/6.4	909	15	5

[a]Average ion coordination distance to water in Å; the first distance corresponds to d(O…O), the second to d(O…S) or d(O…P); *see* Fig. 1

[b]pK_a values in water; http://www.sigmaaldrich.com/france.html

[c]Number of PDB structures with resolution ≤3.0 Å that contain a given ion

[d]Number of nucleic acid (NA) containing structures from the PDB with resolution ≤3.0 Å associated with a given ion

[e]Number of ions establishing direct nucleotide contacts (ion…Ow cutoffs are derived from Fig. 1). Ions with occupancies <1.0 are considered. Note that some ions and especially polyatomic ones are involved in multiple contacts

[f]Fluoride anions close to Ow atoms seem to be all Mg^{2+} mediated in deposited fluoride riboswitch structures

Further, the pH of cacodylate buffers is not sensitive to temperature changes in contrast to TRIS buffers. Although, as sometimes claimed, the fact that crystallization of nucleic acids may not be very sensitive to pH conditions has to be considered with caution. Reports stating that nucleotides (especially adenines and cytosines) can be protonated under specific conditions have been published [26, 27]. Such protonation events are currently not well understood but might be affected by the buffer composition. Indeed, different experimental pH conditions can alter the structure of a particular molecular fragment. For example, the structure of a $tRNA^{Lys}$ [28] and $tRNA_f^{Met}$ [29] anticodon stem-loop obtained at acidic pH by NMR and crystallography, respectively, is different from those extracted from other tRNA crystal structures, essentially because of the formation of a wobble A^+-C pair instead of a more standard (in this context) bifurcated A-C pair [30]. Furthermore, nucleic acid crystallization can be very sensitive to the chemical nature of the buffer ions, i.e. succinate and citrate can chelate divalent ions very effectively and alter the free ions concentration [5]. Popular crystallization screens are described in a detailed review [8].

Halides have no pH regulating functions (although it has been established that all added salts lead to an alteration in measured pH values [31]). Consequently, Cl^- ions are essentially added to buffers as co-salts. In rare cases, other halides were introduced in crystallization solutions. KF has been added to the crystallization solution of a fluoride riboswitch [32] and NaI as a resolution enhancer [33] while some crystals were soaked in KBr [34].

In opposition to NMR buffers [19], phosphate ions are generally avoided in crystallization buffers given their strong tendency to precipitate in presence of Mg^{2+}/Ca^{2+} ions. Nevertheless, phosphate ions are found to contact nucleotides in 24 PDB structures leading to a total of 51 contacts (Table 1).

2.3 Which Anions Are Observed in Crystallographic Structures?

Although most certainly present in crystals, not all buffer anions are detected in PDB structures. Table 1 lists the main anions found in nucleic acid structures and their complexes. This list shows that Cl^- and SO_4^{2-} ions are common in nucleic acid structures. Phosphate ions, despite their biological significance are underrepresented and no example of nucleic acid structures with carbonate ions has been reported. Note that not all anions observed in nucleic acid structures establish direct contacts with nucleotides. For example, a F^- ion in a fluoride riboswitch is directly coordinated to Mg^{2+} cations ($F^-...Mg^{2+} \approx 1.9$ Å) but located at an average distance of ≈3.0 Å from nucleotidic anionic oxygen atoms of phosphate groups [32]. Overall, these data clearly establish the existence of direct contacts between nucleic acids and buffer anions. The issues related to the structural impact of these anions on nucleic acid structure and reactivity remain open. For proteins, NMR evidence has been collected, indicating that buffer anions are, contrary to common believes, not neutral to important biomolecular processes [35–37]. It has also been reported that large hydrophobic groups attached to cationic or anionic parts (like tetraalkyl carboxylate) have a greater ordering effect on the aqueous solvent and promote formation of the less hydrated Z-DNA conformers [38].

Noteworthy, some anions are rarely found in the PDB like perchlorate (ClO_4^-; PDBID: LCP), beryllium tetrafluoride

(BF_4^-; PDBID: BF4) or hexafluorophosphate (PF_6^-) ions with 6, 2 and no entries, respectively. These anions, belonging to the non-coordinating class, are overwhelmingly used for the crystallization of small molecules as observed in the CSD (Cambridge Structural Database; version 5.35; update November 2013) [39–41]. ClO_4^- is found in 9302, BF_4^- in 4791 and PF_6^- in 7126 structures, respectively, while SO_4^{2-} is found in "only" 819 structures. Interestingly, although their ionic radii and geometry are similar, their hydration energies ($-\Delta H_{hyd}$) differ greatly (246 and 1,035 kJ/mol for ClO_4^- and SO_4^{2-}, respectively) [16].

2.4 Anions as Phasing Agents

Heavy atoms are often soaked into biomolecular crystals for the purpose of solving the phase problem. These techniques make use of heavy monoatomic (Cs^+, Tl^+, Hg^{2+}, Sm^{3+}, Yb^{3+}, Lu^{3+}, ...) and/or polyatomic cations ($Os(NH_3)_6^{3+}$ or other hexammine cations, ...) [11, 42, 43].

In their quest for new phasing techniques, crystallographers, with a few exceptions, have rarely considered heavy anions. It has been established that bromide or iodide can rapidly be incorporated into ordered solvent regions around biomolecules [42, 44]. Yet, this technique often results in poorly occupied sites and should be used in combination with other soaking techniques. It has been noted that: "*Because of the close proximity in energies of the selenium and bromine K edges, these methods could conceivably be combined in a single MAD experiment using five (or more) wavelengths around these edges*" [42]. Indeed, the selenate (SeO_4^{2-}) ion, that can be considered as a heavy derivative of the SO_4^{2-} ion and an analog to phosphate anions, has also been considered as a useful tool to obtain phases to solve nucleic acid crystal structures [45].

NaI has been introduced in the crystallization conditions of an U2 small nucleolar RNA system. Three iodide ions were observed in the crystal structure. In this case, the I^- ion acted as a resolution enhancer [33]. On the opposite, replacement of SO_4^{2-} by the heavier SeO_4^{2-} ions led to poorer resolutions albeit unambiguous identification of the SO_4^{2-} binding sites [45].

"Free" bromide ions are rare in nucleic acid structures, although crystal soaking with KBr was attempted [34]. Br^- ions result generally from the dissociation of brominated residues like 8BrG, BrC or BrU due to synchrotron radiation damage and are characterized by low occupancy factors [46, 47].

2.5 Anion Coordination Details

Coordination details of Cl^- and sulfate ions to nucleic acids have been reported in a previous publication along with a nomenclature for anion binding regions around nucleotides [12]. Ion/water coordination distances for most anions and some cations are summarized in Table 1 and Fig. 1. They were derived from a statistical analysis of small crystallographic structures deposited in the CSD and are in agreement with those derived from earlier studies [48].

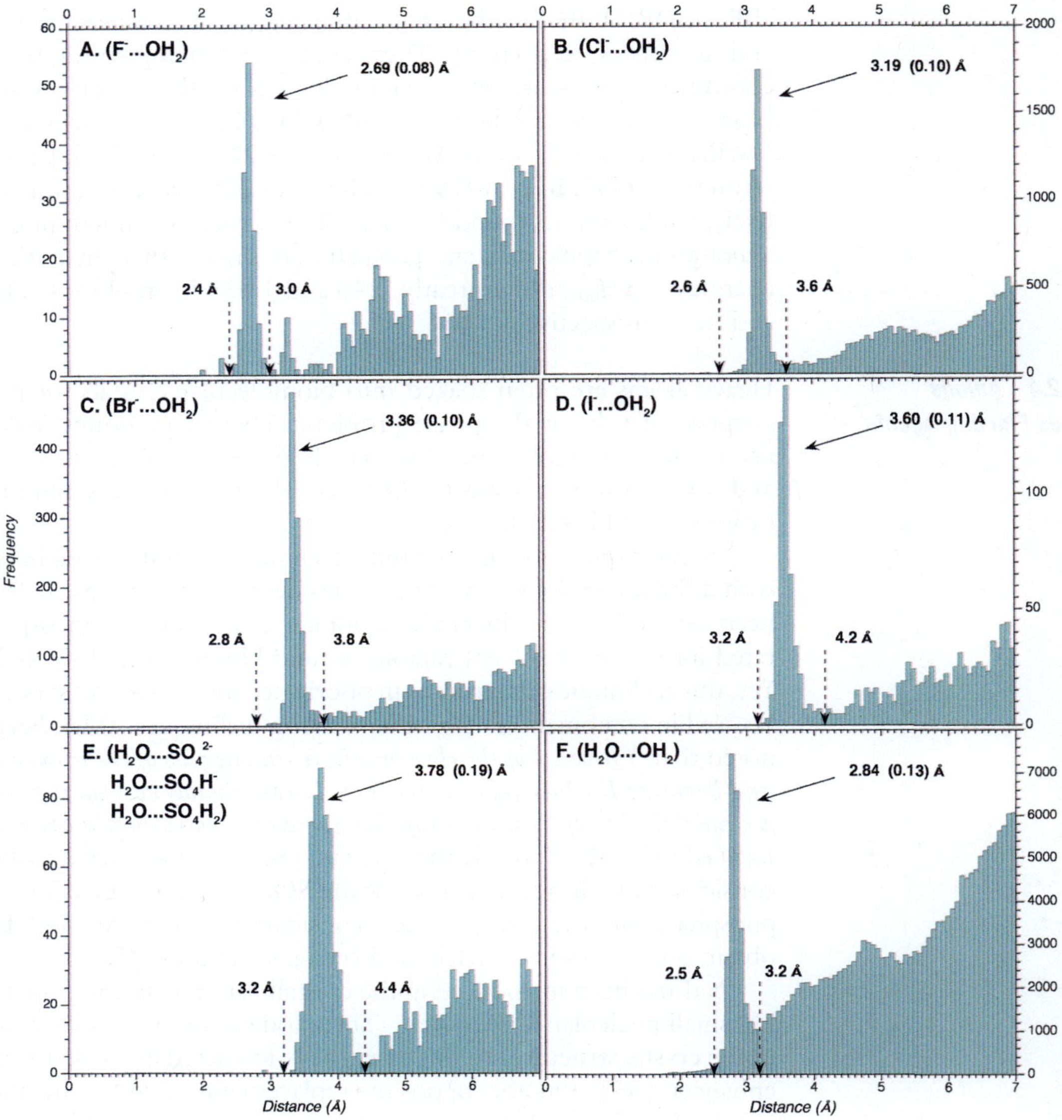

Fig. 1 Distributions of ion…OH_2 and H_2O…OH_2 distances in the CSD (version 5.35; update November 2013; $R \leq 0.05$ unless otherwise specified). Mean coordination distance to water and associated standard deviation (*plain arrow*) are derived from the data points situated between a reasonable lower coordination distance limit and the first minima of the histogram (*doted arrows*). Disordered, error-containing, polymeric and powder structures were excluded from the search. Following coordination distance distributions are shown: (**a**) F^-…OH_2 with $R \leq 0.1$; (**b**) Cl^-…OH_2; (**c**) Br^-…OH_2; (**d**) I^-…OH_2; (**e**) Sulfur…OH_2 (this distribution has been calculated for the three H_2SO_4, HSO_4^- and SO_4^{2-} forms); and (**f**) H_2O…OH_2

Fully ionized anions bind preferentially to electropositive amino and/or imino nitrogen atoms and were also reported close to 2′-OH groups in RNA. Not surprisingly, the coordination distances of these anions to water are very similar to their coordination distances to nucleic acid O or N atoms; ≈3.2 Å for Cl^- and ≈2.8/3.8 Å for oxygen and sulfur atoms, respectively.

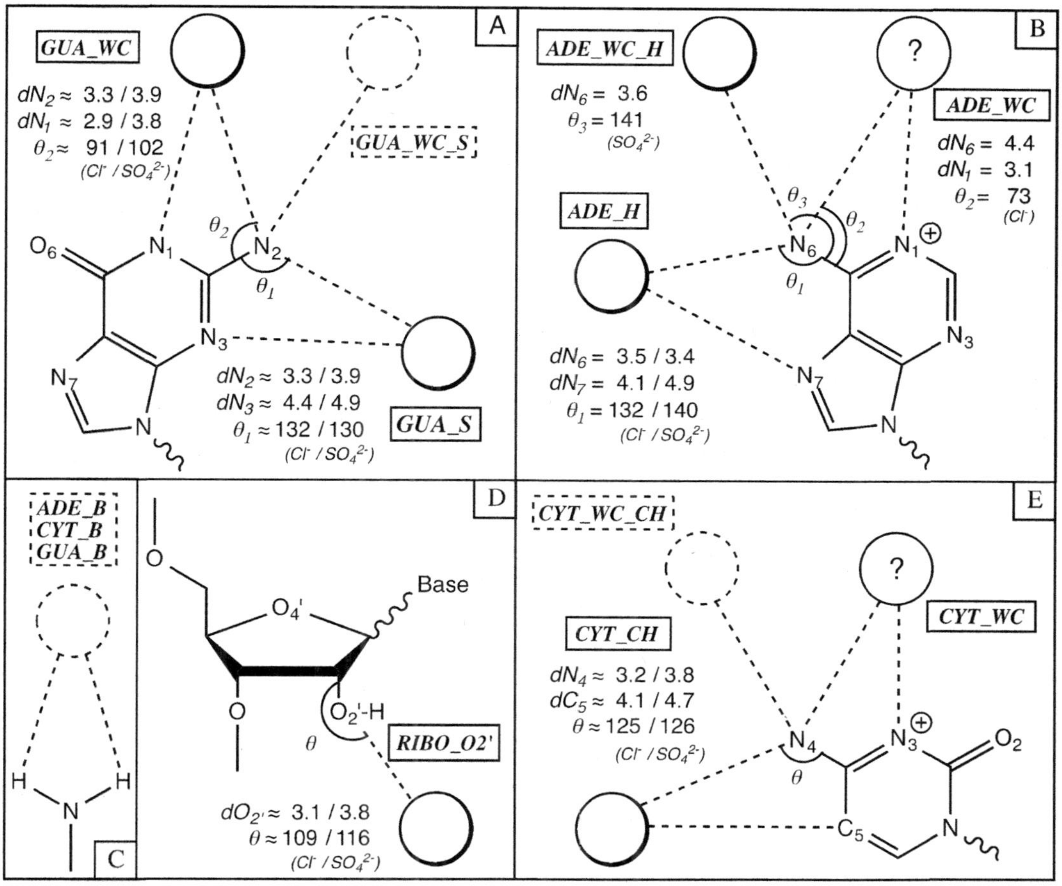

Fig. 2 Anion binding site maps derived from a 2004 PDB survey. Solid spheres represent sites observed in crystal structures. *Dashed spheres* correspond to potential anion binding sites inferred from structures extracted from the PDB and the CSD. *Gray spheres* with a *question mark* (ADE_WC and CYT_WC) correspond to chloride binding sites of small reliability and statistical significance. (Note that occupation of the ADE__WC and CYT_WC sites would imply a protonation of the corresponding bases.) Average coordination distances and angles are given for Cl^- and sulfate ions. (Adapted from ref. 12 with permission from Elsevier)

Coordination sites as observed in the PDB and in the CSD (Fig. 2) involve the Watson-Crick and sugar edges of guanine (GUA_WC and GUA_S), the Hoogsten edge of adenine (ADE_H) and the CH edge of cytosine (CYT_CH). Coordination to possible protonated bases is discussed below.

An example of a sulfate ion bound to a GUA_WC site of a G·G pair [49] is shown Fig. 3. Cl^- binding to two minor groove GUA_H sites of G=C pairs [50] is derived from a 1980 Z-DNA crystal structure account (Fig. 4). Chloride binding to two major groove CYT_CH sites of the same G=C pairs, with smaller occupancy factors, was also reported. This study represents the first account of anion binding to nucleic acid. Yet, in the corresponding PDB file (1ZNA; resolution: 1.5 Å), water molecules replace these

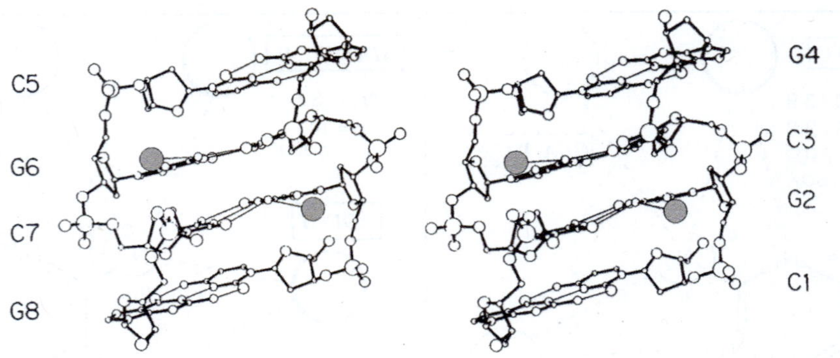

Fig. 3 Stereo pair drawings of the 1.5 Å resolution high-salt Z-DNA CGCG double helix crystal structure. Two chloride ions occupying GUA_H sites of G=C pairs are shown as *grey spheres*. (Adapted from ref. 50 with permission from Nature Publishing Group)

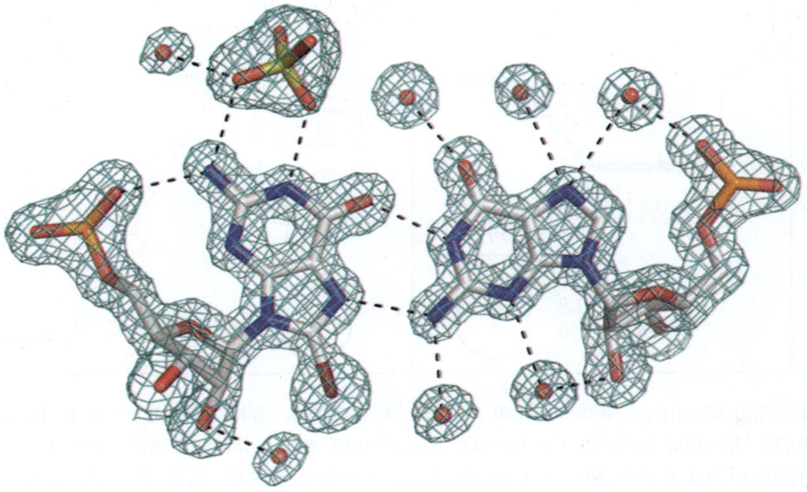

Fig. 4 Sulfate binding and hydration of a G·G pair in a 1.45 Å resolution RNA crystal structure (3RID). The SO_4^{2-} ion occupies a GUA_WC binding site. (Reproduced from ref. 49 with permission from Oxford University Press)

identified Cl^- ions, with surprisingly high occupancy factors that largely exceed 1.0.

Some exceptions, however, can be noted as in the 4FS5/6 structures (resolution: 1.3 Å) where the Cl^- ions bound to GUA_S sites display coordination distances around 3.5–3.7 Å [51]. The latter effect is probably related to the presence of Mg^{2+}/Ca^{2+} cations in the vicinity of the Cl^- ions. Indeed, the close proximity of the divalent cations may stretch the common 3.2 Å ion coordination distance to the nucleobase amino group.

We further noted that Cl^- ions lie mainly in the nucleobase plane and establish contacts with a single nucleotide at most. Sulfate ions instead display a propensity for bridging adjacent nucleotides, especially when bound to the major or minor grooves of G=C pairs.

The number of anion binding events in the PDB increased significantly since our 2004 study [12]. The high number of Cl^- ions to nucleotide contacts (931), noted in Table 1, results from the fact that a large subset of Cl^- ion contacts were identified in the first crystal structures of the 50S ribosome (PDB code: 1JJ2; resolution 2.4 Å) [52] and were kept in the numerous crystallized variants of this structure, including the recent 4V9F revisited 50S ribosomal structure [53]. Interestingly, this structure contains 30 Cl^- ions versus 22 for the original 1JJ2 structure. In the former, 11 Cl^- ions establish direct contacts with nucleotides versus 13 for the latter. The structural redundancy is much smaller for sulfate binding sites where the high number of contacts (311) results essentially from the fact that this large polyatomic anion can easily contact up to three or even four neighboring nucleotides as noted above.

2.6 Anion Occupation Factors

On the average, ≈90 % of anions establishing contacts with nucleotides in the PDB have occupancy factors of one. Yet, evidence has been gathered that in numerous instances ion occupancy factors are fractional, their binding site being possibly occupied by another ion or a water molecule. Further, by using anomalous diffraction data, it was estimated that 90 % of the PDB structures contain unidentified ions. This estimate is considerably larger than the 15 % number of PDB-deposited ion-containing structures [54]. Furthermore, it is probable that a large number of observed ions display non-identified partial occupancies and that ion-binding sites were conservatively associated with water molecules.

Other instances of partial occupancies are related to structures containing polyatomic anions such as SO_4^{2-}, where disorder for the oxygen atomic positions occur although the position of the sulfur atom is precisely defined. An example is provided by a structure of a nonameric RNA duplex with a resolution of 0.97 Å where three symmetry-related G•U pairs form a well defined sulfate binding pocket [55].

2.7 Protonation Issues

2.7.1 Anion Binding to Protonated Nucleotides: Possible Interpretation Errors?

Anions display a propensity to interact with electropositive groups in proteins but also nucleic acids as exemplified in the above sections. In some instances and possibly due to local pH changes, anions could theoretically bind to protonated A/C nucleotides. Indeed, a Cl^- contacting the protonated ADE_WC and CYT_WC sites are reported in the ADOSHC and CYTOSC structures from the CSD with coordination distances in the 3.2 Å range (*see* [12]).

Two examples of possible binding of Cl^- ions to protonated A/C nucleotides can be found in 50S ribosomal structures. First, in the 1FFK and related ribosomal 50S structures, a Cl^- binding event to a protonated adenine (ADE_WC) was identified. Subsequent refinements leading to the 4V9F revised structure [53] stressed an A/G sequence inversion at this position leading to a more reasonable Cl^- binding event to a GUA_WC site. Secondly, a Na^+ that was proposed to bind to a CYT_WC site was replaced by a Cl^- ion in 4V9F. Hence, the attribution of one or the other ion to the observed electron densities seems risky and does not support the existence of an anion binding to a protonated A or C base (Fig. 2).

2.7.2 Anion Protonation: Is It More Likely to Occur?

A surprising outcome of this PDB survey suggests that some polyatomic anions display alternative ionic states as inferred from their proximity to electronegative atoms. An exhaustive account of these events is out of the scope of this review. Nevertheless, we will provide two examples of such interactions. In each case, note that (1) these examples, although rare, are not unique in the current set of PDB structures; (2) their density maps were inspected.

The first example involves interactions of phosphate ions with nucleotides. Such anions can commonly adopt the $H_2PO_4^-$/HPO_4^{2-} forms around neutral pH implying that hydrogen bonding to nucleic acid electronegative sites is possible. Figure 5a shows the binding of a phosphate ion to the ADE_H site expanding the number of possible anion binding sites displayed (Fig. 2). Note that the PDB distinguishes, through their ligand IDs, between the PO_4^{3-} (PO4), HPO_4^{2-} (PI) and $H_2PO_4^-$ (2HP) protonation forms. Yet, since crystallographic evidence for the "correct" protonation state is at best ambiguous, it is generally difficult to decide which ligand ID to use. Similarly, PDB distinguishes between the neutral pyrophosphate (PPV) and the pyrophosphate di-anion (POP) but not between the different sulfate protonation states (Table 1). These protonated phosphate and pyrophosphate ligand IDs will probably become obsolete in future PDB releases as well as the ACT/ACY and FLC/CIT double IDs for acetate and citrate, respectively.

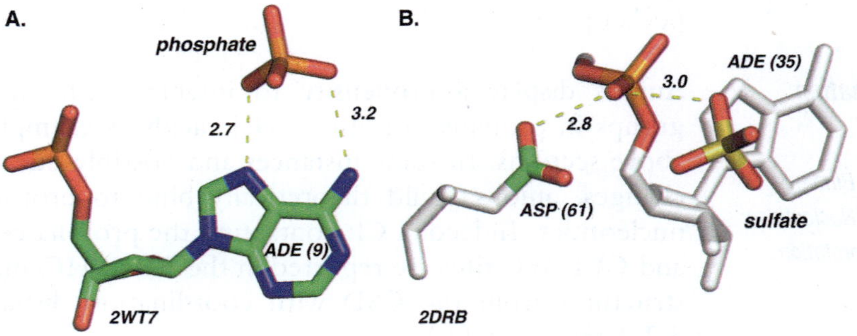

Fig. 5 Binding of protonated phosphate and sulfate ions (electron densities are not ambiguous). (**a**) A phosphate anion, probably adopting the HPO_4^{2-} (or the $H_2PO_4^-$) protonation state, binds to an ADE_H site (2WT7; res: 2.3 Å; pH: 7.5) [68]. (**b**) A sulfate ion, probably adopting the HSO_4^- protonation state, binds counter-intuitively to a negatively charged phosphate group (2DRB; res: 2.8 Å; pH: 7.5) [69]. Note the inferred hydrogen bond to a neutral Asp residue

The second example involves the interaction of a sulfate with a negatively charged phosphate group (Fig. 5b). Given the measured hydrogen bond distances, such proximity between two negatively charged groups can only be explained by the formation of at least one hydrogen bond. Finding a sulfate in the HSO_4^- form is quite surprising since its first pK_a is close to 2.0 (Table 1). Moreover, this structure shows a probably neutral Asp residue that is at hydrogen bond distance from a negatively charged phosphate group.

These two cases, with small statistical significance, stress that the interpretation of experimental electronic densities and associated crystallographic structures must be conducted with great care given our still unsufficient knowledge of how local pH changes affect biomolecular structures in in vitro, in vivo and in cristallo states without even addressing in silico states [56, 57].

3 Summary

Given its rather non-intuitive character, anion binding to nucleic acids has been much less frequently explored than cation binding. Furthermore, since anions are essential constituents of buffers and act as co-ions or pH regulators, they are generally considered as not interfering with nucleic acid or protein systems despite the fact that evidence of buffer anions interacting with proteins [35–37] and anion binding to nucleic acids was reported (Table 1). Clearly these matters need more thorough investigations.

The binding of anions to biomolecular systems obeys rules that are often more complex than just charge compensation rules and render binding predictions quite difficult [58]. For instance, the crystallographic structure of a sulfate binding protein displays no charged residues in its binding pocket. The sulfate ion is solely held by a network of regular hydrogen bonds [59]. Similarly, a large number of phosphate binding sites in proteins rely on hydrogen bond networks [60]. These observations led to the development of new classes of phosphate and sulfate binding neutral synthetic receptors [60, 61].

To summarize, we provide some guidelines related to the detection of nucleic acid anion binding sites. Some of them may appear trivial but are certainly worth some considerations:

1. Certain solvent electron densities might belong to anions present in the crystallized system even if nucleic acids are polyanions.
2. The diffraction power of Cl^-, phosphate or sulfate ions (without mentioning Br^- or I^- ions) is significantly larger than that of H_2O or NH_4^+ molecules or Na^+ and Mg^{2+} cations for full occupancy and reasonable temperature factors.

3. Solvent molecules are sometimes highly diffusive and their occupancy is not always well defined leading to blurred densities [54].
4. When resolution is sufficient (generally ≤2.0 Å) and disorder not too important, sulfates and phosphates can be identified from their tetrahedral geometry.
5. The average coordination distances for halides (F^-, Cl^-, Br^-, I^-) and penta-atomic ions given in Fig. 1 can be used to provide clues on the ion identity; all anion coordination distances differ from the average N/O atoms to water distances that are close to 2.8 Å.
6. As an exception to the preceding rule, the proximity of divalent cations to anion-binding sites may significantly stretch anion coordination distances as observed for Cl^- ions [51].
7. Some anions might adopt counterintuitive protonation states that can only be inferred from distances to electropositive and electronegative atoms.
8. Anomalous signals for sulfur atoms have been successfully used to solve the phase problem, but can also be used to differentiate densities originating from Cl^- ions, water molecule and other ions [54, 62].
9. Sulfate by selenate ion substitution is a possible strategy for identifying anion binding sites by using the Se anomalous signal [45].
10. Assigning solvent electronic densities is a difficult task and some misidentification of anion binding sites occurred in the past [12]. We still believe along with others [54, 63] that a subsequent number of anion-binding sites were left unidentified in PDB structures and that, sometimes, anion electronic densities were misassigned to water molecules or even Mg^{2+} cations [12].

We hope herewith that, thanks to more precise data, a larger number of anion binding sites will be characterized, leading to a better understanding of their characteristics and of the overall solvent structure around nucleic acids. Altogether, this should result into enhanced refinement of forthcoming crystallographic structures and eventual reinterpretation and re-refinement of existing ones [63–67].

Acknowledgements

P.A. wishes to thank Prof. Eric Westhof for ongoing support. We thank Dr Eric Ennifar, Dr Bernard Lorber, Prof. Richard Giegé and Prof. Neena Grover for stimulating discussions. This work has been published under the framework of the LABEX: ANR-10-LABX0036_NETRNA and benefits from a funding from the state managed by the French National Research Agency as part of the program "Investments for the future."

References

1. Dock AC, Lorber B, Moras D, Pixa G, Thierry JC, Giegé R (1984) Crystallization of transfer ribonucleic acids. Biochimie 66:179–201
2. Ducruix A, Giegé R (1999) Crystallization of nucleic acids and proteins, vol 210, A practical approach. The practical approach series. Oxford University Press, Oxford
3. Ferre-D'Amare AR (2010) Use of the spliceosomal protein U1A to facilitate crystallization and structure determination of complex RNAs. Methods 52:159–167
4. Golden BL (2007) Preparation and crystallization of RNA. Methods Mol Biol 317:239–257
5. Golden BL, Kundrot CE (2003) RNA crystallization. J Struct Biol 142:98–107
6. Holbrook SR, Holbrook EL, Walukiewicz HE (2001) Crystallization of RNA. Cell Mol Life Sci 58:234–243
7. Ke A, Doudna JA (2004) Crystallization of RNA and RNA-protein complexes. Methods 34:408–414
8. Mooers BH (2009) Crystallographic studies of DNA and RNA. Methods 47:168–176
9. Reyes FE, Garst AD, Batey RT (2009) Strategies in RNA crystallography. Methods Enzymol 469:119–139
10. Russo Krauss I, Merlino A, Vergara A, Sica F (2013) An overview of biological macromolecule crystallization. Int J Mol Sci 14: 11643–11691
11. Auffinger P, Grover N, Westhof E (2011) Metal ion binding to RNA. Met Ions Life Sci 9:1–35
12. Auffinger P, Bielecki L, Westhof E (2004) Anion binding to nucleic acids. Structure 12: 379–388
13. Giege R (2013) A historical perspective on protein crystallization from 1840 to the present day. FEBS J 280:6456–6497
14. Lo Nostro P, Ninham BW (2012) Hofmeister phenomena: an update on ion specificity in biology. Chem Rev 112:2286–2322
15. Kunz W, Henle J, Ninham BW (2004) 'Zur Lehre von der Wirkung der Salze' (about the science of the effect of salts): Franz Hofmeister's historical papers. Curr Opin Colloid Interface Sci 9:19–37
16. Marcus Y (2012) Ions in water and biophysical implications – from chaos to cosmos. Springer, Berlin
17. Kondo J, Westhof E (2011) Classification of pseudo pairs between nucleotide bases and amino acids by analysis of nucleotide-protein complexes. Nucleic Acids Res 39:8628–8637
18. Zirbel CL, Sponer JE, Sponer J, Stombaugh J, Leontis NB (2009) Classification and energetics of the base-phosphate interactions in RNA. Nucleic Acids Res 37:4898–4918
19. Freund J, Kalbitzer HR (1995) Physiological buffers for NMR spectroscopy. J Biomol NMR 5:321–322
20. Jalilehvand F (2006) Sulfur: not a "silent" element any more. Chem Soc Rev 35:1256–1268
21. Pickering IJ, Prince RC, Divers T, George GN (1998) Sulfur K-edge X-ray absorption spectroscopy for determining the chemical speciation of sulfur in biological systems. FEBS Lett 441:11–14
22. van Eunen K, Bouwman J, Daran-Lapujade P, Postmus J, Canelas AB, Mensonides FI, Orij R, Tuzun I, van den Brink J, Smits GJ, van Gulik WM, Brul S, Heijnen JJ, de Winde JH, de Mattos MJ, Kettner C, Nielsen J, Westerhoff HV, Bakker BM (2010) Measuring enzyme activities under standardized in vivo-like conditions for systems biology. FEBS J 277: 749–760
23. Li S, Smith KD, Davis JH, Gordon PB, Breaker RR, Strobel SA (2013) Eukaryotic resistance to fluoride toxicity mediated by a widespread family of fluoride export proteins. Proc Natl Acad Sci USA 110:19018–19023
24. Godt RE, Maughan DW (1988) On the composition of the cytosol of relaxed skeletal muscle of the frog. Am J Physiol 254:C591–C604
25. Canelas AB, Ras C, ten Pierick A, van Dam JC, Heijnen JJ, van Gulik WM (2008) Leakage-free rapid quenching technique for yeast metabolomics. Metabolomics 4:226–239
26. Wilcox JL, Ahluwalia AK, Bevilacqua PC (2011) Charged nucleobases and their potential for RNA catalysis. Acc Chem Res 44:1270–1279
27. Wilcox JL, Bevilacqua PC (2013) A simple fluorescence method for pK(a) determination in RNA and DNA reveals highly shifted pK(a)'s. J Am Chem Soc 135:7390–7393

28. Durant PC, Davis DR (1999) Stabilization of the anticodon stem-loop of $tRNA^{Lys,3}$ by an A^+-C base pair and by pseudouridine. J Mol Biol 1999:115–131
29. Barraud P, Schmitt E, Mechulam Y, Dardel F, Tisne C (2008) A unique conformation of the anticodon stem-loop is associated with the capacity of $tRNA_f^{Met}$ to initiate protein synthesis. Nucleic Acids Res 36:4894–4901
30. Auffinger P, Westhof E (1999) Singly and bifurcated hydrogen-bonded base-pairs in tRNA anticodon hairpins and ribozymes. J Mol Biol 292:467–483
31. Bostrom M, Lonetti B, Fratini E, Baglioni P, Ninham BW (2006) Why pH titration in protein solutions follows a Hofmeister series. J Phys Chem B 110:7563–7566
32. Ren A, Rajashankar KR, Patel DJ (2012) Fluoride ion encapsulation by Mg^{2+} ions and phosphates in a fluoride riboswitch. Nature 486:85–89
33. Lin Y, Kielkopf CL (2008) X-ray structures of U2 snRNA-branchpoint duplexes containing conserved pseudouridines. Biochemistry 47: 5503–5514
34. Delfosse V, Bouchard P, Bonneau E, Dagenais P, Lemay JF, Lafontaine DA, Legault P (2010) Riboswitch structure: an internal residue mimicking the purine ligand. Nucleic Acids Res 38:2057–2068
35. Long D, Yang D (2009) Buffer interference with protein dynamics: a case study on human liver fatty acid binding protein. Biophys J 96: 1482–1488
36. Quan L, Wei D, Jiang X, Liu Y, Li Z, Li N, Li K, Liu F, Lai L (2008) Resurveying the Tris buffer solution: the specific interaction between tris(hydroxymethyl)aminomethane and lysozyme. Anal Biochem 378:144–150
37. Wong M, Khirich G, Loria JP (2013) What's in your buffer? Solute altered millisecond motions detected by solution NMR. Biochemistry 52:6548–6558
38. McDonnell NB, Preisler RS (1989) Hydrophobic moieties in cations, anions, and alcohols promote the B-to-Z transition in poly[d(G-C)] and poly[d(G-m5C)]. Biochem Biophys Res Commun 164:426–433
39. Allen FH (2002) The Cambridge Structural Database: a quarter of a million crystal structures and rising. Acta Cryst B58:380–388
40. Allen FH, Taylor R (2004) Research applications of the Cambridge Structural Database (CSD). Chem Soc Rev 33:463–475
41. Groom CR, Allen FH (2014) The Cambridge Structural Database in retrospect and prospect. Angew Chem Int Ed Engl 53:662–671
42. Boggon TJ, Shapiro L (2000) Screening for phasing atoms in protein crystallography. Structure 8:R143–R149
43. Brodersen DE, Clemons WM Jr, Carter AP, Wimberly BT, Ramakrishnan V (2003) Phasing the 30S ribosomal subunit structure. Acta Crst B59:2044–2050
44. Dauter Z, Dauter M, Rajashankar KR (2000) Novel approach to phasing proteins: derivatization by short cryo-soaking with halides. Acta Cryst D56:232–237
45. Kieft JS, Chase E, Costantino DA, Golden BL (2010) Identification and characterization of anion binding sites in RNA. RNA 16: 1118–1123
46. Ennifar E, Carpentier P, Ferrer JL, Walter P, Dumas P (2002) X-ray-induced debromination of nucleic acids at the Br K absorption edge and implications for MAD phasing. Acta Cryst D58:1262–1268
47. Olieric V, Ennifar E, Meents A, Fleurant M, Besnard C, Pattison P, Schiltz M, Schulze-Briese C, Dumas P (2007) Using X-ray absorption spectra to monitor specific radiation damage to anomalously scattering atoms in macromolecular crystallography. Acta Cryst D63:759–768
48. Marcus Y (1988) Ionic radius in aqueous solutions. Chem Rev 88:1475–1498
49. Kiliszek A, Kierzek R, Krzyzosiak WJ, Rypniewski W (2011) Crystal structures of CGG RNA repeats with implications for fragile X-associated tremor ataxia syndrome. Nucleic Acids Res 39:7308–7315
50. Drew H, Takano T, Tanaka S, Itakura K, Dickerson RE (1980) High-salt d(CpGpCpG), a left-handed Z′ DNA double helix. Nature 286:567–573
51. Chatake T, Sunami T (2013) Direct interactions between Z-DNA and alkaline earth cations, discovered in the presence of high concentrations of $MgCl_2$ and $CaCl_2$. J Inorg Biochem 124:15–25
52. Klein DJ, Schmeing TM, Moore PB, Steitz TA (2001) The kink-turn: a new RNA secondary structure motif. EMBO J 20:4214–4221
53. Gabdulkhakov A, Nikonov S, Garber M (2013) Revisiting the Haloarcula marismortui 50S ribosomal subunit model. Acta Cryst D69:997–1004
54. Mueller-Dieckmann C, Panjikar S, Schmidt A, Mueller S, Kuper J, Geerlof A, Wilmanns M, Singh RK, Tucker PA, Weiss MS (2007) On the routine use of soft X-rays in macromolecular crystallography. Part IV. Efficient determination of anomalous substructures in biomacromolecules using longer X-ray wavelengths. Acta Cryst D63:366–380

55. Masquida B, Sauter C, Westhof E (1999) A sulfate pocket formed by three GoU pairs in the 0.97 Å resolution x-ray structure of a nonameric RNA. RNA 5:1384–1395
56. Goh GB, Knight JL, Brooks CL 3rd (2013) pH-dependent dynamics of complex RNA macromolecules. J Chem Theor Comput 9: 935–943
57. Goh GB, Knight JL, Brooks CL 3rd (2013) Towards accurate prediction of protonation equilibrium of nucleic acids. J Phys Chem Lett 4:760–766
58. Chakrabarti P (1993) Anion binding sites in protein structures. J Mol Biol 234:463–482
59. Pflugrath JW, Quiocho FA (1988) The 2 Å resolution structure of the sulfate-binding protein involved in active transport in Salmonella typhimurium. J Mol Biol 200:163–180
60. Hirsch AK, Fischer FR, Diederich F (2007) Phosphate recognition in structural biology. Angew Chem Int Ed Engl 46:338–352
61. Ravikumar I, Ghosh P (2012) Recognition and separation of sulfate anions. Chem Soc Rev 41:3077–3098
62. Dauter Z, Dauter M, de La Fortelle E, Bricogne G, Sheldrick GM (1999) Can anomalous signal of sulfur become a tool for solving protein crystal structures? J Mol Biol 289:83–92
63. Echols N, Morshed N, Afonine PV, McCoy AJ, Miller MD, Read RJ, Richardson JS, Terwilliger TC, Adams PD (2014) Automated identification of elemental ions in macromolecular crystal structures. Acta Cryst D70:1104–1114
64. Joosten RP, Joosten K, Cohen SX, Vriend G, Perrakis A (2011) Automatic rebuilding and optimization of crystallographic structures in the Protein Data Bank. Bioinformatics 27: 3392–3398
65. Joosten RP, Salzemann J, Bloch V, Stockinger H, Berglund AC, Blanchet C, Bongcam-Rudloff E, Combet C, Da Costa AL, Deleage G, Diarena M, Fabbretti R, Fettahi G, Flegel V, Gisel A, Kasam V, Kervinen T, Korpelainen E, Mattila K, Pagni M, Reichstadt M, Breton V, Tickle IJ, Vriend G (2009) PDB_REDO: automated re-refinement of X-ray structure models in the PDB. J Appl Crystallogr 42:376–384
66. Kleywegt GJ (2009) On vital aid: the why, what and how of validation. Acta Cryst D65:134–139
67. Williams LD (2005) Between objectivity and whim: nucleic acid structural biology. Top Curr Chem 253:77–88
68. Pogenberg V, Consani Textor L, Vanhille L, Holton SJ, Sieweke MH, Wilmanns M (2014) Design of a bZip transcription factor with homo/heterodimer-induced DNA-binding preference. Structure 22:466–477
69. Tomita K, Ishitani R, Fukai S, Nureki O (2006) Complete crystallographic analysis of the dynamics of CCA sequence addition. Nature 443:956–960

Index

Eric Ennifar (ed.), *Nucleic Acid Crystallography: Methods and Protocols*, Methods in Molecular Biology, vol. 1320, DOI 10.1007/978-1-4939-2763-0,

© Springer Science+Business Media New York 2016

F

G

H

I

M

N

O

P

X

Z

9781493927623